*Text Book Of*



# DESIGN OF MACHINE ELEMENTS

**(Subject Code MEC 604)**

For

## Semester – VI

THIRD YEAR DIPLOMA COURSES IN MECHANICAL ENGINEERING GROUP

***As Per New Syllabus of SBTE, Jharkhand***

**VINOD THOMBRE-PATIL**
*B.E. (Mech.) M. Tech. (Energy Technology)*
Former Vice-Principal and HOD,
S. H. Jondhale Polytechnic, Dombivli,
Dist. Thane, Maharashtra

**APEKSHA THOMBRE-PATIL**
*B.E., M.E. (Electronics and Telecommunication),*
Assistant Professor,
G.V. Acharya College of Engineering and Technology,
Shelu, Dist. Thane, Maharashtra.

NIRALI PRAKASHAN
ADVANCEMENT OF KNOWLEDGE

N4737

**DESIGN OF MACHINE ELEMENTS (SBTE-SEM. VI)** **ISBN 978-93-89825-98-5**

**First Edition : January 2020**



---

**Published By :**

**NIRALI PRAKASHAN**

Abhyudaya Pragati, 1312, Shivaji Nagar,
Off J.M. Road, PUNE – 411005
Tel - (020) 25512336/37/39, Fax - (020) 25511379
Email : niralipune@pragationline.com

## ➢ DISTRIBUTION CENTRES

**PUNE**

**Nirali Prakashan** (For orders within Pune) : 119, Budhwar Peth, Jogeshwari Mandir Lane, Pune 411002, Maharashtra
Tel : (020) 2445 2044, Mobile : 9657703145
Email : niralilocal@pragationline.com

**Nirali Prakashan** (For orders outside Pune) : S. No. 28/27, Dhayari, Near Asian College Pune 411041
Tel : (020) 24690204 Fax : (020) 24690316; Mobile : 9657703143
Email : bookorder@pragationline.com

**MUMBAI**

**Nirali Prakashan** : 385, S.V.P. Road, Rasdhara Co-op. Hsg. Society Ltd.,
Girgaum, Mumbai 400004, Maharashtra; Mobile : 9320129587
Tel : (022) 2385 6339 / 2386 9976, Fax : (022) 2386 9976
Email : niralimumbai@pragationline.com

## ➢ DISTRIBUTION BRANCHES

**JALGAON**

**Nirali Prakashan** : 34, V. V. Golani Market, Navi Peth, Jalgaon 425001, Maharashtra,
Tel : (0257) 222 0395, Mob : 94234 91860;
Email : niralijalgaon@pragationline.com

**KOLHAPUR**

**Nirali Prakashan** : New Mahadvar Road, Kedar Plaza, 1st Floor Opp. IDBI Bank, Kolhapur 416 012
Maharashtra. Mob : 9850046155; Email : niralikolhapur@pragationline.com

**NAGPUR**

**Nirali Prakashan** : Above Maratha Mandir, Shop No. 3, First Floor,
Rani Jhanshi Square, Sitabuldi, Nagpur 440012, Maharashtra
Tel : (0712) 254 7129; Email : niralinagpur@pragationline.com

**DELHI**

**Nirali Prakashan** : 4593/15, Basement, Agarwal Lane, Ansari Road, Daryaganj
Near Times of India Building, New Delhi 110002 Mob : 08505972553
Email : niralidelhi@pragationline.com

**BENGALURU**

**Nirali Prakashan** : Maitri Ground Floor, Jaya Apartments, No. 99, 6th Cross, 6th Main,
Malleswaram, Bengaluru 560003, Karnataka; Mob : 9449043034
Email: niralibangalore@pragationline.com

**Other Branches : Hyderabad, Chennai**

niralipune@pragationline.com | www.pragationline.com

Also find us on www.facebook.com/niralibooks

## Dedication ...

*In the loving memory of my Father,*

**Late Mr. S. T. Thombre–Patil,**
**Hon'ble Advocate, Jalna**

*Thanks,*

*To Our Beloved Son,*

**"Abhishek Vinod Thombre-Patil"**

# Preface ...

We are glad to present the 1st edition of book entitled **"Design of Machine Elements"** for **III$^{rd}$ Year Diploma, Semester VI** in Diploma in Mechanical Engineering Group as per the syllabus prescribed by SBTE.

We have observed the students facing extreme difficulties in understanding the basic principles and fundamental concepts without adequate solved problems along with the text. To meet this basic requirement of students, sincere efforts have been made to present the subject matter with frequent use of figures and lots of numerical examples.

We are thankful to Mr. Dineshbhai Furia, Mr. Pradeepbhai Furia, Mr. Jigneshbhai Furia, Mr. Shashikant Patel, Mr. Santosh Bare, Mrs. Manasi Pingle, Mrs. Prachi Sawant, Mr. Ravindra Walodare of Nirali Prakashan along with their staff members for bringing this publication timely to the Market.

We must acknowledge special gratitude to Mr. Raosaheb Patil Danve (Minister of State, Govt. of India), Er. Rajesh Tope (MLA and President of Matsyodari Shikshan Sanstha, Maharashtra), Er. Sagar Jondhale (Director, Jondhale Educational Group), Dr. Manoj Shete ( President, College of Engg and Tech., Kasara), Dr. Ram Reddy, (Principal, LRTCOE, Mumbai), Dr. R. G. Tated, Prof. Shinde N. N. (HOD – Energy Technology, DOT, SUK), Prof. Sharad Pawar (Principal, Fabtech Polytechnic), Dr. Nehte (Principal, Father Agnel), Dr. Pable M. J. (G. P. Bandra), Prof. R. B. Sholapurkar, (Principal, Virar Polytechnic), Prof. Taskar (Principal, G.P., Jintur), Prof. Kantute (Principal, SBM, Mumbai), Prof. Bavaskar (Principal, Muchala Polytechnic, Thane), Prof. Tandon (Principal, BVIT, Mumbai), Prof. Mange Amar (Principal, Pillai Polytechnic, Rasayani), Prof. Tambe (Principal, Shantiniketan Polytechnic, Panvel), Prof. S. N. Mahajan (H.O.D., V.B.V.P., Vasai), Prof. Anoop Kumar and Prof. Sawant (H.O.D., S.S.J.P., Ambernath), Prof. Mrs. Nehal Muchala (HOD, Thakur Polytechnic), Prof. Sable (HOD, Tasgaonkar Polytechnic, Karjat), Prof. Kamble (HOD, Navjeevan Polytechnic), Prof. S. S. Bhadane (HOD, SSJ Polytechnic, Asangaon), Prof. A. S. Dayma (HOD, SSJCOE, Asangaon), Prof. K. A. Jawalkar (HOD, SSJ Polytechnic, Asangaon), Prof. M. G. Bagale (HOD, SPM Polytechnic, Kunthe, Dist. Solapur) Prof. Killedar (G.P., Solapur), Prof. Lonkar, Prof. Salunkhe, Prof. Sanghvikar, Prof. P. P. Chaudhari, Prof. Shele, Prof. Pandit, (G.P., Thane), Mr. Anish Gunjal (M. Tech., I.I.T., Mumbai) and many more, who have extended their co-operation and suggestions time to time.

We specially thank our students for motivating and encouraging us from time to time.

Any errors or suggestions for the improvement of this book brought to notice will be thankfully acknowledged and incorporated in the next edition.

Write: vinodthombrepatil@gmail.com

**Vinod Thombre-Patil**

**Apeksha Thombre-Patil**

# Syllabus ...

## 1. INTRODUCTION TO DESIGN (08 Hours)

1.1 Fundamentals : Types of loads, Concepts of stress, Strain, Strain-strain diagram for ductile and brittle materials, Types of stresses such as tension, Compression, Shear, Bearing pressure intensity, Crushing, Bending and torsion, Principle stresses (Simple Numerical)

1.2 Fatigue, Creep, S-N curve, Endurance limit. Factor of safety and factors governing selection of factor of safety. Stress concentration - Causes and remedies.

1.3 Converting actual load or torque into design load or torque using design factors like velocity factor, Factor of safety and service factor.

Properties of engineering materials, Designation of materials as per IS and introduction to international standards and advantages of standardization, Use of design data book.

## 2. DESIGN OF SIMPLE MACHINE PARTS (06 Hours)

2.1 Cotter joint, Kunckle joint.

2.2 Design of Levers ; Hand/foot lever and bell crack lever.

2.3 Design of C-clamp, Off-set links, Over hang crank, Arm of pulley.

## 3. DESIGN OF SHAFTS, KEYS AND COUPLINGS (08 Hours)

3.1 Types of shafts, Shaft materials, Standard sizes, Design of shafts (Hollow and Solid) using strength and rigidity criteria, ASME code of design for line shafts supported between bearings with one or two pulleys in between or one over hung pulley.

3.2 Design of sunk keys, Effect of key ways on strength of shaft.

3.3 Design of couplings - Muff coupling, Flange coupling, Bush-pin type flexible coupling.

## 4. DESIGN OF POWER SCREWS AND SPUR GEAR (08 Hours)

4.1 Thread profiles used for power screws, Relative merits and demerits of each, Self-locking and over hauling property.

Torque required to overcome thread friction, Efficiency of power screws, Types of stresses induced.

4.2 Design of screw jack, Toggle jack.

4.3 Spur gear design considerations. Lewis equation for static beam strength of spur gear teeth.

## 5. DESIGN OF SPRINGS (05 Hours)

5.1 Classification and applications of springs, Spring – Terminology, Materials and specifications. Stresses in springs, Wahl's correction factor, Deflection of springs, Energy stored in springs.

5.2 Design of helical springs subjected to uniformly applied loads.

5.3 Leaf springs ; Construction and applications.

## 6. DESIGN OF FASTENERS (05 Hours)

6.1 Riveted joints : Design of riveted joints, Efficiency and frictional resistance of riveted joints.

6.2 Welded joints : Represented of welds, Design of welded joints for static loads, Strength of welds at varying load.

6.3 Stresses in Screwed fasteners, Bolts of uniform strength. Design of bolts subjected to fatigue loading.

## 7. ERGONOMICS AND AESTHETIC CONSIDERATION IN DESIGN (02 Hours)

7.1 Ergonomics of design - man - machine relationship.

Design of equipment for control, Environment and safety.

Aesthetic considerations regarding shape, size, colour and surface finish.

❑❑❑

# Contents ...

❑❑❑

Chapter 1

# INTRODUCTION TO DESIGN

## 1.1 INTRODUCTION TO MACHINE DESIGN PHILOSOPHY

- **Machine design** is defined as *"the use of scientific principles, technical information and imagination in the description of a machine or mechanical system to perform specific functions"*.
- Machine design refers to create new and better machines alongwith improvement and updation of existing machines.

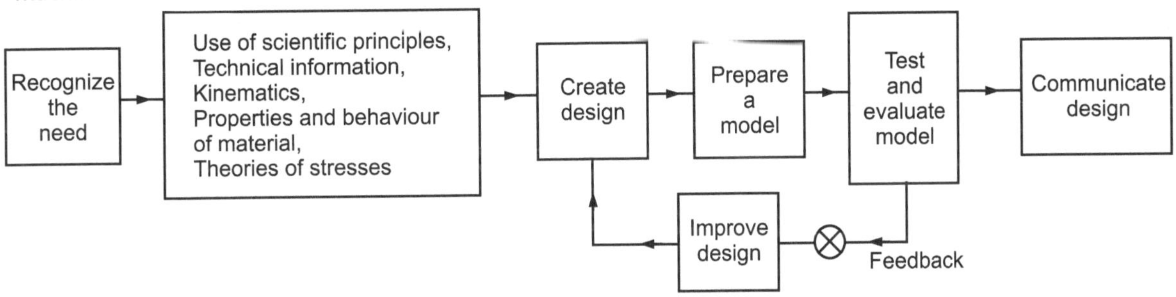


**A flow chart showing process of machine design**

- As per the needs and demands of consumers, small scale units, industries etc., machines are invented or modified by creating an idea in mind.
- This idea is then reformed and translated in terms of dimensions and suitable drawings.

## 1.2 GENERAL PROCEDURE OR STEPS INVOLVED IN DESIGNING A MACHINE COMPONENT OR PHASES IN DESIGN

**Steps Involved in the Process of Designing a Machine Component :**

1. **Recognition of need :** First of all, make a complete statement of problem, indicate the need, aim or purpose, for which, the machine is to be designed.
2. **Mechanism :** Select the possible mechanism or group of mechanisms, which will give the desired motion.
3. **Analysis of forces :** Find the forces acting on each member of machine.
4. **Material selection :** Select most suitable material for each member.
5. **Design of machine elements :** Find size of each member by considering forces acting on the member and permissible stress value for the selected material from design data book.
6. **Modification :** Modify size of member to agree with past experience and judgment to reduce the overall production cost. This will benefit the manufacturer.
7. **Detailed drawings :** Draw detailed drawings of each component and assembly of machine with complete specifications for the manufacturing process suggested.
8. **Production :** Each component is manufactured according to drawings and specifications of design in production department.

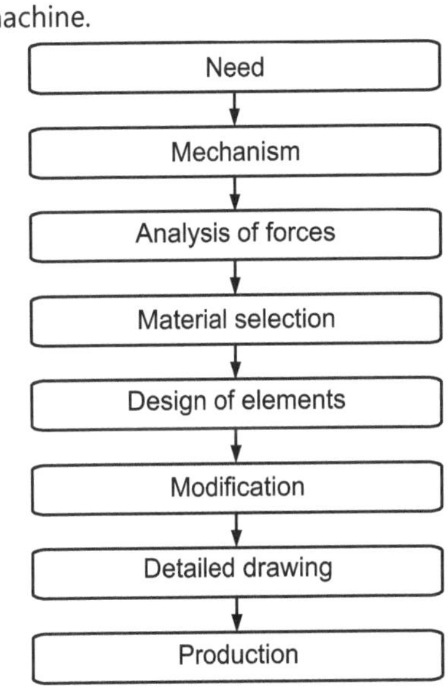


**Chart : Steps involved in design process**

## 1.3 GENERAL CONSIDERATIONS IN MACHINE DESIGN

(1) **Type of externally applied loads and stresses induced in machine component :** Externally applied loads on a machine component may act in several ways. Component tries to resist these loads, due to which, internal stresses are induced in component.

(2) **Mechanism (Motion of the parts or kinematics of machine) :** Successful operation of any machine depends largely upon the simplest arrangement of parts, which will give the required motion.

(3) **Selection of material :** Designer should have a deep knowledge of properties of materials and their behaviour under working conditions, i.e. strength, durability, flexibility, weight, resistance to heat and corrosion, castability, weldability, hardenability, machinability etc.

(4) **Convenient and economical features :** Machine designed must be convenient to operate and cost-wise economical for the customer to purchase.

(5) **Use of standard part :** Use of standard parts reduces the overall cost.

(6) **Safety of operation :** To avoid accidental hazards, care should be taken by designer, against all possibilities of dangerous situations.

(7) **Workshop facilities :** A design engineer should be familiar with limitations of his employer's workshop, in order to avoid the necessity of vendors.

(8) **Number of machines to be manufactured.**

(9) **Cost of production :** Design engineer should always try to design a part, product or machine, in such a way that, production cost will be less. This benefits manufacturer as well as customer.

(10) **Ease in assembly :** All the components should be easy to assemble for creating a machine. Simplicity in design should be top priority for any design engineer.

(11) **Frictional resistance and lubrication :** Designer should provide necessary lubrication to all moving parts of machine. Movements may be sliding, rolling, rotary, reciprocating etc.

## 1.4 BASIC DESIGN REQUIREMENTS

- Basic design requirements in machine design are explained as follows.

| Sr. No. | Design requirement | Description |
|---|---|---|
| 1. | **Strength** | A machine part should not fail under the effect of externally applied loads or forces acting on it. It should have sufficient strength to avoid failure either due to fracture or due to general yielding. |
| 2. | **Rigidity** | A machine component should be rigid, i.e. it should not deflect or bend too much due to forces or bending moments acting on it. |
| 3. | **Wear resistance** | Machine parts worn out very early as compared to their expected life due to wear. Wear also leads to the loss of accuracy of machine tool. Wear resistance of any machine part can be increased by increasing its hardness. **For example :** Case hardening of gears and cams. |
| 4. | **Manufacturability** | Manufacturability is the ease of production, fabrication and assembly. Shape and material of machine part should be selected in such a way that, it can be produced with minimum labour and material costs. |
| 5. | **Safety** | Shape and dimensions of machine parts should ensure safety to the operators of machine. |
| 6. | **Standardization** | A machine part should conform to national or international standards covering its profile, dimensions, grade and material used. |
| 7. | **Reliability** | Reliability is defined as, 'the probability of a machine part, that it will give satisfactory performance over sufficient time period'. Any machine part can be called as reliable, if it performs its intended functions under desired operating conditions over a specified period of time. |
| 8. | **Maintainability** | A machine part should be maintainable. Maintainability is the ease, with which, a machine part can be serviced or repaired. Maintainability is defined as, "the probability that, a product will be restored to its operational effectiveness within the given period, when maintenance action is carried out as per the prescribed procedure mentioned in service manual". |

- In addition to above, other desired requirements are Durability, Size and shape, Efficiency, Lubrication, Cost etc.

## 1.5 TYPES OF LOADS

- **Load acting on a body** is defined as, "*the combined effect of all externally applied forces on the body at a time*".
- Load acting on a member can be classified as,

**(a) Dead or Steady Load :**

- It is the gradually applied load, which does not change in magnitude, direction or point of application with respect to time.
- It can be a force, torque or twisting moment, bending moment or combination of these.
- **Example :** Load acting on a column or strut.

**(b) Variable or Fluctuating Load :**

- It varies in magnitude, direction or point of application with respect to time.
- It is also called as **fatigue load**.
- Types of fatigue loads are, (i) Reversed (ii) Repeated, and (iii) Completely reversed.
- **Examples :** Force in I.C. engine valve spring, Bending moment on rotating shafts.

**(c) Impact Load :**

- It is suddenly applied with initial velocity.
- **Example :** Blows of a hammer.

**(d) Shock Load :**

- It is applied or removed suddenly.
- **Example :** Leaf spring of automobile.

## 1.6 STRESS AND STRAIN

**Stress :**

- Stress is defined as, "*the internal resistive force per unit cross-sectional area of body, when subjected to external load*".
- Stress is denoted by $\sigma$.
- Mathematically, Stress = $\sigma = \frac{W}{A}$

Where, W = Load acting on a body in 'N'

A = Cross-sectional area of body in 'mm$^2$'

**SI unit of stress :** N/mm$^2$.

**Note that :** 1 MPa = 1 N/mm$^2$

$$1 \text{ kPa} = 1 \times 10^3 \text{ N/m}^2 = \frac{1 \times 10^3 \text{ N}}{10^6 \text{ mm}^2} = 1 \times 10^{-3} \text{ N/mm}^2 = 1 \times 10^{-3} \text{ MPa}$$

$$1 \text{ GPa} = 1 \times 10^9 \text{ N/m}^2 = \frac{1 \times 10^9 \text{ N}}{10^6 \text{ mm}^2} = 1 \times 10^3 \text{ N/mm}^2 = 1 \times 10^3 \text{ MPa}$$

**Strain :**

- Strain is defined as, "*the ratio of change in dimension to original dimension of a body subjected to external loads*".
- Linear strain is denoted by e, whereas, lateral strain is denoted by $e_L$.

Mathematically, Strain = $\frac{\text{Change in dimension}}{\text{Original dimension}}$

**S.I. unit of strain :** It is an unitless quantity.

## 1.6.1 Stress-Strain Diagram for Ductile Material (Mild Steel)

- Fig. 1.1 or Graph shows relationship between stress and strain, when a mild steel specimen (ductile material) having uniform cross-sectional area is subjected to a gradually increasing tensile load up to fracture.
- This tensile test is carried out on "Universal testing machine (U.T.M.)".

1. **Proportional Limit (O to A) :**

- In stress-strain diagram, distance "O" to "A" represents a straight line graph. It means that, **stress is directly proportional to strain** and it obeys **Hooke's law**.
- Therefore, this region is called as "**Proportional limit"** and point "A" is called as **"Limit of proportionality".**

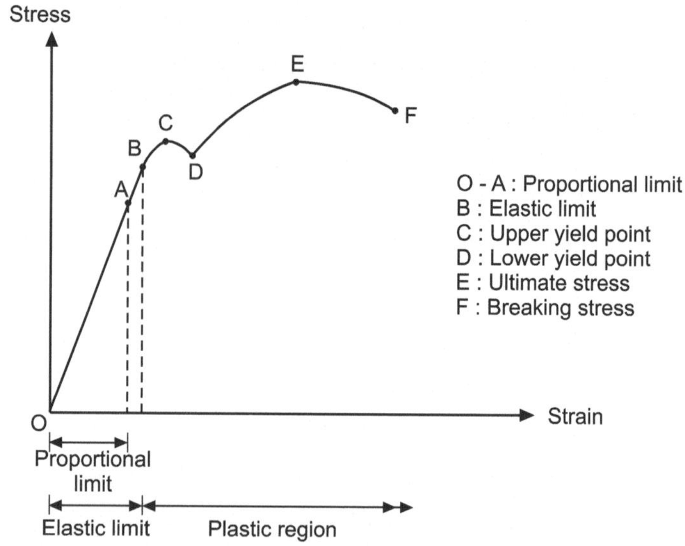


**Fig. 1.1 : Stress-strain diagram for ductile material**

2. **Elastic Limit (B) :**

- If the tensile load applied is gradually increased from point 'A' upto point 'B', then we observe that, beyond point A, stress-strain relationship no longer remains straight (linear). It deviates from straight line and shows non-linearity from point A to point B.
- If the tensile load applied on specimen is suddenly removed on or before reaching point 'B', the whole deformation of mild steel specimen will vanish and specimen will regain its original shape and size.
- We observe that, during gradual increase in load from point 'A' to point 'B', specimen behaves in elastic manner. It means, specimen possesses elasticity upto point B. Therefore, point 'B' is referred as **Elastic limit**.
- But, if the specimen is loaded beyond elastic limit, i.e. point 'B', the specimen will not return back to its original shape or position. It will be permanently deformed. This is called as **permanent set**.
- Thus, the material will be in its elastic range upto point B, but beyond point B, it will reach plastic stage.

3. **Upper Yield Point Stress (C) :**

- Increase in tensile load beyond point B will lead to formation of creep making the specimen to be plastic.
- During this stage, the strain increases at a faster rate even without a small increase in load and hence, stress. This process continues till point "C" is reached.
- Therefore, **point "C"** is called as **upper yield point** and stress corresponding to point 'C' is called as **upper yield stress**.
- **Yield stress** is defined as, "*a unit stress causing increase in length without increase in load*".

4. **Lower Yield Point Stress (D) :**

- If the tensile load is further increased beyond point C, by a very small amount, then the stress-strain curve drops to point "D" indicating the process of yielding.
- Therefore, **point "D"** is called as **lower yield point** and stress corresponding to point 'D' is called as **lower yield stress**.

5. **Ultimate Stress (E) :**

- After increasing the load beyond lower yield point D, there will be increase in stress along with increase in strain upto point E. Therefore, a rise in stress-strain curve is observed between points D and point E.
- **Point 'E'** is called as **ultimate (or maximum) load point** in the stress-strain diagram.
- Value of stress at point E is maximum and therefore, stress at point 'E' is called as **ultimate stress**.
- Due to increase in strain between points D and E, cross-sectional area of specimen decreases.

6. **Breaking Stress (F) :**

- During points D and E, cross-sectional area of specimen decreases uniformly and at point E, we observe **neck formation** in specimen.
- At this point, stress required to break away the specimen is very low, and even less than maximum stress. Therefore, stress-strain curve falls down from point E to point F.
- As the elongation continues, cross-sectional area of specimen becomes smaller and smaller and ultimately, the specimen is broken at point "F" into two pieces showing cup cone type of ductile fracture.
- **Point 'F'** is called as **breaking point** and **stress corresponding to point 'F'** is called as **breaking stress**.

**Formulae**

$$\text{Yield stress} = \frac{W_{yt}}{A}$$

$$\text{Ultimate Stress} = \frac{W_{ut}}{A}$$

$$\text{Percentage reduction in area} = \frac{\text{Original area} - \text{Reduced area}}{\text{Original area}} \times 100$$

$$\text{Percentage elongation} = \frac{\text{Increased length} - \text{Original length}}{\text{Original length}} \times 100$$

**Numerical Type No. 1 : "Stress-Strain Diagram"**

**Problem 1.1 :** *A mild steel rod of 12 mm diameter was tested for tensile strength with gauge length of 600 mm. Following observations were recorded. Final length = 800 mm, Final diameter = 7 mm, Yield load = 4 kN, Ultimate load = 6 kN. Calculate :*

*(a) Yield stress. (b) Ultimate stress. (c) Percentage reduction in area and percentage elongation.*

**Solution : Given data :** $d = 12$ mm, $L = 600$ mm, $L_1 = 800$ mm, $d_1 = 7$ mm

$$\text{Yield load} = \mathbf{W_{yt}} = 4 \text{ kN} = \mathbf{4 \times 10^3 \text{ N}}$$

$$\text{Ultimate load} = \mathbf{W_{ut}} = 6 \text{ kN} = \mathbf{6 \times 10^3 \text{ N}}$$

**Procedure :** Original area = $\mathbf{A} = \frac{\pi}{4} d^2 = \frac{\pi}{4} \times (12)^2 = \mathbf{113.09 \text{ mm}^2}$

(a) $$\textbf{Yield stress, } \sigma_{yt} = \frac{W_{yt}}{A} = \frac{4 \times 10^3}{113.09} = \mathbf{35.37 \text{ N/mm}^2}$$

(b) $$\textbf{Ultimate stress, } \sigma_{ut} = \frac{W_{ut}}{A} = \frac{6 \times 10^3}{113.09} = \mathbf{53.05 \text{ N/mm}^2}$$

(c) (i) **% Reduction in area :**

$$\text{Reduced area} = \mathbf{A_1} = \frac{\pi}{4} \cdot (d_1)^2 = \frac{\pi}{4} \times (7)^2 = \mathbf{38.48 \text{ mm}^2}$$

$$\therefore \quad \text{\% Reduction in area} = \frac{A - A_1}{A} \times 100 = \frac{113.09 - 38.48}{113.09} \times 100 = \mathbf{65.97\%}$$

(ii) $$\textbf{\% Elongation} = \frac{L_1 - L}{L_1} \times 100 = \frac{800 - 600}{800} \times 100 = \mathbf{25\%}$$

### 1.6.2 Stress-Strain Diagram for Brittle Material

- Figure 1.2 shows relationship between stress and strain, when a brittle material is subjected to tensile test.
- Brittle materials fail at relatively low strain.

- Consider a specimen of harder material like cast iron subjected to load. It will be observed that, stress is directly proportional to strain from point 'O' upto point 'A'. It means that, there will be continuous elongation of bar, if it is loaded within this limit. This is called as **proportional or proportionality limit**.
    - In such materials, the elastic limit lies above proportionality limit. As shown in Fig. 1.2, **elastic limit** is from point 'A' to point 'Y'.
    - After elastic limit, the curve shows permanent deformation, i.e. the material reaches plastic stage.
    - Point 'E' is the **breaking point**, at which, the specimen breaks or fails. Here, **breaking stress** is almost equal to **ultimate stress** i.e. point 'F'.
    - Therefore, design of parts made of brittle material is based on ultimate stress.

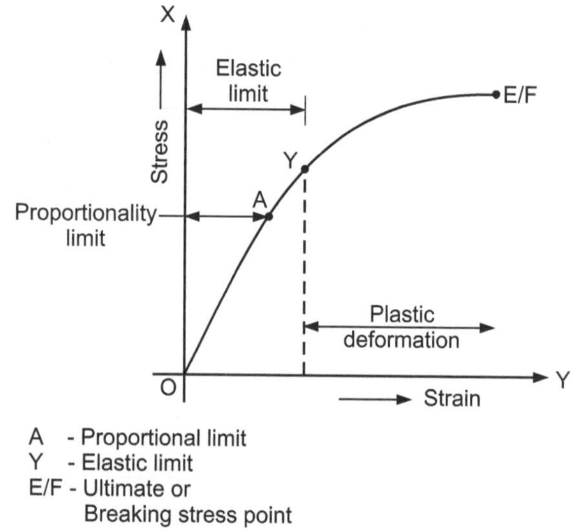


**Fig. 1.2 : Stress-strain diagram for brittle material**

### 1.6.3 Difference between Stress-Strain Diagram for Brittle and Ductile Materials

- Stress-strain diagram for ductile materials obeys the Hooke's law within elastic limit, whereas in case of stress-strain diagram for brittle materials, deviations from Hooke's law begins very early.
- Stress-strain diagram for ductile materials shows that, there is always sufficient plastic deformation prior to failure, which is a warning bell in advance, whereas, stress-strain diagram for brittle materials shows that, fatigue cracks are not visible, till they reach the surface and by that time, the failure has occurred and the fracture is said to be sudden.

## 1.7 TYPES OF STRESSES

The different types of stresses are :

**(1) Tensile Stress :**

- *"When a body is subjected to equal and opposite axial pull forces, the stress produced or induced in the body is called as **tensile stress**".*
- Fig. 1.3 shows a circular rod subjected to axial tensile force 'W'. Tensile stress induced is given by,

$$\sigma_t = \frac{W}{A}$$

where,

W = Axial tensile force in 'N'

A = Cross-sectional area of rod

$= \frac{\pi}{4} d^2$ in 'mm²'

d = Diameter of rod in 'mm'

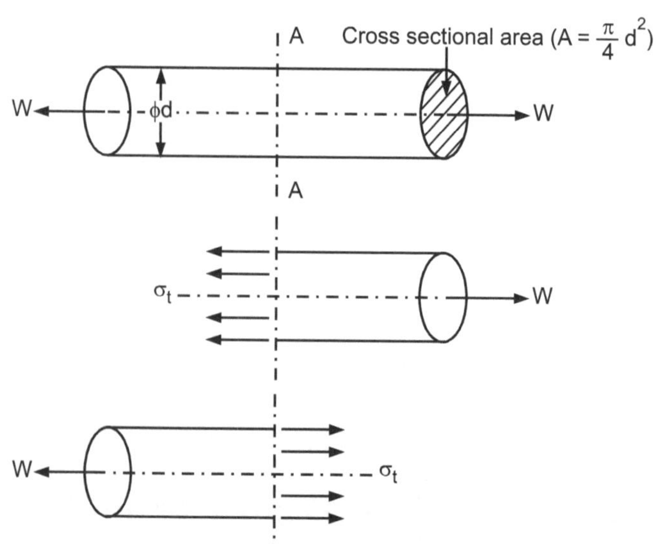


**Fig. 1.3 : Tensile stress**

### (2) Compressive Stress :

- *"When a body is subjected to equal and opposite axial push forces, the stress produced or induced in the body is called as **compressive stress**".*
- Fig. 1.4 shows a circular rod subjected to axial compressive force 'W'. Compressive stress induced is given by,

$$\sigma_c = \frac{W}{A}$$

where,

$W$ = Axial compressive force in 'N'

$A$ = Cross-sectional area of rod = $\frac{\pi}{4} d^2$ in '$mm^2$'

$d$ = Diameter of rod in 'mm'

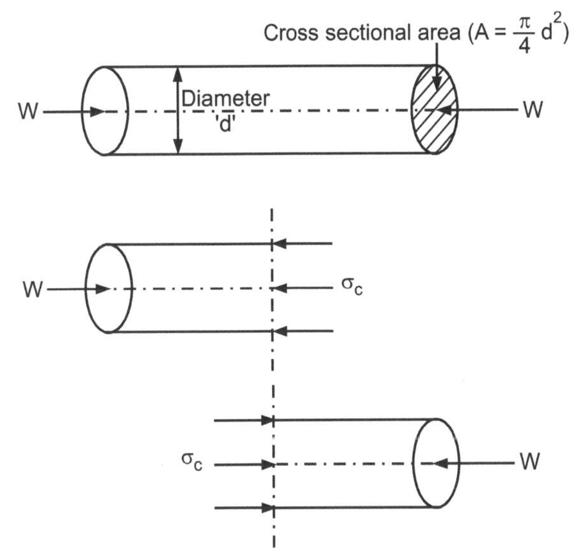


**Fig. 1.4 : Compressive stress**

### (3) Bending Stress :

- *"When a beam is subjected to bending moment (M), the stress induced in it is called as **bending stress**".*
- Consider a straight beam subjected to bending moment as shown in Fig. 1.5. It is clear that, the upper fibres will be shortened due to compression, while the lower fibres will be elongated due to tension. Also, there is a surface, called as *neutral axis (N.A.)*, where fibres are neither shortened nor elongated.
- Thus, bending stress is nothing but, a tensile or compressive stress. Fibres above the neutral axis are subjected to compressive stress, whereas, fibres below the neutral axis are subjected to tensile stress.
- Bending stress induced at extreme fibre i.e. at greatest distance from *neutral axis* is given by,

$$\sigma_{b\,max} = \frac{M}{I} \cdot y_{max} = \frac{M}{\left(\frac{I}{y_{max}}\right)} = \frac{M}{Z}$$

where,

$Z$ = Section modulus = $\frac{I}{y_{max}}$ in '$mm^3$'

$I$ = Moment of inertia of cross-section in '$mm^4$'

$M$ = Bending moment acting on the beam in 'N.mm'

$y_{max}$ = Distance of extreme (outermost) fibre from Neutral axis (N.A.). in 'mm'

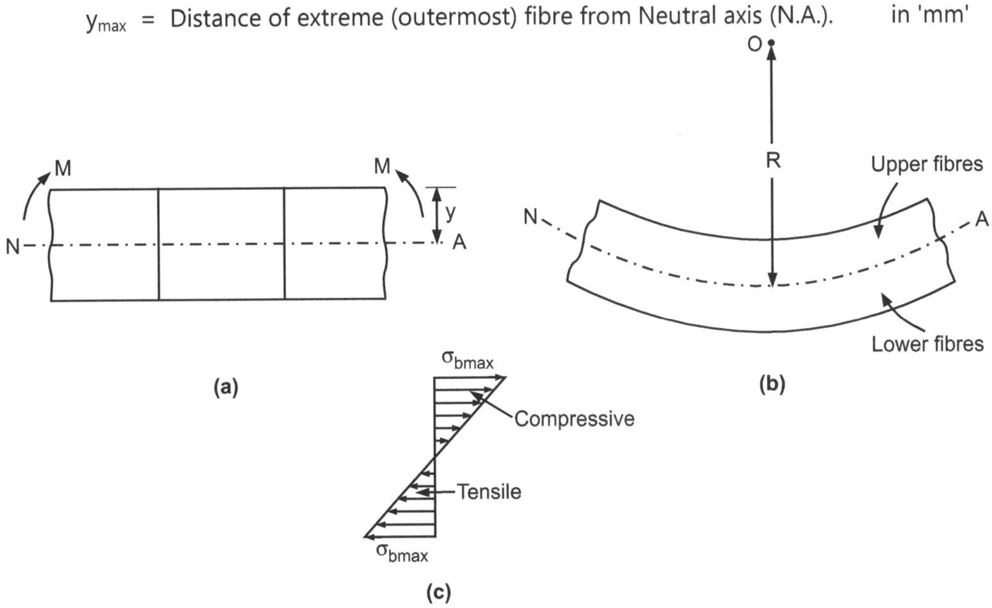


**Fig. 1.5 : Bending stress**

**(4) Transverse Shear Stress :**

- *"When a section is subjected to two equal and opposite forces acting tangentially, it may lead to shear off the section. In such a case, the stress produced or induced in the section is called as* ***transverse shear stress****"*.
- Fig. 1.6 shows a riveted joint for two plates resting one above another, subjected to single shear.
- Transverse shear stress is given by, $\tau = \dfrac{W}{A}$

where, W = Force applied tangentially in 'N'

A = Cross-sectional area of rivet under single shear = $\frac{\pi}{4}d^2$ in 'mm$^2$'

d = Diameter of rivet in 'mm'

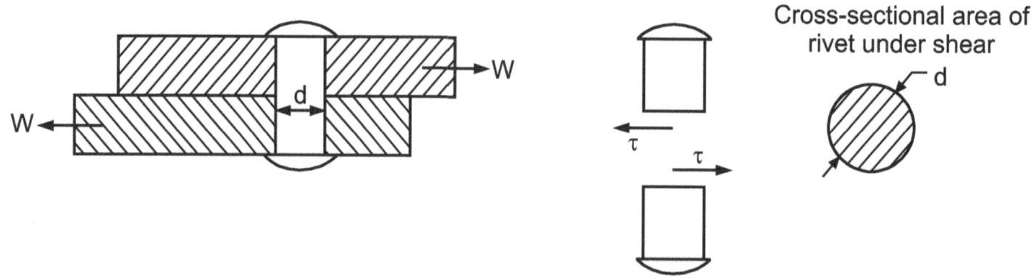


**Fig. 1.6 : Single transverse shear stress**

- Fig. 1.7 shows a riveted joint for three plates resting one above other, subjected to double shear.

Here, $\tau = \dfrac{W}{A} = \dfrac{W}{2 \times \left(\frac{\pi}{4}d^2\right)}$

where, W = Force applied tangentially in 'N'

A = Cross-sectional area of rivet under double shear

∴ $A = 2 \times \frac{\pi}{4}d^2$ in 'mm$^2$' (∵ Double shear)

d = Diameter of rivet in 'mm'

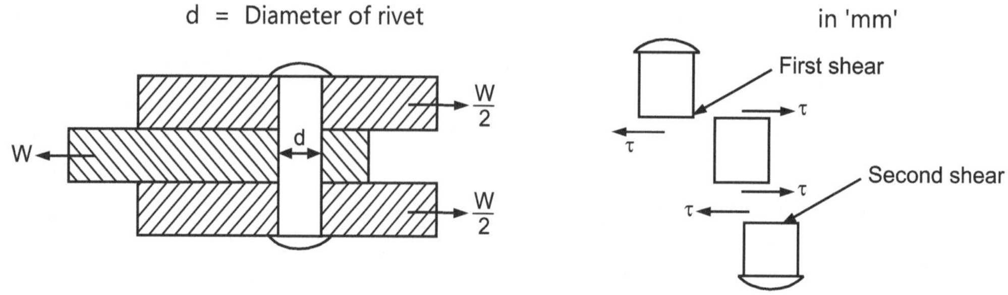


**Fig. 1.7 : Double transverse shear stress**

**(5) Torsional Shear Stress :**

- *"When a machine component is under the action of two equal and opposite couples, i.e. a twisting moment or torque, then component is said to be in torsion and the stress induced due to torsion is called as* ***torsional shear stress****"*.
- It is zero at neutral axis and maximum at outer fibres of machine component, such as, shaft subjected to torque.

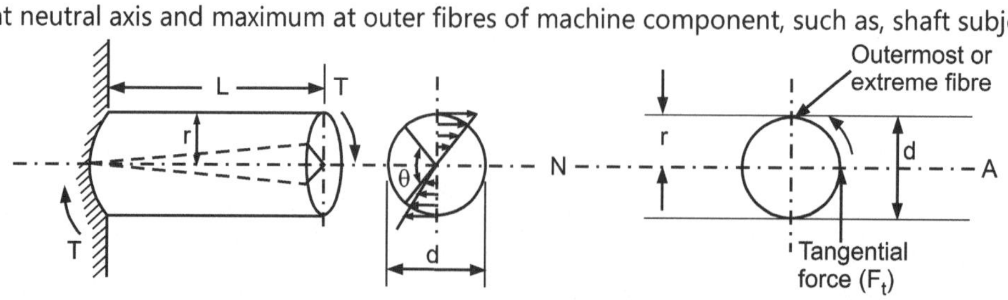


**Fig. 1.8 : Torsional shear stress**

- Consider a component of circular cross-section having diameter 'd'.
- Torsional shear stress is induced in the component, when torque 'T' is applied on component.

- Torsional shear stress is calculated by using basic torsion equation given below.

$$\frac{T}{J} = \frac{\tau}{r} = \frac{G\theta}{L}$$

where, $r$ = Distance of outermost fibre from neutral axis = $\frac{d}{2}$ in 'mm'

$J$ = Polar moment of inertia of circular cross-section

$\therefore$ $J = I_{XX} + I_{YY} = \frac{\pi}{64} d^4 + \frac{\pi}{64} d^4 = \frac{\pi}{32} d^4$ in 'mm$^4$'

$G$ = Modulus of rigidity of component material in 'N/mm$^2$'

$\theta$ = Angle of twist in 'radians'

$L$ = Length of component (shaft) in 'mm'

**(6) Crushing Stress :**

- Crushing stress is defined as, "*a localized compressive stress induced at the surface of contact between two components of a machine, which are relatively at rest*".
- Here, there is no relative motion between the two components.
- Consider a riveted joint subjected to load 'W' as shown in Fig. 1.9.
- Crushing stress induced is calculated as,

$$\sigma_{ck} = \frac{\text{Load}}{\text{Projected area of contact}} = \frac{W}{d \cdot t \cdot n}$$

where, $d$ = Diameter of rivet, $t$ = Thickness of plate

$n$ = Number of rivets

- Other examples of crushing stresses are cotter joints, knuckle joint etc.

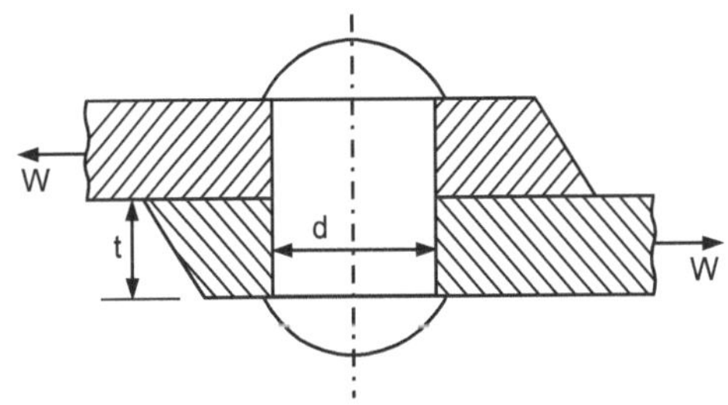


**Fig. 1.9 : Crushing stress**

**(7) Bearing Stress :**

- Bearing stress is defined as, "*a localized compressive stress at the surface of contact between two components of a machine, which have relative motion between them*'.
- Bearing stress is taken into account in design of machine elements like,
  (i) Screwed joint,
  (ii) Crank pin and large end bearings in I.C. engines, and
  (iii) Small end of connecting rod and piston pin etc.
- Let us consider a journal (or shaft) supported in a bearing as shown in Fig. 1.10, subjected to load 'W'.
  Bearing stress (also called as bearing pressure) is given by,
- Bearing stress (pressure), $P_b = \frac{\text{Load}}{\text{Projected area of contact}} = \frac{W}{l \times d}$
- Usually length $l$ is taken as 1.25 times the diameter. $\therefore$ $l = 1.25$ d

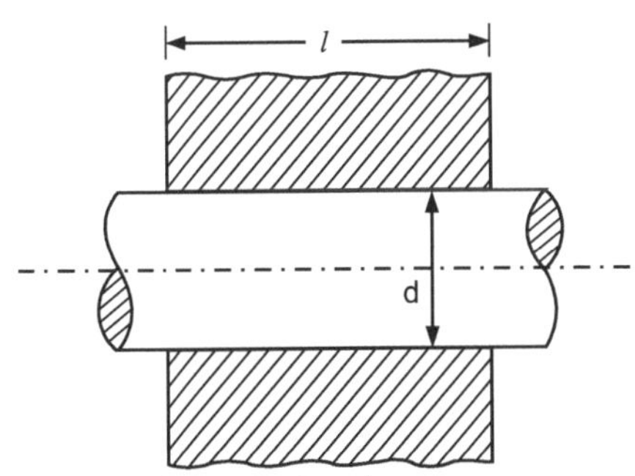


**Fig. 1.10 : Bearing stress**

## 1.7.1 Difference between Crushing and Bearing Stress

| Sr. No. | Crushing Stress | Bearing Stress |
|---|---|---|
| 1. | Two machine components are in contact with each other. | Two machine components are in contact with a small lubricating film. |
| 2. | There is no relative motion between two machine components. | There is relative motion between two machine components. |
| 3. | Examples:<br>(a) Knuckle pin inserted in a circular hole in knuckle joint.<br>(b) Cotter inserted in a rectangular slot in cotter joint. | Examples :<br>(a) Cross head and guide.<br>(b) Screwed joint.<br>(c) Crank pin in bearing.<br>(d) Fulcrum pin of lever. |

**Numerical Type No. 2 : "Tensile Stress and Compressive Stress"**

**Problem 1.2 :** *A load of 5 kN is to be raised by means of steel wire. Find minimum wire diameter required for safe stress of 100 MPa.*

**Solution : Given data : W** = 5 kN = $\mathbf{5 \times 10^3}$ **N**, $\sigma_t$ = 100 MPa = **100 N/mm²**

**Procedure :** Let d be diameter of wire.

We have, $$\sigma_t = \frac{W}{A}$$

$$\therefore \quad 100 = \frac{5 \times 10^3}{\frac{\pi}{4} \cdot d^2}$$

$$\therefore \quad d = 7.97 \text{ mm} \cong \mathbf{8 \text{ mm (say)}}$$

---

**Problem 1.3 :** *A press exerts a load of 2 MN. This load is carried by two rods of 2 metre length and supports the upper head of press. Find : (i) Diameter of rods, and (ii) Change in length of rods. Assume safe stress = 80 MPa, E = 210 kN/mm².*

**Solution : Given data :** $\mathbf{W_1}$ = 2 MN = $\mathbf{2 \times 10^6}$ **N**, **L** = 2 m = $\mathbf{2 \times 10^3}$ **mm**,

E = $210 \times 10^3$ N/mm², $\sigma_t$ = 80 MPa = **80 N/mm²**.

**Procedure :** As the load is carried by two rods, therefore, load taken by each rod is,

$$\mathbf{W} = \frac{W_1}{2} = \frac{2 \times 10^6}{2} = \mathbf{1 \times 10^6 \text{ N}}$$

**(i) Diameter of rods :**

We have, $$\sigma_t = \frac{W}{\frac{\pi}{4} d^2}$$

$$\therefore \quad 80 = \frac{1 \times 10^6 \times 4}{\pi \times d^2}$$

$$\therefore \quad \mathbf{d = 126.15 \text{ mm} \cong 128 \text{ mm (say)}}$$

**(ii) Change in length of rods :**

We have, $$\delta L = \frac{WL}{AE} = \frac{1 \times 10^6 \times 2 \times 10^3}{\frac{\pi}{4} \times d^2 \times 210 \times 10^3} = \frac{1 \times 10^6 \times 2 \times 10^3 \times 4}{\pi \times (128)^2 \times 210 \times 10^3} = \mathbf{0.74 \text{ mm}}$$

---

**Problem 1.4 :** *A cast iron link as shown in Fig. 1.11 is required to transmit a steady tensile load of 45 kN. Find tensile stress induced in linked material at section A-A and B-B. All dimensions given are in mm.*

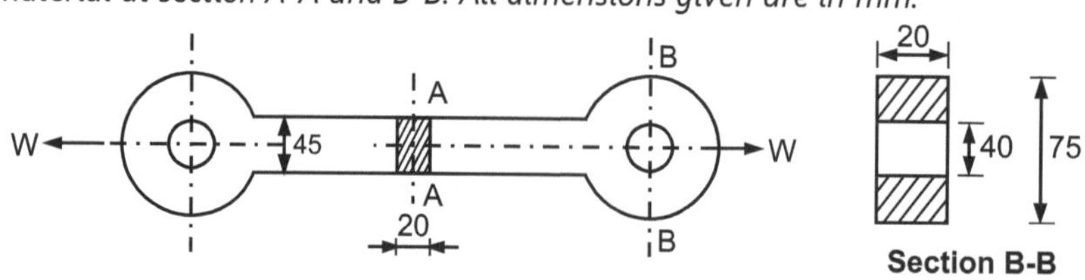


**Fig. 1.11**

**Solution : Given data :** **W** = 45 kN = $\mathbf{45 \times 10^3}$ **N**

**Procedure :** Tensile stress induced at section A-A,

$$\sigma_{t\,(A-A)} = \frac{W}{A} = \frac{45 \times 10^3}{45 \times 20} = \mathbf{50 \text{ N/mm}^2}$$

Tensile stress induced at section B-B, $\sigma_{t\,(B-B)} = \frac{W}{A} = \frac{45 \times 10^3}{(75 - 40) \times 20} = \mathbf{64.28 \text{ N/mm}^2}$

---

**Problem 1.5 :** *Connecting rod of a reciprocating steam engine is subjected to maximum load of 65 kN. Find diameter of connecting rod at thinnest part, if permissible tensile stress is 35 MPa.*

**Solution : Given data :** **W** = 65 kN = $\mathbf{65 \times 10^3}$ **N**, $\sigma_t$ = 35 MPa = **35 N/mm²**

**Procedure :** Let d be diameter of connecting rod.

We know, $\sigma_t = \frac{W}{A}$

$\therefore \quad 35 = \frac{65 \times 10^3}{\frac{\pi}{4} \cdot d^2}$

$\therefore \quad$ **d** $= 48.63$ mm $\cong$ **50 mm (say)**

**Problem 1.6 :** *For a reciprocating steam engine, diameter of piston is 300 mm and maximum steam pressure is 0.7 N/mm². If compressive stress is 40 MPa, find diameter of piston rod.*

**Solution : Given data :** D = 300 mm, p = 0.7 N/mm², $\sigma_c$ = 40 MPa

**Procedure :** We have, $\text{Pressure} = \frac{\text{Force (Load)}}{\text{Area}}$

$\therefore \quad p = \frac{W}{\frac{\pi}{4} D^2}$

$\therefore$ Load on piston, **W** $= \frac{\pi}{4} D^2 \times p = \frac{\pi}{4} (300)^2 \times 0.7 = \mathbf{49.48 \times 10^3\ N}$

Let, d = diameter of piston rod.

We have, $\sigma_c = \frac{W}{\frac{\pi}{4} \cdot d^2}$

$\therefore \quad 40 = \frac{49.48 \times 10^3}{\frac{\pi}{4} \cdot d^2}$

$\therefore$ **d** $= 39.68$ mm $\cong$ **40 mm (say)**

## Numerical Type No. 3 : "Transverse Shear Stress"

**Problem 1.7 :** *Determine the maximum thickness of plate, that can be punched in a hole of diameter of 20 mm, having ultimate shear stress of 0.40 kN/mm². Maximum load on punch is 0.25 MN.*

**Solution : Given data : W** = 0.25 MN = $\mathbf{0.25 \times 10^6\ N}$, d = 20 mm, $\tau_{ut}$ = 0.4 kN/mm² = **400 N/mm²**.

**Procedure :** Value of ultimate shear stress is given in the problem. Therefore, to calculate value of permissible shear stress, we should use factor of safety (F.O.S..).

**Let us assume that, F.O.S. = 5.**

Permissible shear stress $= \tau = \frac{\tau_{ut}}{\text{F.O.S.}} = \frac{400}{5} = \mathbf{80\ N/mm^2}$.

We know that, $\tau = \frac{W}{\pi \times d \times t}$

$\therefore$ **t** $= \frac{W}{\pi \times d \times \tau} = \frac{0.25 \times 10^6}{\pi \times 20 \times 80} = \mathbf{49.73\ mm \cong 50\ mm\ (say)}$

**Problem 1.8 :** *Calculate force required to punch a circular blank of 60 mm diameter in a plate of 5 mm thickness. Ultimate shear stress of plate is 350 N/mm².*

**Solution : Given data :** d = 60 mm, t = 5 mm, $\tau$ = 350 N/mm²

**Procedure :** We have, $\tau = \frac{W}{A} = \frac{W}{\pi \times d \times t}$

$\therefore \quad 350 = \frac{W}{\pi \times 60 \times 5}$

$\therefore \quad W = \mathbf{329.86 \times 10^3\ N}$

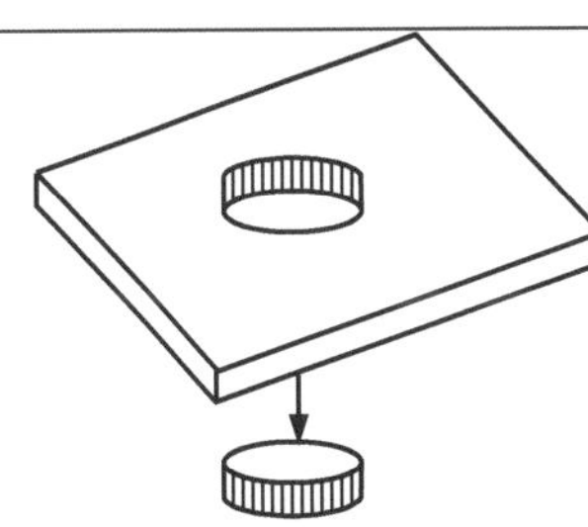

**Fig. 1.12**

**Problem 1.9 :** *Circular rods of diameter 50 mm are connected by knuckle joint with a pin of 40 mm diameter. If pull of 120 kN acts at each end, find tensile stress induced in rod and shear stress induced in the pin.*

**Solution : Given data :** D = 50 mm, d = 40 mm, **W** = 120 kN = $\mathbf{120 \times 10^3\ N}$

**Procedure :** Tensile stress induced in the rod is calculated as,

$$\sigma_t = \frac{W}{A} = \frac{W}{\frac{\pi}{4} \cdot D^2} = \frac{120 \times 10^3}{\frac{\pi}{4}(50)^2} = \mathbf{61.11\ N/mm^2}$$

Shear stress induced in pin is calculated as,

$$\tau = \frac{\text{Load}}{\text{Cross-sectional area of kunckle pin under shear}}$$

$\therefore \quad \tau = \frac{W}{2 \times A}$ [Since pin undergoes double shear, Area under shear is doubled]

$$\therefore \quad \tau = \frac{W}{2 \times \frac{\pi}{4} d^2} = \frac{120 \times 10^3}{2 \times \frac{\pi}{4} \times (40)^2} = \mathbf{47.74\ N/mm^2}$$

---

**Numerical Type No. 4 : "Crushing Stress"**

**Problem 1.10 :** *Two plates of 16 mm thickness are joined by a double riveted lap joint. Rivets are 25 mm in diameter. Find the crushing stress induced, if maximum tensile load on joint is 48 kN.*

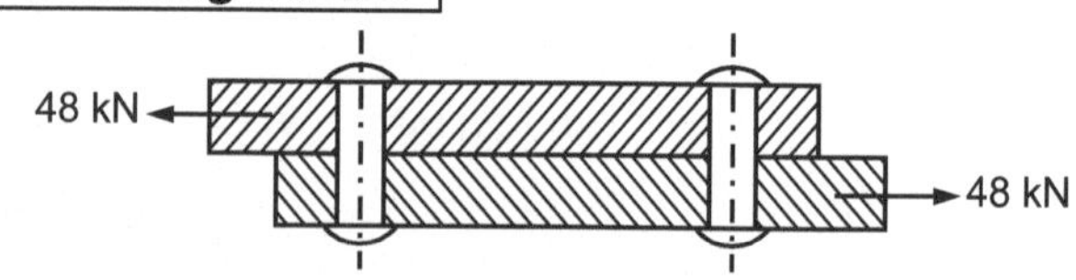


**Fig. 1.13**

**Solution : Given data :** n = 2, t = 16 mm, d = 25 mm, **W** = 48 kN = $\mathbf{48 \times 10^3\ N}$

**Procedure :** We have, $\sigma_{ck} = \frac{W}{\pi \times d \times t \times n} = \frac{48 \times 10^3}{\pi \times 25 \times 16 \times 2} = \mathbf{19.09\ N/mm^2}$

---

**Problem 1.11 :** *Find the maximum diameter of hole that can be punched in a M.S. plate 10 mm thick having ultimate shear stress of 220 N/mm². Allowable crushing stress in punch material is 350 N/mm².*

**Solution : Given data :** t = 10 mm, $\tau_{ut}$ = 220 N/mm², $\sigma_{ck}$ = 350 N/mm².

**Procedure :** Value of ultimate shear stress is given in the problem. Therefore, let us **assume value of factor of safety as 5** to calculate permissible shear stress.

Permissible shear stress = $\tau = \frac{\tau_{ut}}{\text{F.O.S.}} = \frac{220}{5} = \mathbf{44\ N/mm^2}$

We know that, $W = \text{Area under shear} \times \tau$

$W = (\pi \times d \times t) \times \tau = (\pi \times d \times 10) \times 44 = \mathbf{1382.30\ d}$ ... (1)

Also, $W = \text{Area under crushing} \times \sigma_{ck} = \left(\frac{\pi}{4} d^2\right) \times \sigma_{ck} = \frac{\pi}{4} d^2 \times 350 = \mathbf{274.88\ d^2}$ ... (2)

Equating equations (1) and (2), we get,

$1382.30\ d = 274.88\ d^2$

$\therefore \quad \mathbf{d = 5.02\ mm \cong 6\ mm\ (say)}$

---

**Numerical Type No. 5 : "Bearing Stress"**

**Problem 1.12 :** *A journal 25 mm in diameter and supported in sliding bearing has maximum end reaction of 2.5 kN. Assuming allowable pressure 5 N/mm², find length of sliding bearing.*

**Solution : Given data :** d = 25 mm, **W** = 2.5 kN = $\mathbf{2.5 \times 10^3\ N}$, $\sigma_{bearing}$ = 5 N/mm²

**Procedure :** We have, bearing pressure or bearing stress as,

$$P_b \text{ or } \sigma_{bearing} = \frac{W}{d \times l}$$

$$\therefore \quad 5 = \frac{2.5 \times 10^3}{25 \times l}$$

$$\therefore \quad \mathbf{l = 20\ mm}$$

## 1.8 PRINCIPAL STRESS

- When a component is subjected to several types of loads simultaneously, it is necessary to determine the state of stresses.

  **For example :** A transmission shaft subjected to bending moment as well as twisting moment (torque) at same time.

- **Principal plane** is defined as, *"a plane, on which, only normal stresses act and no shear stress acts"*.
- **Principal stress** is defined as, *"the magnitude of normal stress acting on principal plane"*.

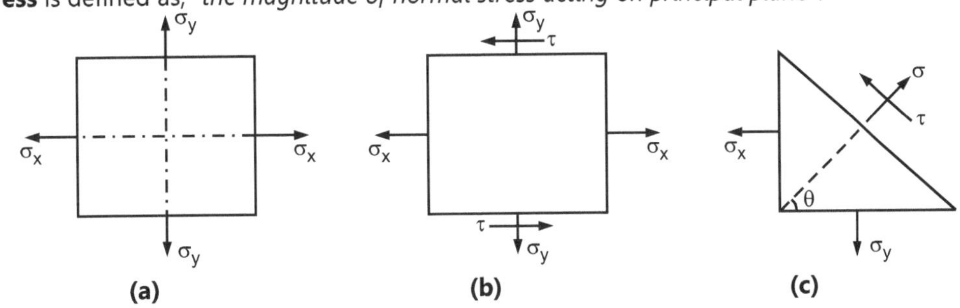


**Fig. 1.14 : Principal stresses**

- *"**Major principal stress** is the maximum value of normal stress acting on the principal plane, whereas, **Minor principal stress** is the minimum value of normal stress acting on principal plane"*.

  Maximum principal stress, $\sigma_{t_1} = \dfrac{\sigma_x + \sigma_y}{2} + \dfrac{1}{2}\sqrt{(\sigma_x - \sigma_y)^2 + 4\tau^2}$

  Minimum principal stress, $\sigma_{t_2} = \dfrac{\sigma_x + \sigma_y}{2} - \dfrac{1}{2}\sqrt{(\sigma_x - \sigma_y)^2 + 4\tau^2}$

- Planes of maximum shear stress are at right angles to each other and they are inclined at 45° to principal planes. Maximum shear stress is given by one half of the algebraic difference between the principal stresses (i.e. $\sigma_{t_1}$ and $\sigma_{t_2}$).

$$\therefore \quad \tau_{max} = \frac{\sigma_{t_1} - \sigma_{t_2}}{2} = \frac{1}{2}\left[\sqrt{(\sigma_x - \sigma_y)^2 + 4\tau^2}\right]$$

## 1.9 CREEP AND CREEP CURVE

- "When a component is subjected to constant stress at high temperature over a long period of time, it undergoes a slow and permanent deformation, which is called as ***creep***".
  - **Creep** is defined as, *"slow and progressive deformation of material with time under constant stress"*.
- Creep is a function of stress and temperature. Therefore, it becomes an essential and important factor to be considered in design procedure of components operating at elevated temperature.

  **For example :** Bolts and pipes in thermal power plant.

**Creep Curve :**

- Fig. 1.15 shows an idealized creep curve.
- When load is applied at the beginning of creep test, instantaneous elastic deformation OA occurs.
- This elastic deformation is followed by creep curve 'ABCD'.

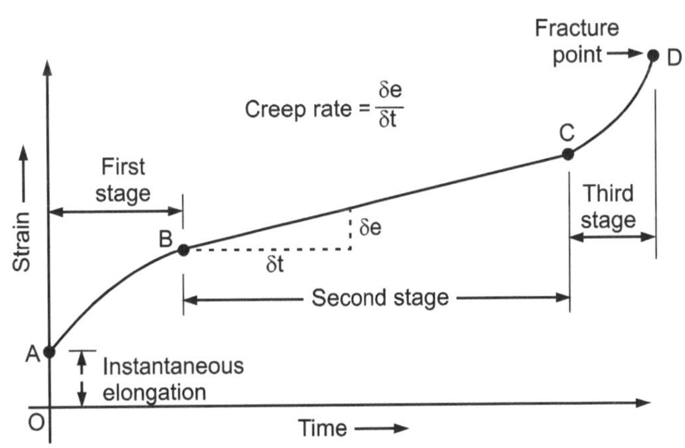


**Fig. 1.15 : Creep curve**

- Creep occurs in three stages.

1. **First Stage :**
   - It is called as **'primary creep'** shown by AB on the creep curve.
   - During this, creep rate decreases i.e. slope of creep curve increases from A to B progressively with time. Hence, the metal strain hardens to support the external load.
2. **Second Stage :**
   - It is called as **'secondary creep'** shown by BC on the creep curve.
   - Here, the creep rate is almost constant. This stage occupies a major portion of the life of component.
3. **Third Stage :**
   - It is called as **'tertiary creep'** shown by CD on the creep curve.
   - During this, creep rate is accelerated or gets increased due to necking (neck formation) and finally results into fracture at point D.

## 1.10 FATIGUE

- When a material is subjected to repeated stresses, it fails at stresses below yield point stresses. Such a type of failure is called as *fatigue failure.*
- This failure is caused due to progressive crack formations. Actually, these cracks are very very small in size., i.e. fine and microscopic size. Therefore, failure may occur even without any prior indication.
- Fatigue failure begins with a crack at some point in the component.
- Crack is more likely to occur in the following regions :
  - (i) Regions of discontinuity, such as, oil holes or keyways.
  - (ii) Regions of abrupt change in cross-section, such as, shoulder or steps.
  - (iii) Regions of irregularities during machining operations, such as, machining scratches, stamp mark or inspection marks.
  - (iv) Internal cracks in component material such as blow holes.
- These regions are subjected to stress concentration due to presence of crack.
- Fig. 1.16 shows a component subjected to alternate tensile and compressive stresses. This component has a crack. Refer Fig. 1.16 (a).
- During first half of stress cycle, crack opens due to tensile stress [Refer Fig. 1.16 (b)] and during second or next half of stress cycle, crack closes due to compressive stress [Refer Fig. 1.16 (c)]. Thus, one stress cycle gets completed with opening of crack followed by closing of crack.

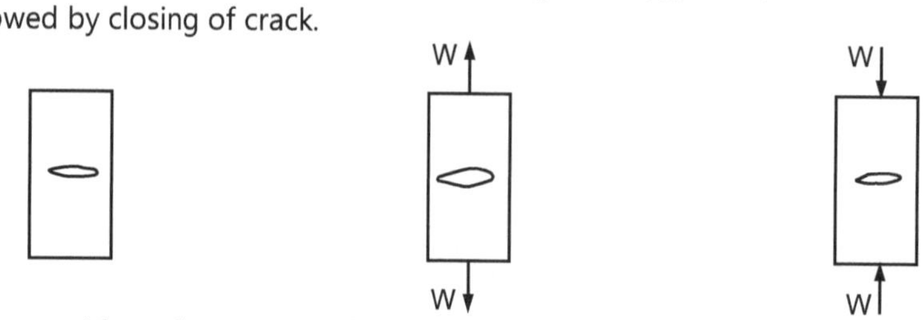


**(a) Component with crack** **(b) Opening of crack (Tensile half)** **(c) Closing of crack (Compressive half)**

**Fig. 1.16 : Crack propagation**

- Opening and closing of crack during each cycle continues. Therefore, crack propagates with increase in number of stress cycles and this phenomenon is termed as, **"Crack Propagation"**.
- Finally, cross-sectional area of component subjected to external load gets reduced to such a minimum value, at which, the remaining cross-sectional area of component cannot sustain against externally applied load and it results into sudden fracture.
- Fatigue failure depends upon number of factors, such as number of cycles, mean stress, stress amplitude, stress concentration, residual stresses, corrosion and creep.

## 1.11 NOTCH SENSITIVITY

- **Notch sensitivity** is defined as, *"the degree to which, theoretical expected effect of stress concentration is reached"*.
- It is given by,

$$\text{Notch sensitivity} = \frac{\text{Increase of actual stress over nominal stress}}{\text{Increase of theoretical stress over nominal stess}}$$

## 1.12 ENDURANCE LIMIT

- Consider a standard mirror polished specimen rotating in a fatigue-testing machine and loaded in bending as shown in Fig. 1.17 (a).
- When the specimen rotates, bending stresses at the upper fibers vary from maximum compressive to maximum tensile, whereas, bending stresses at the bottom fibers vary from maximum tensile to maximum compressive.
- In other words, specimen is subjected to completely reversed stresses. This is represented by time-stress graph shown in Fig. 1.17 (b).
- Number of cycles (N) required to produce a failure at given stress ($\sigma$) are noted and recorded. Then, these results are plotted in stress-cycle curve. Refer Fig. 1.17 (c).
- A little consideration will show that, if design stress is kept lower than a certain value as shown by dotted line, the material will not fail, whatever may be the number of cycles. This stress (shown by dotted line) is known as **endurance or fatigue limit.**

- **Endurance or fatigue limit** is defined as '*maximum value of completely reversed bending stress, which a standard specimen can withstand without failure, for infinite number of cycles of loads*'.

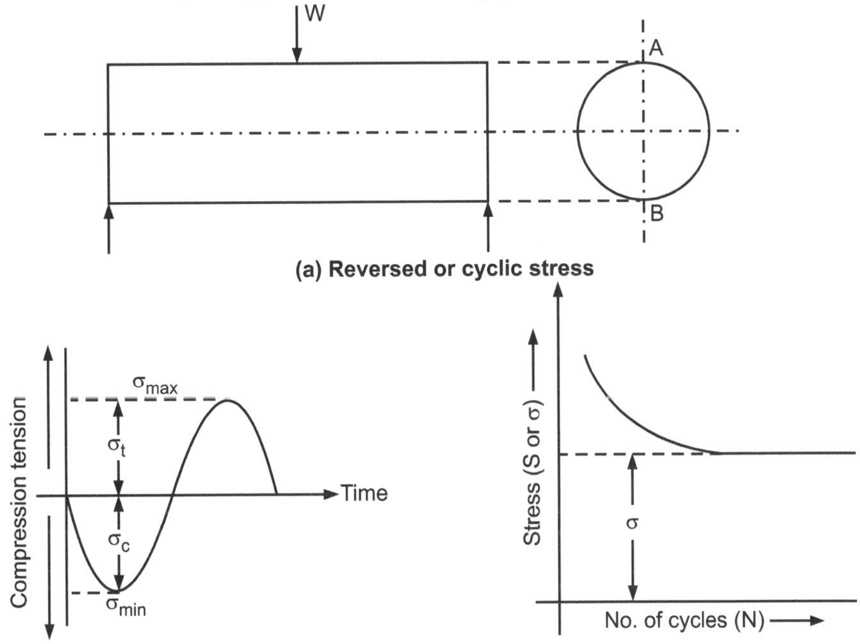


**(a) Reversed or cyclic stress**

**(b) Time-stress graph** **(c) Stress-cycle graph or S-N curve**

**Fig. 1.17 : Endurance limit**

## 1.13 DESIGN STRESS

- While designing machine parts, it is desirable to keep the stress lower than the maximum stress, at which, failure takes place. This stress is known as ***working stress*** or ***design stress***.
- Mathematically, Design stress = $\dfrac{\text{Maximum stress}}{\text{Factor of safety}}$

## 1.14 FACTOR OF SAFETY

- There are a number of parameters, which are difficult to evaluate accurately in design analysis, such as,
  - (a) Uncertainty in magnitude of external forces acting on the component.
  - (b) Variations in the properties of material like yield strength or ultimate strength.
  - (c) Variations in the dimensions of component due to imperfect workmanship.
- In order to ensure the safety against such circumstances, factor of safety is used.
- **Factor of safety** is defined as "*ratio of maximum stress to working stress*".

  Mathematically, Factor of safety (F.O.S.) = $\dfrac{\text{Maximum stress}}{\text{Working or design stress}}$
- **In case of ductile materials**, e.g. mild steel, where yield point is clearly defined, factor of safety is based on yield point stress.

  Factor of safety = $\dfrac{\text{Yield point stress}}{\text{Working or Design stress}} = \dfrac{\sigma_{yt}}{\sigma_t}$
- **In case of brittle materials**, e.g. cast iron, where yield point is not well defined, factor of safety is based on ultimate stress.

  Factor of safety = $\dfrac{\text{Ultimate stress}}{\text{Working or Design stress}} = \dfrac{\sigma_{ut}}{\sigma_t}$

### 1.14.1 Factors to be Considered While Selecting Factor of Safety

1. Reliability of properties of material and change in these properties during service.
2. Reliability of applied load.
3. No doubt about exact type of failure.
4. Extent of stress concentration.
5. Extent of initial stresses set up during manufacture.
6. Extent of loss of life, if failure occurs.
7. Extent of simplifying assumptions.
8. Reliability of test results carried out on actual machine parts.

### 1.14.2 In What Cases, Factor of Safety is Taken High ?

1. Analysis of loads with respect to magnitude and nature is difficult to estimate.
2. Impact loading.
3. Chances of accidental loading.
4. Non-uniformity in material.
5. High temperature applications.
6. Reliability of parts.
7. Materials likely to corrode (corrosion).
8. In case of possibilities of accidents, leading to damage property, risk towards lives of workers etc.

### 1.14.3 In What Cases, Factor of Safety May be Taken Low ?

1. No damage to property and lives of workers in case of design failure.
2. Less frequently used parts.
3. Analysis of loads with respect to magnitude and nature is easy to estimate.
4. Past knowledge of working conditions and strength of similar parts is known.
5. In case of failure, little time and low money is needed for rectification.

### 1.14.4 Why Factor of Safety is High in Design Subjected to Uncertain Environments ?

- There may be different environmental conditions at the time of use than those assumed at design stage.
- **For example :** A machine component is designed and manufactured with reference to standard atmospheric conditions. When this machine component is actually put to operate, then it may be subjected to high temperature surrounding and corrosive atmosphere.
- In such circumstances, factor of safety chosen for machine component should be increased considering the difference between assumed or standard environmental conditions and actual environmental conditions, otherwise, it will reduce the strength of component.

### 1.14.5 Factor of Safety According to Type of Load

- Following table shows range of factor of safety to be taken or assumed according to type or nature of load.

**Table 1.1**

| Type of load | Factor of safety |
|---|---|
| Static load | 1.5 to 3 |
| Variable load | 3 to 6 |
| Impact/shock load | 6 to 12 |

## 1.15 STRESS CONCENTRATION AND ITS CAUSES

- Whenever a machine component changes the shape of its cross-section, the simple stress distribution no longer holds good. This irregularity in the stress distribution caused by abrupt changes of form is called as *stress concentration*.
- **Stress concentration** is defined as, *"the irregularity in stress distribution caused by abrupt changes of form"*.
- **Stress concentration may occur due to,**
  1. Change in cross-section such as stepped axle, key-ways, grooves, threaded holes.
  2. Concentrated loads applied at minimum areas of machine parts, such as contact between gear teeth, a beam and its supports.
  3. Variation in mechanical properties of materials from point to point due to cavities, cracks or air pockets.
  4. Surface irregularities or poor surface finish.
- Stress concentration occurs for all kinds of stresses in presence of fillets, notches, holes, keyways etc.
- Consider a member with different cross-section subjected to tensile load as shown in Fig. 1.18. A little consideration will show that, the nominal stress is uniform in the right and left-hand sides, but in the region, where cross-section is changing, a redistribution of forces is essential to avoid stress concentration.

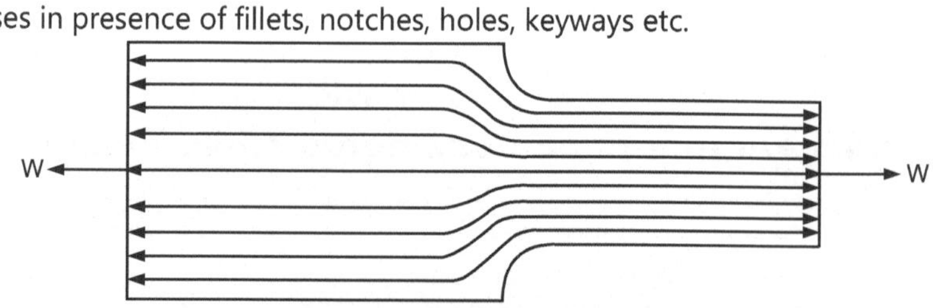

**Fig. 1.18 : Stress concentration**

## 1.16 METHODS OF REDUCING STRESS CONCENTRATION

- Presence of stress concentration cannot be totally eliminated, but it can be reduced.
- **To reduce stress concentration, various methods adopted are,**
  1. By providing fillets, undercutting and notches.
  2. By providing additional notches and holes.
  3. By reducing stress concentration in case of threaded members (For example : Screw and Nut).
- Fig. 1.19 (a) shows that, **stress lines tend to bunch up and cut very close to the edges or sharp corners.** For improvement, fillets may be provided as shown in Fig. 1.19 (b) and (c).

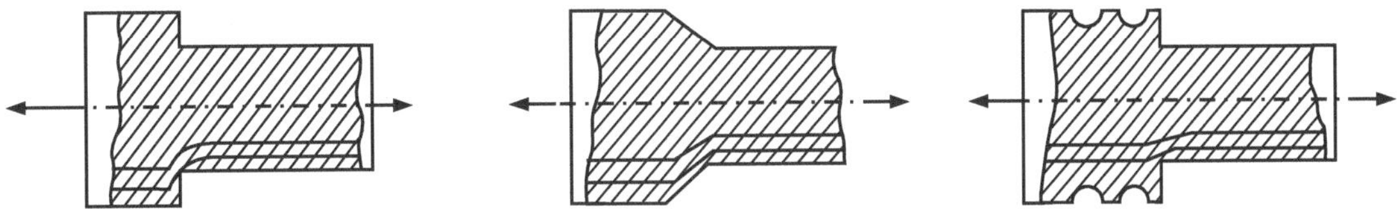

(a) Poor (b) Good (c) Preferred

Fig. 1.19

- Fig. 1.19 (d), (e) and (f) show methods of **reducing stress concentration in case of cylindrical members with shoulders.**

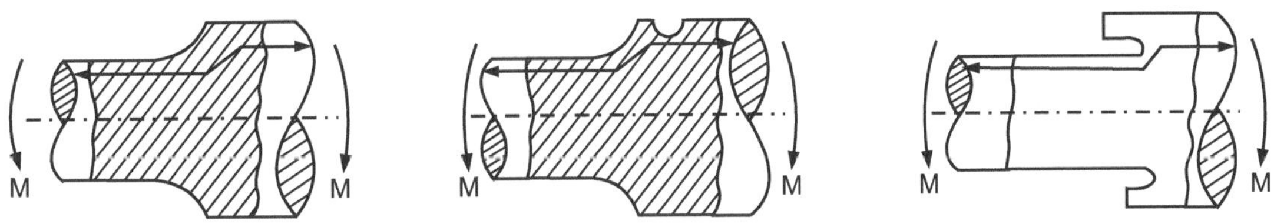


(d) Poor (e) Good (f) Preferred

Fig. 1.19

- Fig. 1.19 (g) and (h) show methods of **reducing stress concentration in case of cylindrical members with holes.**

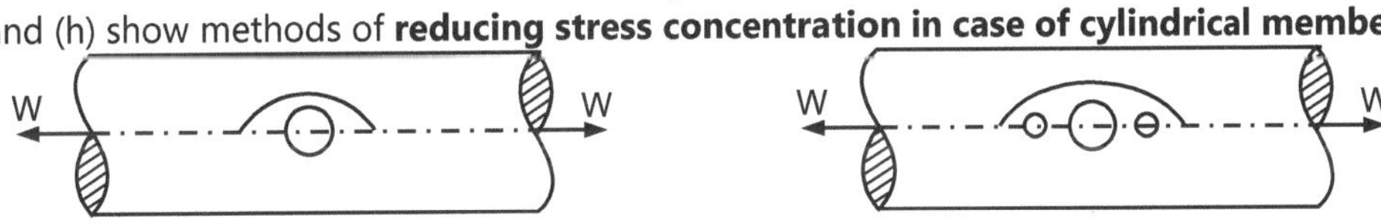


(g) Poor (h) Preferred

Fig. 1.19

- Fig. 1.19 (i), (j) and (k) show methods of **reducing stress concentration in case of cylindrical members with threads.**

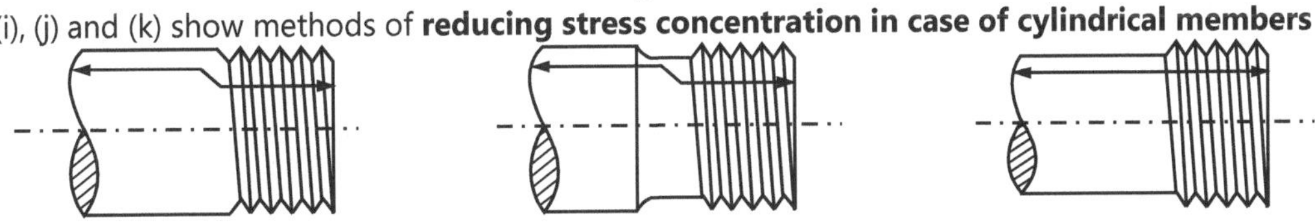

(i) Poor (j) Good (k) Preferred

Fig. 1.19

### 1.16.1 Theoretical Stress Concentration Factor

- **Theoretical stress concentration factor** is defined as *'the ratio of maximum stress in member (i.e., stress at a notch or fillet) to the nominal stress at the same section based upon net area".*
- Mathematically, Theoretical stress concentration factor, $K_t = \dfrac{\text{Maximum or Ultimate sterss}}{\text{Nominal stress}}$
- When a component made up of brittle material is subjected to static loading, theoretical stress concentration factor is used for safe design.
- Value of $K_t$ may vary due to,
  1. Type of loading i.e. dead load or live load.
  2. Type of force such as tensile, compressive or torque or combined loading.
  3. Material used.

### 1.16.2 Fatigue Stress Concentration Factor

- Stress concentration factors may vary according to type of material, type of loads (tensile, compressive, bending, combined of them), nature of loading (steady or cyclic) etc.
- When a material is subjected to cyclic or fatigue loading, the value of fatigue stress concentration factor is applied instead of theoretical stress concentration factor.
- For example, if a component made up of ductile material is subjected to variable loading (also called as cyclic loading or fatigue loading or fluctuating loads), then fatigue stress concentration factor is used, because endurance strength of component is greatly reduced.
- This is due to the reason that, in fatigue loading, the effect of stress concentration is usually less than that predicted by theoretical stress concentration factor. Therefore, value of fatigue stress concentration factor ($K_f$) is less than theoretical stress concentration factor ($K_t$).
- **Fatigue stress concentration factor** is defined as, *"the ratio of endurance limit without stress concentration to the nominal stress"*.
- Mathematically, Fatigue stress concentration factor, $K_f = \dfrac{\text{Endurance limit without stress concentration}}{\text{Nominal stress}}$

## 1.17 CONVERTING ACTUAL LOAD OR TORQUE INTO DESIGN LOAD OR TORQUE USING DESIGN FACTORS LIKE VELOCITY FACTOR, FACTOR OF SAFETY AND SERVICE FACTOR

- Design factor is not the factor of safety. It is used to take into account unknown contingencies and emergencies.
- Design engineer has to compensate for the unknown or unidentified conditions such as, unreliable material, the probable effects of unknown or accidental overloads, temperature effect, stresses encountered during machining, assembly, transportation, intentional overloading, chances of failure leading to injury, loss of life, costly shutdowns, expensive repairs etc.
- Therefore, for a safety design, design engineer uses design factor, which is also called margin of safety.

**Various design factors are,**

1. **Velocity factor :**

- This factor is considered in design of gears. It depends on velocity of operating gears. It is represented by '$C_v$'.
- Permissible working stress is obtained as follows :

$$\sigma_{wp} = \sigma_p \times C_v$$

Here,
$\sigma_{wp}$ = Permissible working stress
$\sigma_p$ = Permissible stress
$C_v$ = Velocity factor

2. **Factor of safety :**

- It is already discussed in article 1.14.

3. **Service factor :**

- Whenever power is transmitted from a prime mover (such as an electric motor) to a machine through one or more intermediate mechanical elements like: shaft, coupling, clutch, gears, chain, or belt etc., then there is a possibility of momentary shock on these intermediate elements either due to prime mover (driver) or machine to be driven. This leads to a momentary high load (or torque) acting on these elements. This momentary load (or torque) is higher than the average load acting on these elements.
- Consider an example of electric motor driving the hoisting drum through a reduction gear box.
- At the beginning of operation, the torque required is much higher than the average torque required. In a design process, the torque (or load) calculated from the formula of power $\left(P = \dfrac{2\pi NT}{60}\right)$ is an average torque (or load), whereas for a shorter duration, the mechanical elements are subjected to high momentary torque or load. This high momentary load is accounted in design process by a factor known as *service factor* or *application factor* or *overload factor.*
- Thus, the service factor or application factor is defined as, *"the ratio of maximum load (or maximum torque) to the average load (or average torque)".*
- If service factor is given in the problem, then find maximum torque as follows :

$$T_{max} = \text{Service factor} \times T_{average}$$

- All the above factors are considered for protecting machine parts in overloading condition and to increase their useful life.

## 1.18 PROPERTIES OF ENGINEERING MATERIAL

**Mechanical Property :**

- It is concerned with the ability of a material to resist various types of loads.
- Various mechanical properties are,

**(a) Strength :**

- Strength is defined as *"the ability of a material to resist, without rupture, under the action of external force causing various types of stresses"*.
- Strength is measured by different quantities. Depending upon the type of stresses induced due to externally applied loads, strength is expressed as tensile strength, compressive strength or shear strength.
- Tensile strength is the ability of a material to resist external load causing tensile stress, without failure.
- Compressive strength is the ability of a material to resist external load causing compressive stress, without failure.

**(b) Stiffness :**

- Stiffness can be defined as *"the property of material to resist deformation under stress"*. All materials deform to more or less extent, under stress.
- For a given stress within elastic limit, if deformation of a material is very very less, then we say, the material is stiffest.

**(c) Elasticity :**

- *"When a body regains its original shape on removal of external load, then the body is said to be elastic"* and its *material is said to possess elasticity.*
- All engineering materials are elastic, but degree of elasticity varies. Steel is perfectly elastic within elastic limit.

**(d) Plasticity :**

- *"When a body is permanently deformed under the action of externally applied load and does not regain its original shape on removal of external load, then the body is said to be plastic and its material is said to possess plasticity.*

**(e) Ductility :**

- *"The property of a material, which enables it to be drawn into wire under the action of tensile load, is called as ductility"*.
- Ductility is also defined as *"the ability of material to deform to a greater extent before the sign (signal or indication) of crack, when it is subjected to tensile force"*.
- **Examples :** Mild steel, copper, aluminium.
- Ductility is a desirable property in machine components, which are subjected to unexpected overloads or impact loads.

**(f) Brittleness :**

- Brittleness is defined as, *"the property of a material of breaking down under the action of externally applied load"*. It is opposite to that of ductility.
- A brittle material undergoes little plastic deformation prior to fracture, when it is subjected to tensile force (in tension test).
- Components of ductile material fail by yielding, whereas, components of brittle material fail by sudden fracture.
- **Example :** Cast iron.

**(g) Malleability :**

- *"The property of a material, which enables it to be rolled into thin sheets, is called as malleability"*.
- It may be considered as a special case of ductility.
- **Malleability** is also defined as *"the ability of a material to deform to a greater extent before the sign (signal or indication) of crack, when it is subjected to compressive force"*.
- The term 'Malleability' was originated from a word meaning 'hammer'. In a narrow sense, it means the ability of materials to be hammered out into thin sections.
- **Examples :** Low carbon steels, copper, aluminium.

**(h) Toughness :**

- *"The property, which resists the fracture under the action of impact loading, is called as toughness".*
- In other words, toughness is energy for failure by fracture. This property is essential for machine components, which are required to withstand impact loads.
- Tough materials have ability to bend, twist or stretch before failure takes place.
- **Example :** All structural steels.

**(i) Hardness :**

- **Hardness** is defined as *'the ability of a metal to cut another metal'.*
- It is usually expressed in numbers. i.e. Brinell Hardness Number (BHN) and Rockwell Hardness Number ($R_a$).
- Following tests are carried out to determine hardness value of any metal.
  1. Brinell hardness test.
  2. Rockwell hardness test.
- Hardness indicates resistance to abrasion, scratching, cutting or shaping.
- Hardness is an important property in selection of material for mating parts, which rub on one another such as pinion and gear, cam and follower, rail and wheel and parts of ball bearing. Wear resistance of these parts is improved by increasing surface hardness using the method of case hardening.

### 1.18.1 Difference between Malleability and Ductility

| Malleability | Ductility |
|---|---|
| 1. It is the ability to deform under compressive load. | 1. It is the ability to deform under tensile load. |
| 2. All malleable materials cannot be ductile. | 2. All ductile materials are malleable. |
| 3. It is desirable, when component is forged, rolled or extruded. | 3. It is desirable, when component is formed or drawn. |
| 4. It increases with increase in temperature. | 4. It decreases with increase in temperature. |

### 1.18.2 Difference between Ductility and Brittleness

| Ductility | Brittleness |
|---|---|
| 1. Ductile materials deform to a greater extent before fracture in tension test. | 1. Brittle materials show negligible plastic deformation prior to fracture. |
| 2. Failure of ductile materials takes place by yielding. | 2. Failure of brittle materials takes place by sudden fracture. |
| 3. More energy is absorbed by ductile specimen before fracture in a tension test. | 3. Energy absorbed is negligible. |
| 4. Examples : Steel, copper, aluminium. | 4. Example : Cast iron. |

## 1.19 FACTORS TO BE CONSIDERED FOR SELECTION OF MATERIAL IN DESIGNING MACHINE ELEMENT

Following considerations are to be made, while selecting the material :

1. **Availability :** Material should be easily available in the market.
2. **Cost :** The material should be available at cheaper rates.
3. **Physical properties :** Like colour, density, thermal conductivity, melting point, magnetic properties and specific heat.
4. **Mechanical properties :** Such as strength, ductility, malleability, toughness and hardness.
5. **Manufacturing ease :** From the manufacturing point of view, various abilities such as, cast ability, weldability, machinability, forgeability, surface properties are considered.
6. **Corrosion resistance :** Material should be corrosion resistant.
7. Properties of material should not change or deteriorate with time.

### 1.19.1 Materials Possessing Specific Mechanical Property

**Table 1.2**

| Mechanical property | Material |
|---|---|
| Elasticity | Steel, Rubber |
| Ductility | Mild steel, Aluminium, Copper, Tin |
| Brittleness | Cast iron |
| Malleability | Lead, Soft steel, Aluminium |

### 1.19.2 List of Machine Parts with Suggested Material

**Table 1.3**

| Sr. No. | Machine parts | Suitable material |
|---|---|---|
| 1. | **Crank shaft** | Alloy steel like 35Mn2Mo28 |
| 2. | **Helical spring** | Oil tempered carbon steel/ chromium vanadium alloy steel |
| 3. | **Bushes for knuckle pin** | Phosphor bronze, Grey C.I. |
| 4. | **Lathe bed** | Grey cast iron like FG150 |
| 5. | **Spring for I.C. engine valve** | Carbon steel |
| 6. | **Screw of screw jack** | Carbon steel, low alloy steel |
| 7. | **Nut of screw jack** | Phosphor bronze |
| 8. | **Bearing bushes** | Bronze (Alloy steel) |
| 9. | **Shaft** | Mild steel |
| 10. | **Body or casing of centrifugal pump** | Cast iron |
| 11. | **Balls for ball bearing** | Chrome steel |
| 12. | **Hydraulic cylinder** | Plain carbon steel, steel casting or aluminium alloys. |
| 13. | **Machine tool spindle** | High quality steel. |
| 14. | **Heavy duty gear** | Ferrous metal. |
| 15. | **Connecting rod** | Drop forged heat treated steel. |
| 16. | **Turbine blade** | Chromium steel like 7Cr13 |
| 17. | **Clutch spring** | Plain carbon steel with manganese like 60C4 |
| 18. | **Locomotive carriage or Wagon wheels** | Fe 290 or FeE230, plain carbon steel |

## 1.20 DESIGNATION OF DIFFERENT MATERIALS

1. **Ferrous metals :**

| | |
|---|---|
| Magnetite | $Fe_2O_3$ |
| Hematite | $Fe_3O_4$ |
| Limonite | $FeCO_3$ |
| Siderite | $Fe_2O_3(H_2O)$ |

2. **Cast iron :**

a) **Grey cast iron :**

| | |
|---|---|
| FG200 | Grey C.I. with tensile strength 200 MPa ($N/mm^2$). |
| FG300 | Grey C.I. with tensile strength 300 MPa ($N/mm^2$). |
| Examples | FG220, FG260, FG300, FG350, FG400. |

b) **Malleable cast iron :**

| | |
|---|---|
| WM400 | White malleable cast iron with minimum tensile strength 350 $N/mm^2$ or 400 $N/mm^2$. |
| BM300<br>BM320<br>BM350 | Black malleable C.I. with 300 $N/mm^2$ tensile strength.<br>Black malleable C.I. with 320 $N/mm^2$ tensile strength.<br>Black malleable C.I. with 350 $N/mm^2$ tensile strength. |
| PM450 | Pearlitic malleable C.I. with 450 $N/mm^2$ tensile strength. |

**c) Nodular or spheroidal C.I. :**

| | |
|---|---|
| SG450/10 | Spheroidal graphite cast iron with minimum tensile strength **450 N/mm$^2$** followed by **10% elongation**. |
| Examples | SG800/2, SG400/15, SG350/18. |

**3. Steels designated on the basis of mechanical properties :**

| | |
|---|---|
| Fe290 | Steel with minimum **tensile** strength of 290 N/mm$^2$. |
| FeE230 | Steel with minimum **yield** strength of 230 N/mm$^2$. |
| FeE220 | Steel with minimum **yield** strength of 220 N/mm$^2$. |
| Examples | Fe310, Fe330, Fe360, Fe410, Fe490, Fe540, Fe620, Fe690, FeE250, FeE270, FeE310, FeE370, FeE400 etc. |

**4. Steel designated on the basis of chemical composition :**

| | |
|---|---|
| 35C8 | Steel with 0.35% carbon and 0.8% manganese. |
| 40C8 | Steel with 0.4% carbon and 0.8% manganese. |
| 20C8 | Steel with 0.2% carbon and 0.8% manganese. |
| Examples | 30C8, 45C8, 50C4, 70C6, 75C6. |

**5. Free cutting steels :**

| | |
|---|---|
| 10C8S10 | Steel with 0.1% carbon, 0.8% manganese and 0.1% sulphur |
| Examples | 14C14S14, 28C12S14, 40C10S18, 11C10S25, 40C15S12 |

- Instead of sulphur, Pb i.e. lead can be used.

**6. Alloy steel :**

It is designated as –

- Figure indicating 100 times the percentage of carbon.
- Chemical symbol for alloying elements followed by figure for its average percentage content multiplied by a factor as given below.

| Alloying Element | Multiplying Factor |
|---|---|
| Cr, Co, Ni, Si, W, Mn | 4 |
| Al, Be, V, Pb, Cu, Ti, Zn and Mo | 10 |
| P, S and N | 100 |

**For example :**

| | |
|---|---|
| 40Cr4Mo2 | Alloy steel having 0.4% carbon, 1% chromium and 0.2% molybdenum. |
| 25Cr3Mo55 | Alloy steel having 0.25% carbon, 0.75% chromium and 5.5% molybdenum |
| 35Mn2Mo28 | Alloy steel having 0.35% carbon, 0.5% manganese and 2.8% molybdenum |
| 20Mn2 | Alloy steel having 0.20% carbon and 0.5% manganese. |
| 40Ni2Cr1Mo28 | Alloy steel having 0.4% carbon, 0.5% nickel, 0.25% chromium and 2.8% molybdenum. |
| 30Ni4Cr1 | Alloy steel having 0.30% carbon, 1% Nickel and 0.25% Chromium. |
| 30Ni16Cr5 | Alloy steel having 0.3% carbon, 4% nickel and 1.25% chromium. |

**7. High alloy steel :**

They are designated as –

(a) Letter X.

(b) Figure indicating 100 times the percentage of carbon content.

(c) Chemical symbol for alloying elements followed by a figure for its average % content rounded off to nearest integer.

(d) Chemical symbol to indicate specially added element to provide desired properties.

**Note :**

- If material is not designated by X, then see chromium %. If it is written Cr13, Cr14 (i.e. more than 9), then it is a type of high alloy steel (stainless steel or heat resisting steel).
- Remember that, **if chromium % is more than 9,** then it is a type of **high alloy steel.**

**For example :**

| | |
|---|---|
| 30Cr13 | Alloy steel having carbon 0.3% and chromium 13%. |
| 20Cr18Ni2 | Alloy steel having carbon 0.20%, chromium 18%, and nickel 2%. |
| 15Cr16Ni2 | Alloy steel having average carbon 0.15%, chromium 16%, nickel 2%. |
| 45Cr9Si4 | Alloy steel having carbon 0. 45%, chromium 9% and silicon 4%. |
| X20Cr18Ni2 | High alloy steel having carbon 0.20%, chromium 18%, Nickel 2%. |
| X10Cr18Ni9Mo4Si2 | High alloy steel having 0.1% carbon, 18% chromium, 9% nickel, 4% molybdenum and 2% silicon. |
| X15Cr25Ni12 | High alloy steel having 0.15% carbon, 25% chromium and 12% nickel. |

8) **High speed tool steel :**

They are designated as,

(a) Letter XT.

(b) Figure indicating 100 times the percentage of carbon.

(c) Chemical symbol indicating alloying elements in their average %.

(d) Chemical symbol to indicate specially added element to attain desired properties.

| | C% | Si% | Cr% | W% | V% | Co% | Mo% | Mn% |
|---|---|---|---|---|---|---|---|---|
| **XT72W18Cr4V1** | 0.72 | --- | 4 | 18 | 1 | --- | --- | --- |
| **XT72W18CoCr4V1** | 0.72 | --- | 4 | 18 | 1 | 4 | --- | --- |
| **XT125W10Co10CrMo4** | 1.25 | --- | 4 | 10 | --- | 10 | 4 | --- |
| **XT90W6CoMo5Cr4V2** | 0.90 | | 4 | 6 | 2 | 5 | 5 | --- |

## 1.21 STANDARDIZATION

- **Standardization** is defined as, "*the process of establishing standards so as to minimize the varieties in the characteristics*".
- Standard is a set of specifications, defined by a local, national or international regulatory body or authority, to which various characteristics of a component, a system, or a product should conform. The characteristics include, material's quality, dimensions, shapes, tolerances, surface finish, method of testing, method of use, method of packing and storing etc.
- **Purpose of standardization:** To establish the norms intended to achieve uniformity, specified quality, interchangeability, safety, and to put reasonable limit on the variety.
- In India, BIS (Bureau of Indian standards) is responsible for evolving all types of technical standards.

**Standards used in design are,**

1. Standards for material, their mechanical properties and chemical composition.
2. Standards for dimensions of commonly used machine elements.
3. Standards for fits, tolerances and surface finish of machine elements.
4. Standards for engineering drawing of components.

**Categories of Standards:** Based on the defining body or organization, the standards can be divided into three categories:

1) **Company Standards :**
- These standards are defined or set by a company or a group of companies.

2) **National Standards :**
- National standards of India are defined and set by an apex body, known as Bureau of Indian Standards (BIS).
- It is mandatory (compulsory) for every organisation to follow these standards throughout the country.

3) **International Standards:**
- These standards are defined and set by an international apex body and normally followed all over the world.
- These standards are prepared by :
    - International Standards Organization (ISO)
    - International Bureau of Weights and Measures (IBWM).

### 1.21.1 Advantages OR Benefits of Standardization

1. Standardization helps in manufacturing the components quickly and economically.
2. Standardization saves effort of design engineer to design and manufacture new machines, as standard components are readily designed by experts.
3. Standardization helps in manufacturing the components on mass production.

4. Interchangeability of components is possible.
5. Standardization ensures a minimum specified quality.
6. Effective utilization of available resources is ensured.
7. Easy and quick replacement of components is possible.
8. Standardization ensures safety, protection from hazards, accidents etc.

## 1.22 IMPORTANCE OF DESIGN DATA BOOK FOR A DESIGNER

- To design and develop a product, design engineer requires lot of information such as material specifications, physical and mechanical properties of materials, standards, different manufacturing processes, process specifications, empirical relations, tool data, details of standard machines etc.
- Standard dimensions of nuts, bolts, shafts, bearings, springs, along with factors of safety for specific work, stress concentration factors etc. are readily available in design data book.
- Use of design data book makes easy for the designer to collect the above data.
- Design data book saves the time of design engineers.
- In short, use of design data book makes design procedure as easy and quick.

## 1.23 THEORIES OF ELASTIC FAILURE

- Theories of elastic failure provide a relationship between the strength of machine component subjected to complex state of stresses with the mechanical properties obtained in a tension test.
- Mechanical properties include yield strength, ultimate strength and percentage elongation.
- With the help of these theories, the data obtained in tension test can be used to determine dimensions of any component, irrespective of nature of stresses induced in component due to several types of loads simultaneously.
- Various theories of elastic failure are enlisted below.
  (1) Maximum principal stress theory or Normal stress theory (Rankine's theory).
  (2) Maximum shear stress theory (Guest's theory).
  (3) Maximum distortion energy theory (Von Mises and Henry theory).
  (4) Maximum strain energy theory (Haigb's theory).

**(1) Maximum Principal Stress Theory or Normal Stress Theory or Rankine's THEORY:**

- According to this theory, *"failure of machine part occurs, when maximum principal stress in a biaxial stress system reaches limiting stress value of the material in a simple tension test. Therefore, factor of safety is taken into consideration"*.
- According to this theory, Maximum principal stress is given by,

$$\sigma_{t\,max} = \frac{\sigma_{yt}}{F.O.S.} \quad \text{... (1.1) (For ductile material)}$$

where, $\sigma_{yt}$ = Yield point tensile stress

F.O.S. = Factor of safety

Also,

$$\sigma_{t\,max} = \frac{\sigma_{ut}}{F.O.S.} \quad \text{... (1.2) (For brittle material)}$$

where, $\sigma_{ut}$ = Ultimate tensile stress

- Values of limiting stresses for ductile and brittle material are yield point tensile stress and ultimate tensile stress respectively.

**Applications :**

(i) This theory is preferred for brittle material, as it considers possibility of failures only either in tension or compression. It ignores the possibility of failure due to shearing.
(ii) For machine's spindle design.
(iii) For machine beds design.
(iv) For overhang crank design.

**(2) Maximum Shear Stress Theory or Guest's Theory :**

- According to this theory, *"failure of a machine part occurs, when the maximum shear stress in a biaxial stress system reaches to a value equal to shear stress at yield point in a simple tension test"*.

- According to this theory, Maximum shear stress is given by,

$$\tau_{max} = \frac{\tau_y}{F.O.S.} \quad \ldots (1.3)$$

where, $\tau_y$ = Yield point shear stress

F.O.S. = Factory of safety

But, *yield point shear stress is equal to half of yield point tensile stress.*

Mathematically, $$\boldsymbol{\tau_y = \frac{1}{2}\sigma_{yt}} \quad \ldots (1.4)$$

where, $\sigma_{yt}$ = Yield point tensile stress

Putting the formula of $\tau_y$ from equation (1.4) in equation (1.3), we get,

$$\tau_{max} = \frac{\frac{1}{2}\sigma_{yt}}{F.O.S.} = 0.5 \times \frac{\sigma_{yt}}{F.O.S.} = \mathbf{0.5 \times \sigma_{t\,max}} \quad \left[\text{From equation (1.1), } \sigma_{t\,max} = \frac{\sigma_{yt}}{F.O.S.}\right]$$

**Applications :**

(i) This theory is preferred for ductile materials.
(ii) Crank shaft design.
(iii) Propeller shaft design.
(iv) Design of 'C' frame.

**(3) Maximum Distortion Energy Theory :**

- According to this theory, "*failure of a machine part or member occurs, when the distortion strain energy or shear strain energy per unit volume in a biaxial stress system reaches the value of distortion energy at yield point per unit volume as obtained from a simple tension test*".
- According to this theory, considering the factor of safety into account, we have

$$\sigma_{t\,max}^2 + \sigma_{t\,min}^2 - 2 \cdot \sigma_{t\,max} \cdot \sigma_{t\,min} = \left(\frac{\sigma_{yt}}{F.O.S.}\right)^2$$

where, $\sigma_{t\,max}$ = Maximum principal stress

$\sigma_{t\,min}$ = Minimum principal stress

$\sigma_{yt}$ = Yield point tensile stress

- **Application :** This theory is mostly used for ductile materials only.

## Theory Questions for Practice

1. State important factors to be considered, while selecting material for any part.
2. Give general procedure in machine design.
3. State general considerations in machine design.
4. Define following stresses induced in machine member with suitable example : (i) Crushing stress (ii) Bearing stress (iii) Torsional shear stress (iv)Transverse shear stress.
5. Draw stress-strain diagram for ductile material and brittle material and show clearly 'proportional limit', 'elastic limit', 'yield point' and 'ultimate tensile strength'.
6. Explain the term "stress concentration".
7. Explain four remedies to reduce "stress concentration".
8. What do you mean by 'creep' ? Explain different stages of creep with graph.
9. Define any four desirable properties of material.
10. Suggest suitable material for the following machine parts :
    (i) Hydraulic cylinder (ii) Turbine blades (iii) Helical spring (iv) Bearing bushes.
11. Suggest suitable material for the designing of following machine parts :
    (i) Crank shaft (ii) Helical spring (iii) Bushes for knuckle pin (iv) Lathe bed (v) Spring for I.C. engine valve (vi) Screw for screw jack.
12. Define endurance limit and draw typical S-N curve for steel.
13. State the meaning of (i) 30Ni4Cr1 (ii) SG400/12
14. Define factor of safety for ductile material and brittle material.

15. State any four factors to be considered, while selecting factor of safety.
16. Name different theories of elastic failure and explain any one in brief.
17. State maximum principal stress theory and maximum shear stress theory with their uses.
18. What do you mean by product design ? Explain its importance.
19. Define 'ergonomics'. State areas covered under ergonomics.
20. List ergonomic considerations in design of display and controls.
21. Illustrate the role played by 'colour' in ergonomics and aesthetic design of products.
22. Discuss the aesthetic considerations in machine design regarding shape and colour.
23. Explain the term "aesthetic". Give its important features.

## Numerical Problems for Practice

1. Determine the smallest size of the punch that can be made to punch a 10 mm thick mild steel plate having ultimate shear stress as 0.3 kN/mm$^2$. The permissible crushing stress for the hardened punch is 1.3 kN/mm$^2$.

   (**Ans.** d = 2 mm)

2. Determine the smallest size of a hole that can be punched in 10 mm thick M.S. plate having an permissible shear stress of 39 N/mm$^2$, if permissible crushing stress for punch material is 24 N/mm$^2$. Sketch an arrangement.

   (**Ans.** d = 65 mm)

❑❑❑

*Chapter* **2**

# DESIGN OF JOINTS, LEVERS AND OFFSET LINKS

## 2.1 INTRODUCTION TO MACHINE PARTS SUBJECTED TO DIRECT LOAD

- **Direct load** produces same kind of effect in any part of machine or machine itself all the time.
- Direct load does not change its magnitude as well as direction.
- Direct load may be tensile, compressive or shear load. Therefore, while designing any part or component subjected to direct load, it is essential to know the nature of load and area of machine part sustaining that load.
- **Practical examples of machine parts subjected to direct load are:**
  (1) Bolts used for securing top cover to top end of engine cylinder.
  (2) Cotter joint to connect big end bearing to connecting rod of a steam engine.
  (3) Knuckle joint.
  (4) Screwed end of a piston.
  (5) Turnbuckle etc.

## 2.2 COTTER JOINT

### 2.2.1 Introduction

- A **cotter joint** is used to connect two co-axial rods or bars, which are subjected to axial tensile and/or compressive forces. It is a temporary joint.
- A **cotter** is a flat wedge shaped piece having rectangular cross-section.
- Width of cotter is tapered (either on one side or both sides) from top to bottom for easy adjustments, such as, insertion and withdrawal (removal) of cotter in a rectangular slot made in spigot as well socket.
- **Value of taper :** Taper in cotter varies from 1 in 48 to 1 in 24.
- **Materials used for cotter :** Mild steel and Wrought iron.

### 2.2.2 Applications of Cotter Joint

(1) Connection of piston rod to cross-head of a reciprocating steam engine.
(2) Valve rod and its stem.
(3) Strap end of connecting rod.
(4) Lewis foundation bolt and other cotter foundation bolts.
(5) Piston rod to the tail end (rod) in an air pump.

### 2.2.3 Types of Cotter Joint

There are three main types of cotter joint :
(1) Socket and Spigot cotter joint. (2) Sleeve and cotter joint. (3) Gib and cotter joint.

### 2.2.4 Advantages of Cotter Joint

(1) Quick to assemble and dismantle parts.
(2) Very high tightening force due to the wedge action developed, which prevents loosening of parts in service.
(3) Simple to design.
(4) Easy to manufacture.

### 2.2.5 Design of Socket and Spigot Cotter Joint

- Fig. 2.1 (a) shows a socket and spigot cotter joint used to connect two circular rods 'A' and 'B'.

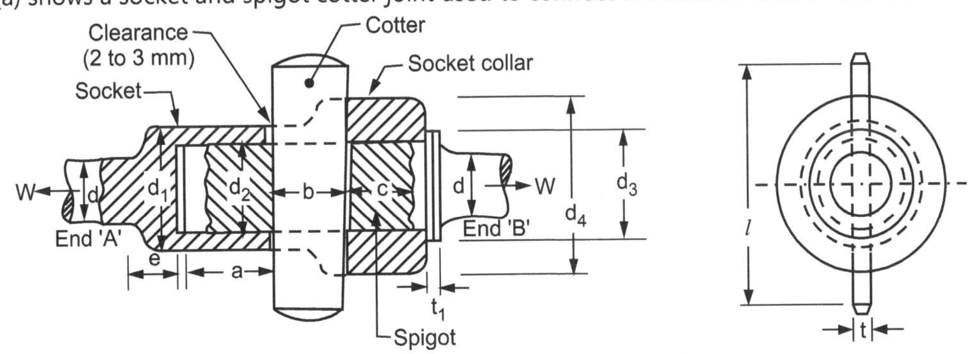


**Fig. 2.1 (a) : Socket and spigot cotter joint**

- In a socket and spigot cotter joint, left side rod (say A) is provided with a socket and right side rod (say B) is modified as spigot end.
- Spigot is inserted into the socket as shown in Fig. 2.1 (b).

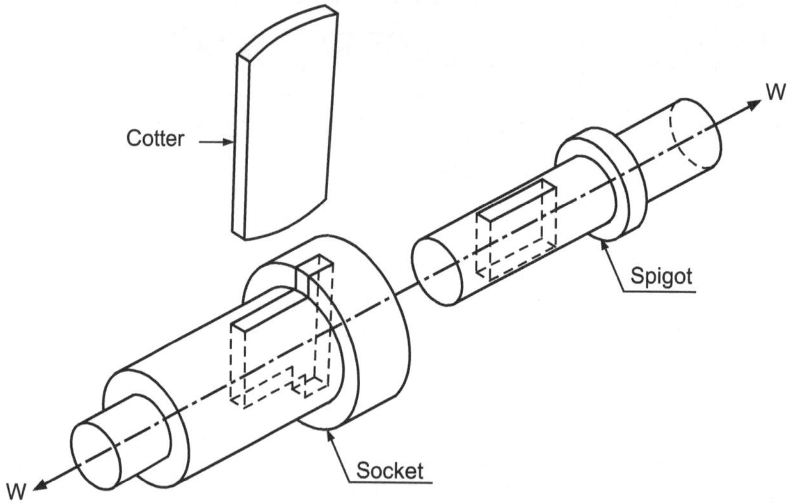


**Fig. 2.1 (b) : Isometric view of socket and spigot cotter joint**

- A rectangular slot is made in socket as well as in spigot.
- A cotter is then driven tightly through this rectangular slot to make a temporary connection between two rods.

Let,

| | | | |
|---|---|---|---|
| W | = | Load carried by each rod | in 'N' |
| d | = | Diameter of each rod | in 'mm' |
| $d_1$ | = | Outside diameter of socket | in 'mm' |
| $d_2$ | = | Diameter of spigot or Inside diameter of socket | in 'mm' |
| $d_3$ | = | Outside diameter of spigot collar | in 'mm' |
| $d_4$ | = | Diameter of socket collar | in 'mm' |
| $t_1$ | = | Thickness of spigot collar | in 'mm' |
| t | = | Thickness of cotter | in 'mm' |
| b | = | Width of cotter | in 'mm' |
| $l$ | = | Length of cotter | in 'mm' |
| a | = | Distance from end of slot to the end of spigot | in 'mm' |
| c | = | Thickness of socket collar beyond slot | in 'mm' |
| $\tau$ | = | Permissible shear stress for cotter material | in 'N/mm$^2$' |
| $\sigma_t$ | = | Permissible tensile stress for rod material | in 'N/mm$^2$' |
| $\sigma_{ck}$ | = | Permissible crushing stress for cotter material | in 'N/mm$^2$' |

**Step I : Failure of rod in tension :** The rod may fail in tension due to tensile load W.

We know that, Area resisting tearing $= \frac{\pi}{4} \cdot d^2$

$$\text{Considering tensile stress, } \sigma_t = \frac{\text{Load}}{\text{Area resisting tearing}} = \frac{W}{\frac{\pi}{4} \cdot d^2}$$

Here, diameter of rod (d) can be determined.

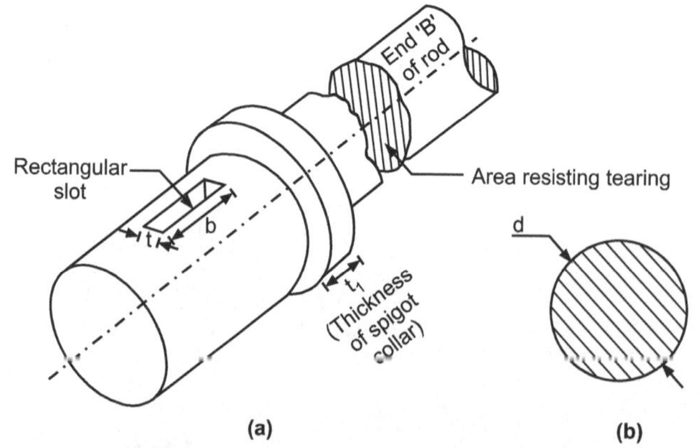


**Fig. 2.2 : Area of rod under tensile failure**

**Step II : Failure of spigot in tension across the slot (weakest section) :**

Area resisting tearing of spigot across the slot

$$= \left[\frac{\pi}{4} \cdot (d_2)^2\right] - [d_2 \cdot t]$$

Considering tensile stress,

$$\sigma_t = \frac{W}{\left[\frac{\pi}{4} \cdot (d_2)^2\right] - [d_2 \cdot t]}$$

Assuming $t = \left(\frac{d_2}{4}\right)$, diameter of spigot ($d_2$) and thickness of cotter (t) can be determined.

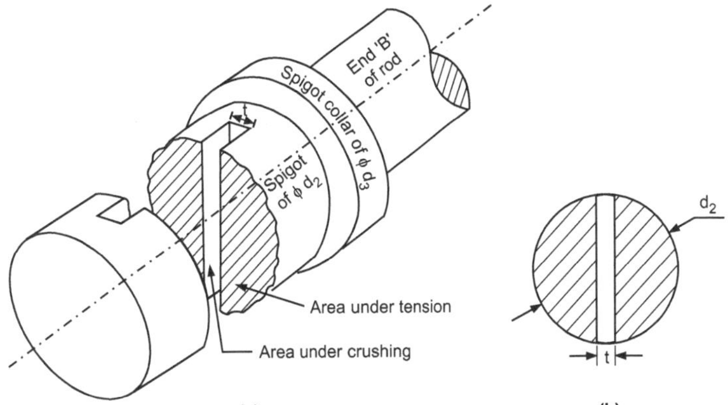


**Fig. 2.3 : Area of spigot under tensile failure**

**Step III : Checking the spigot for crushing stress :** Contact area between rod end 'B' (spigot end) at the slot and cotter inserted into slot shall be under crushing stress.

To check for safe crushing, the induced crushing stress should not exceed the permissible crushing stress.

Area resisting crushing of spigot or cotter = $d_2 \cdot t$

Induced crushing stress, $\sigma_{ck} = \frac{W}{d_2 \cdot t}$

If the induced crushing stress (calculated by using above equation) is less than permissible crushing stress, we say that, design is safe.

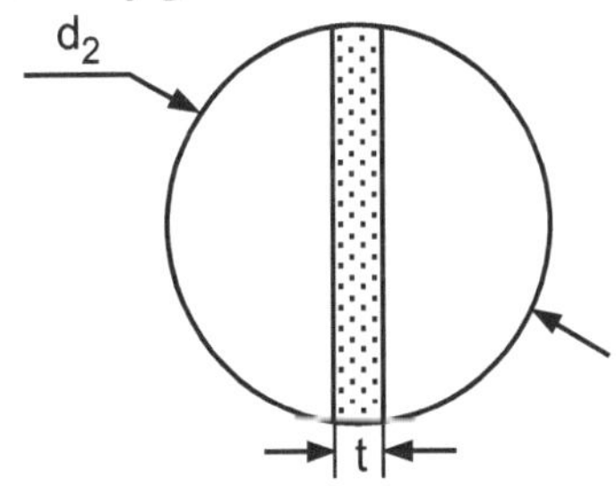


**Fig. 2.4 : Spigot area under crushing**

**Step IV : Failure of spigot end in shear (shear failure of spigot end beyond slot) :** Due to cotter, the spigot end may undergo a double shear.

$\therefore$ Area resisting shear of spigot end = $2 \cdot a \cdot d_2$ [Area ($a \cdot d_2$) is multiplied by 2, due to double shear of spigot end]

Considering transverse shear stress, $\tau = \frac{W}{2 \cdot a \cdot d_2}$

Here, distance 'a' can be determined.

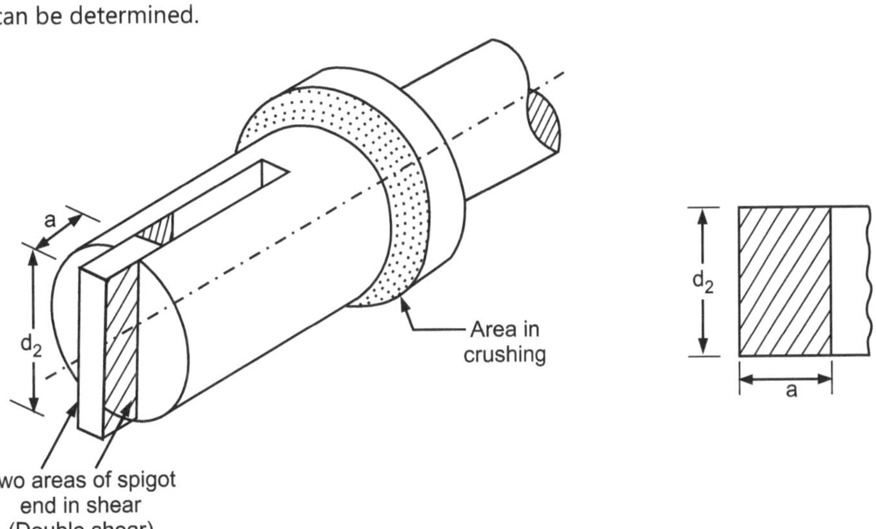


**(a)** **(b) Area of spigot end under double shear**

**Fig. 2.5**

**Step V: Failure of spigot collar in crushing (with socket collar) :**

Area resisting crushing of collar = $\frac{\pi}{4} \cdot [(d_3)^2 - (d_2)^2]$

Considering crushing stress, $\sigma_{ck} = \dfrac{W}{\frac{\pi}{4} \cdot [(d_3)^2 - (d_2)^2]}$

Here, $d_3$ = Outside diameter of spigot collar can be determined.

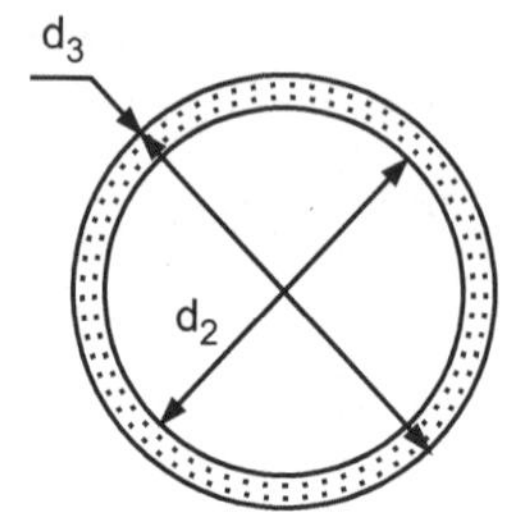


**Fig. 2.6 : Area of spigot collar under crushing**

**Step VI : Failure of spigot collar in shearing :**

Area resisting shearing = $\pi \cdot d_2 \times t_1$

Considering transverse shear stress, $\tau = \dfrac{W}{\pi \cdot d_2 \times t_1}$

Here, thickness of spigot collar '$t_1$' can be determined.

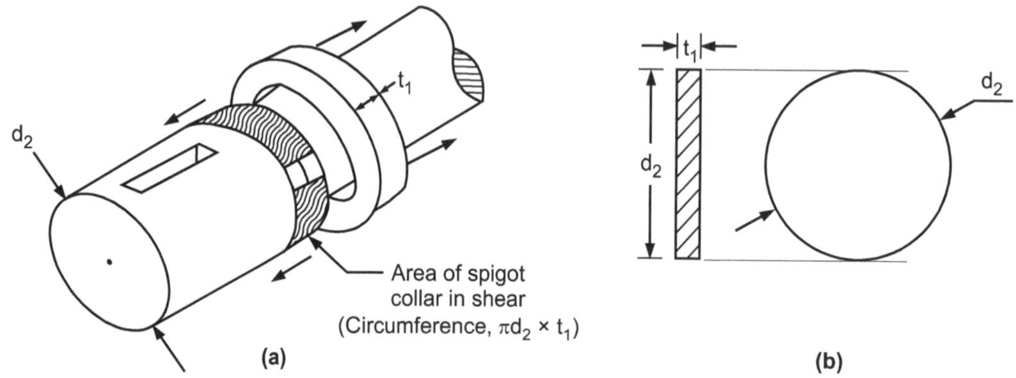


**Fig. 2.7 : Area of spigot collar under shear**

**Step VII : Failure of socket in tension across the slot :**

Area resisting tension across the slot = $\left[\left\{\frac{\pi}{4} \times (d_1^2 - d_2^2)\right\} - \{(d_1 - d_2) \times t\}\right]$

Considering tensile stress, $\sigma_t = \dfrac{W}{\left[\left\{\frac{\pi}{4} \times (d_1^2 - d_2^2)\right\} - \{(d_1 - d_2) \times t\}\right]}$

Hence, $d_1$ = outside diameter of socket can be determined.

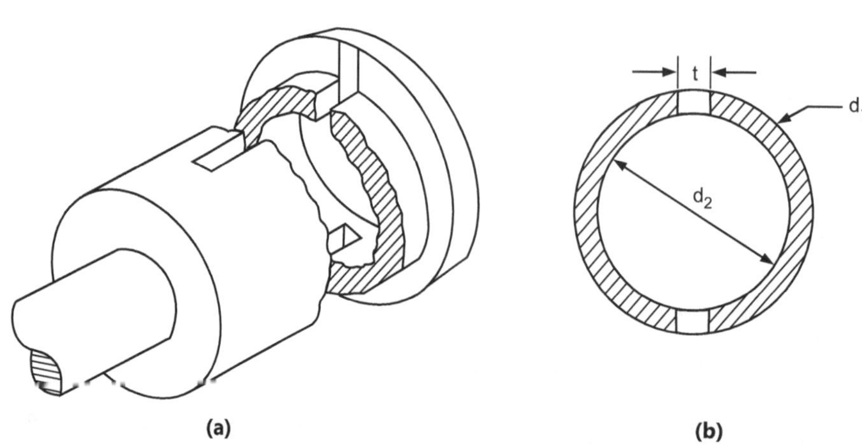


**Fig. 2.8 : Area of socket under tension**

**Step VIII : Failure of socket collar in crushing :**

Area resisting crushing $= (d_4 - d_2) \cdot t$

Considering crushing stress, $\sigma_{ck} = \dfrac{W}{(d_4 - d_2) \cdot t}$

Hence, $d_4$ = diameter of socket collar can be determined.

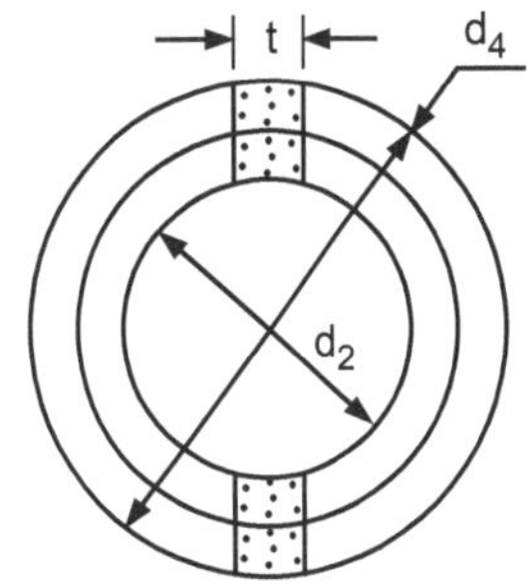


**Fig. 2.9 : Area of socket collar under crushing**

**Step IX : Failure of socket end in shearing :**

Area resisting shear $= 2 \cdot (d_4 - d_2) \cdot c$ [For double shear]

Considering shear stress, $\tau = \dfrac{W}{2 \cdot (d_4 - d_2) \cdot c}$

Hence, distance 'c' = thickness of socket collar beyond slot can be determined.

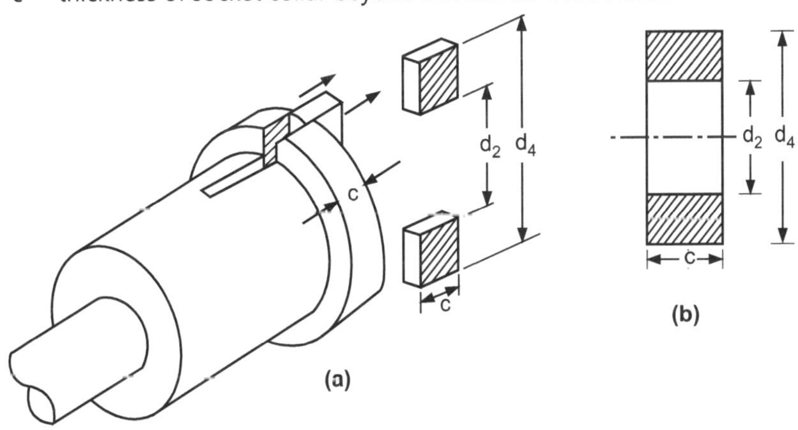


**Fig. 2.10 : Area of socket end in double shear**

**Step X : Failure of cotter in shear :**

Area of cotter resisting shear= $2 \times (b \times t)$

Cross-sectional area $(b \times t)$ is multiplied by 2, because cotter shears off at two places. Refer Fig. 2.11.

Considering shear stress, $\tau = \dfrac{W}{2 \times b \times t}$

Here, we can find the value of 'b' i.e. width of cotter.

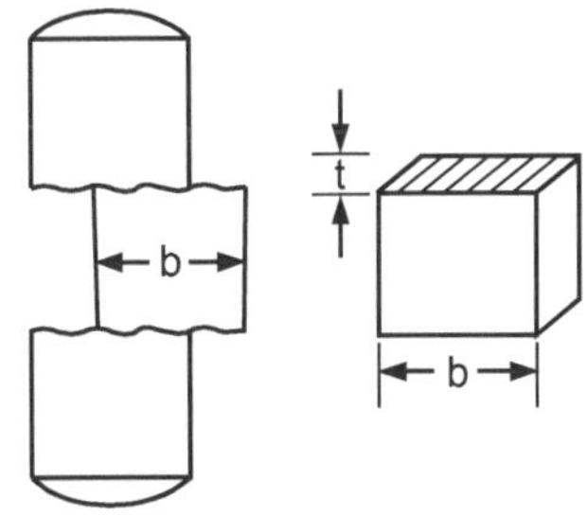


**Fig. 2.11 : Area of cotter in double shear**

**Step XI :** Take length of cotter $= 4 \times d$.

**Numerical Type No. 1 : "Socket and Spigot Cotter Joint"**

**Problem 2.1 :** *Design single cotter joint to transmit 200 kN. Allowable stresses for the material are 75 MPa in tension and 50 MPa in shear.*

**Solution : Given data :**

**W** = Load carried by rods = 200 kN = $\mathbf{200 \times 10^3}$ **N**, $\tau = 50$ N/mm$^2$, $\sigma_t = 75$ N/mm$^2$

**Procedure :** For design of single cotter joint, known as socket and spigot cotter joint, value of permissible crushing stress is required. Since, its value is not given in numerical problem, **let us assume that,** $\boldsymbol{\sigma_{ck} = 75}$ **N/mm$^2$**.

**Step I : Failure of rod in tension :** The rod may fail in tension due to tensile load W.

We know that, Area resisting tearing $= \frac{\pi}{4} \times (d)^2$

Considering tensile strength, $\sigma_t = \dfrac{W}{\frac{\pi}{4} d^2}$

$$\therefore \quad (d)^2 = \frac{4 \times W}{\pi \times \sigma_t} = \frac{4 \times 200 \times 10^3}{\pi \times 75} = 3395.3$$

$\therefore$ Diameter of rod = **d** = 58.269 mm ≅ **60 mm (say)**

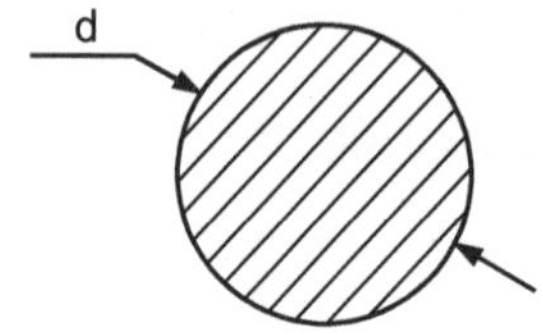


**Fig. 2.12**

**Step II : Failure of spigot in tension across the slot (weakest section) :**

Area resisting tearing of spigot across the slot $= \frac{\pi}{4} \times (d_2)^2 - d_2 \times t$.

Where, '$d_2$' is the diameter of spigot and 't' is thickness of cotter.

Considering tensile strength, $\sigma_t = \dfrac{W}{\left[\frac{\pi}{4} \times (d_2)^2\right] - [d_2 \times t]}$

Using empirical relation $t = \frac{d_2}{4}$, the above equation becomes,

$$75 = \frac{200 \times 10^3}{\left[\frac{\pi}{4} (d_2)^2\right] - \left[d_2 \cdot \frac{d_2}{4}\right]}$$

$$\therefore \quad (d_2)^2 = \frac{4 \times 200 \times 10^3}{(\pi - 1) \times 75} = 4980.716$$

$$\therefore \quad \mathbf{d_2} = 70.57 \text{ mm} \cong \mathbf{72 \text{ mm}} \text{ and}$$

$$\mathbf{t} = \frac{d_2}{4} = \frac{72}{4} = \mathbf{18 \text{ mm}}$$

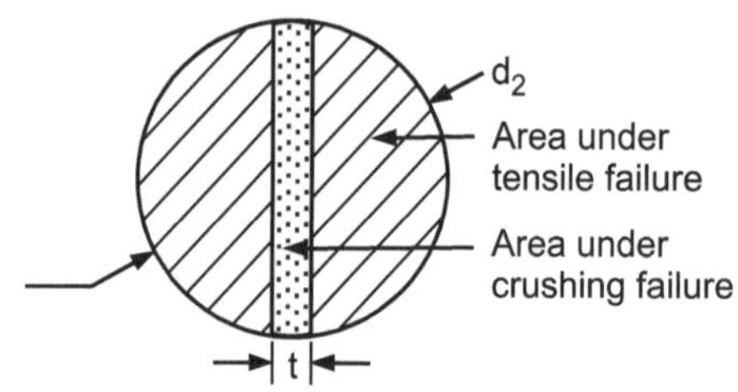


**Fig. 2.13**

**Step III : Checking the spigot for crushing stress :**

Area resisting crushing of rod or cotter $= d_2 \times t$

Induced crushing stress, $\sigma_{ck} = \dfrac{W}{d_2 \cdot t} = \dfrac{200 \times 10^3}{72 \times 18}$ **= 154.32 N/mm² > 75 N/mm² (Permissible stress)**

Since the induced crushing stress (154.32 N/mm²) is more than permissible crushing stress (75 N/mm²), design will not be safe, if values of $d_2$ and t are taken as 72 mm and 18 mm respectively. Therefore, values of $d_2$ and t should be increased to prevent crushing failure. For this, we will put the value of permissible crushing stress in design equation to calculate new dimensions of $d_2$ and t.

We have, $\sigma_{ck} = \dfrac{W}{d_2 \cdot t} = \dfrac{W}{d_2 \times \frac{d_2}{4}}$ $\left[\because t = \frac{d_2}{4}\right]$

$$\therefore \quad 75 = \frac{200 \times 10^3 \times 4}{d_2^2}$$

$$\therefore \quad \mathbf{d_2} = 103.279 \text{ mm} \cong \mathbf{104 \text{ mm (say)}}$$

and $\mathbf{t} = \dfrac{d_2}{4} = \dfrac{104}{4} = \mathbf{26 \text{ mm}}$

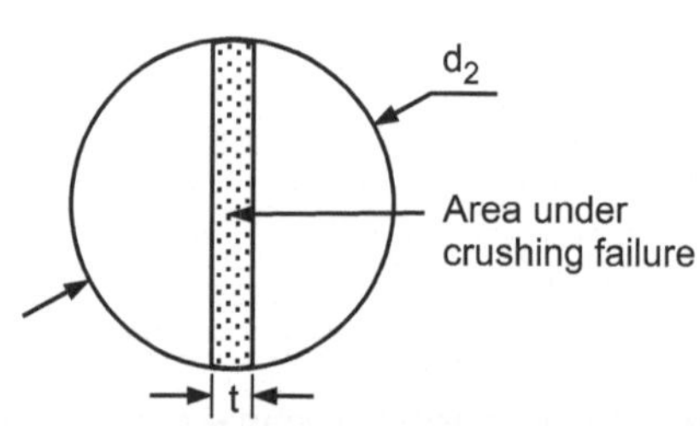


**Fig. 2.14**

**Step IV : Failure of spigot end in shear :** Due to cotter, the spigot end may undergo a double shear.

Area-resisting shear of spigot end = $2 \times a \times d_2$.

Considering shear stress, $\tau = \frac{W}{2 \times a \times d_2}$;

$$\therefore \quad 50 = \frac{200 \times 10^3}{2 \times a \times 104}$$

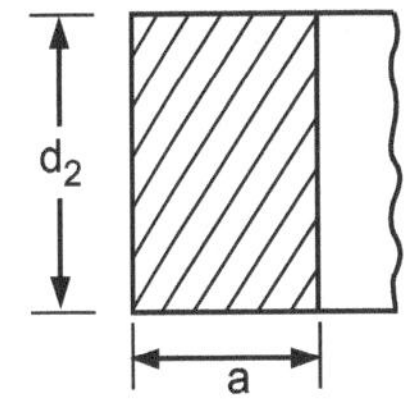


Fig. 2.15

Thus, **a** = distance from end of slot to the end of spigot = **19.23 mm ≅ 20 mm (say)**.

**Step V : Failure of spigot collar in crushing :**

Area resisting crushing of collar = $\frac{\pi}{4}\ [(d_3)^2 - (d_2)^2]$

Considering crushing stress, $\sigma_{ck} = \frac{W}{\frac{\pi}{4}[(d_3)^2 - (d_2)^2]}$

$$\therefore \quad 75 = \frac{4 \times 200 \times 10^3}{\pi \times (d_3^2 - 104^2)}$$

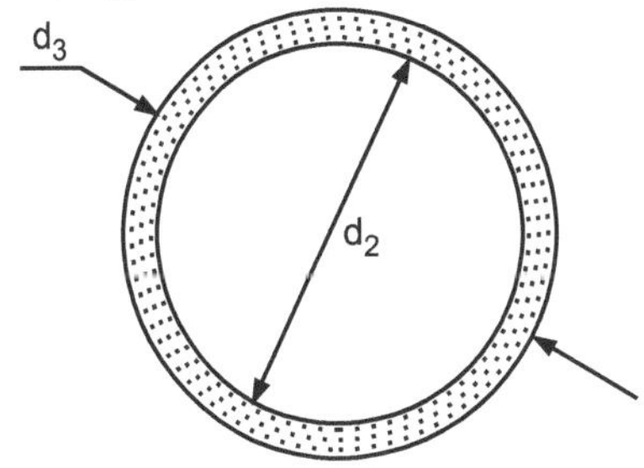


Fig. 2.16

$\therefore$ **$d_3$** = diameter of spigot collar = **119.21 mm ≅ 120 mm (say)**

**Step VI : Failure of spigot collar in shearing :**

Area resisting shearing of spigot collar = $\pi\ d_2 \times t_1$

Considering transverse shear stress,

$$\tau = \frac{W}{\pi\ d_2\ t_1}$$

$$\therefore \quad 50 = \frac{200 \times 10^3}{\pi \times 104 \times t_1}$$

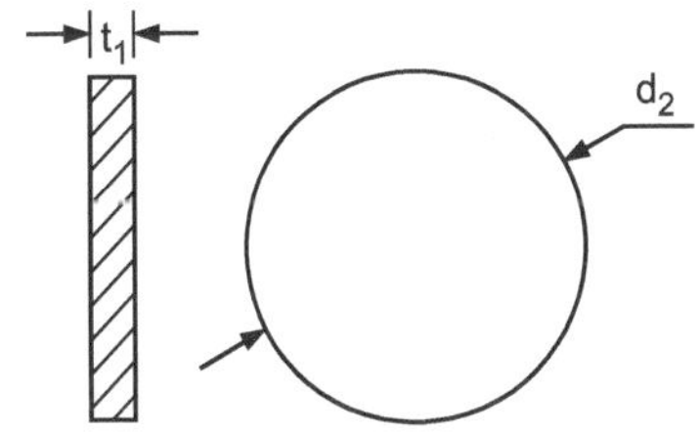


Fig. 2.17

$\therefore$ **$t_1$** = thickness of spigot collar = **12.24 mm ≅ 14 mm (say)**

**Step VII : Failure of socket in tension across the slot :**

Area resisting tension across slot (weakest section)

$$= \left[\left\{\frac{\pi}{4} \times (d_1^2 - d_2^2)\right\} - \{(d_1 - d_2) \times t\}\right]$$

Considering transverse tensile stress,

$$\sigma_t = \frac{W}{\left[\left\{\frac{\pi}{4} \times (d_1^2 - d_2^2)\right\} - \{(d_1 - d_2) \times t\}\right]}$$

$$\therefore \quad 75 = \frac{200 \times 10^3}{[\{0.7853 \times [d_1^2 - 104^2]\} - \{(d_1 - 104) \times 26\}]}$$

$$\therefore \quad 0.7853\ d_1^2 - 26 d_1 - 10778.7 = 0$$

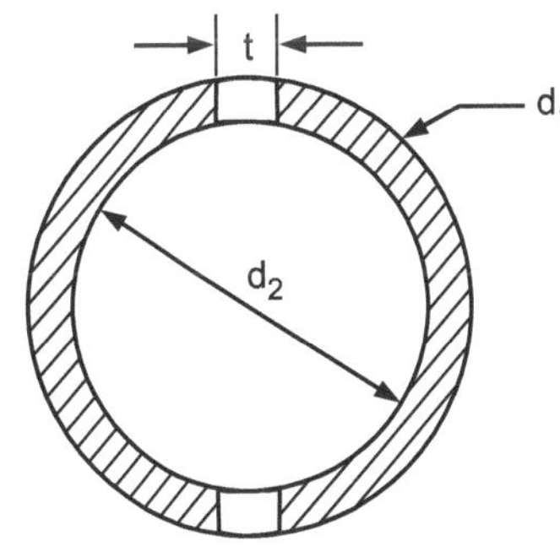


Fig. 2.18

$\therefore$ **$d_1$ = 134.87 mm ≅ 136 mm (say)**

Thus, $d_1$ = outside diameter of socket = **136 mm**

**Step VIII : Failure of socket collar in crushing :**

Area resisting crushing = $(d_4 - d_2) \times t$

Considering crushing stress,

$$\sigma_{ck} = \frac{W}{(d_4 - d_2) \times t}$$

$$\therefore \quad 75 = \frac{200 \times 10^3}{(d_4 - 104) \times 26}$$

$\therefore$ **$d_4$ = 206.56 mm ≅ 208 mm (say)**

Outside diameter of socket collar = $d_4$ = **208 mm**

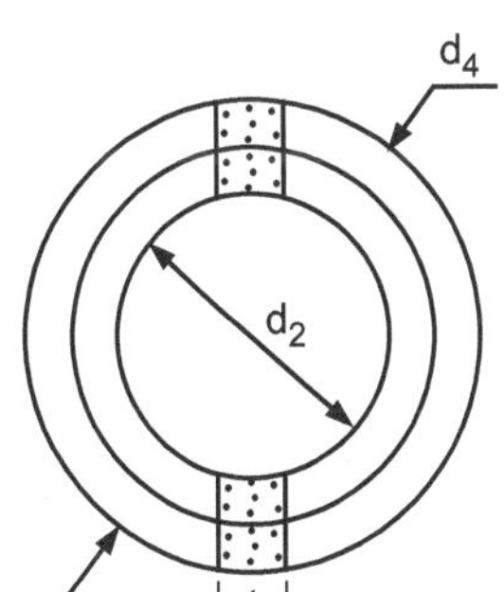


Fig. 2.19

**Step IX : Failure of socket end in shearing :**

Area resisting double shear = $2 \times [(d_4 - d_2) \times c]$

Considering shear failure, $\tau = \dfrac{W}{2 \times [(d_4 - d_2) \times c]}$ [For double shear]

$\therefore \quad 50 = \dfrac{200 \times 10^3}{2 \times (208 - 104) \times c}$

Hence, **c** = thickness of socket collar = **19.23 mm ≅ 20 mm (say)**

**Step X : Failure of cotter in shear :** Cotter is in double shear as shown in Fig. 2.20.

$\therefore$ Area resisting shear = $2 \times (b \times t)$

Considering shear stress, $\tau = \dfrac{W}{2 \times b \times t}$

$\therefore \quad 50 = \dfrac{200 \times 10^3}{2 \times b \times 26}$

$\therefore$ b = width of cotter = **76.32 mm ≅ 78 mm (say)**

**Step XI : Take length of cotter = 4 × d = 4 × 60 = 240 mm.**

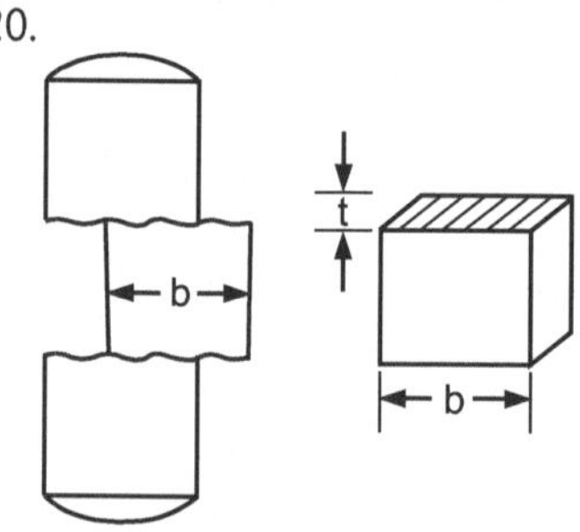


**Fig. 2.20**

## 2.2.6 Failure of Cotter in Bending

- When the cotter is tight in socket and spigot, it is subjected to shear stress.
- But, when it becomes loose, it is subjected to bending. Forces acting on cotter are shown in Fig. 2.21 (a).
- Load 'W' between cotter and spigot is assumed to be **uniformly distributed** over length $d_2$.
- Whereas, the load between socket end and cotter is assumed to be varying linearly from zero to maximum with **triangular distribution**.
- The cotter is treated as a beam as shown in Fig. 2.21 (b).

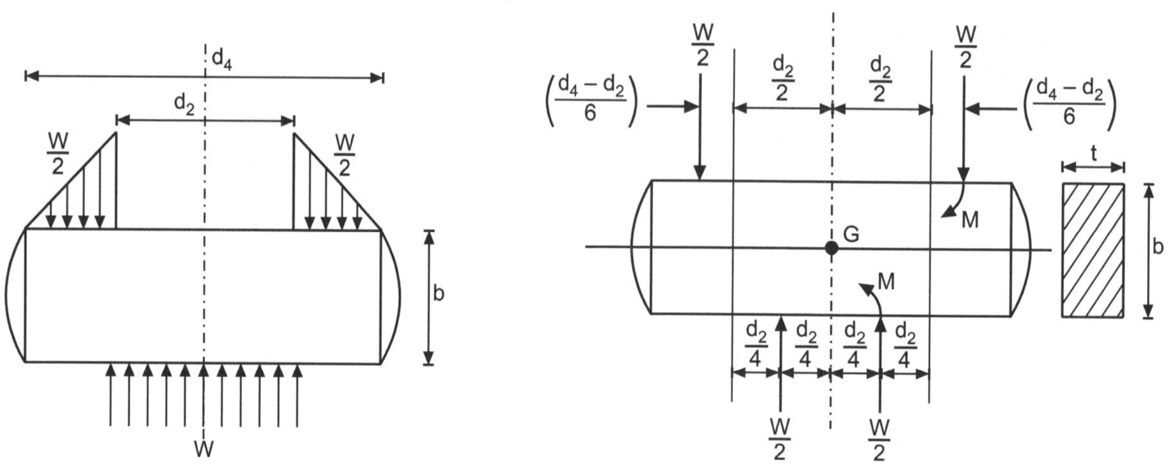


**(a) Actual distribution of forces** **(b) Cotter treated as a beam**

**Fig. 2.21 : Failure of cotter in bending**

Maximum bending moment at centre of cotter 'G' is given by,

$$M = \frac{W}{2} \times \left[\frac{d_4 - d_2}{6} + \frac{d_2}{2}\right] - \frac{W}{2} \times \left[\frac{d_2}{4}\right]$$

$$\therefore \quad M = \frac{W}{2} \times \left[\frac{(d_4 - d_2)}{6} + \frac{d_2}{2} - \frac{d_2}{4}\right]$$

$$\therefore \quad M = \frac{W}{2} \times \left[\frac{(d_4 - d_2)}{6} + \frac{d_2}{4}\right]$$

On re-arranging the terms, we can write,

$$\therefore \quad \mathbf{M} = \frac{\mathbf{W}}{\mathbf{2}}\left[\frac{\mathbf{d_2}}{\mathbf{4}} + \frac{\mathbf{(d_4 - d_2)}}{\mathbf{6}}\right] \qquad \text{... (2.1)}$$

We know that,

$$Z = \frac{I}{y} = \frac{\frac{1}{12}tb^3}{\frac{b}{2}} = \frac{\mathbf{1}}{\mathbf{6}}\mathbf{tb^2} \qquad \text{... (2.2)}$$

Bending stress is given by,

$$\sigma_b = \frac{M}{Z}$$

$$\therefore \quad \sigma_b = \frac{\frac{W}{2}\left[\frac{d_2}{4} + \frac{(d_4 - d_2)}{6}\right]}{\frac{1}{6}tb^2} \qquad \text{[From equations (2.1) and (2.2)]}$$

Assuming suitable proportion like b = 3t or 4t, dimensions of cotter can be calculated.

#### 2.2.6.1 Reasons for Taper Provided on Cotter

- **Cotter** is a flat wedge shape piece of rectangular cross-section and its width is tapered (either on one side or both sides) from one end to another.
- **Reasons for providing taper:**
  1. Taper helps in easy removal of cotter.
  2. Due to taper, cotter remains in its desirable position i.e. after inserting cotter in to rectangular slots of spigot and socket, it will not drop down.
  3. Taper provides maximum friction area.
- Taper provided is usually small, so that, self locking can be achieved.
- Generally, **value of taper** provided lies in the range of **1 in 48** to **1 in 24**.
- **Higher taper** than this range can increase the **risk of slipping back**.
- If taper is more and joint is subjected to variable loading, then, some locking arrangement should be provided with cotter so as to keep it in position.
- Usually taper is provided only on one side, because, machining taper on both sides is rather difficult and expensive.

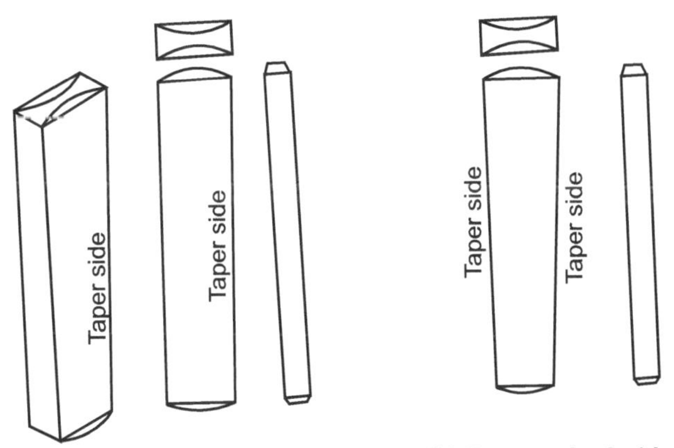


(a) Taper on one side (b) Taper on both sides

**Fig. 2.22 : Cotter showing taper**

#### 2.2.6.2 Purpose of Providing Clearance in Cotter Joint

- A clearance of 2 to 3 mm is provided between the slots and cotter.
- When cotter is driven in slots, the two rods are drawn together, until the spigot rests on the socket collar.

## 2.3 KNUCKLE JOINT

### 2.3.1 Introduction

- A knuckle joint is used to connect two circular rods, which are subjected to tensile loads. A knuckle joint can be easily disconnected for adjustments or repairs.
- Knuckle joint is also called as **'forked pin joint'**.
- In knuckle joint, one end of one rod is made into shape of an eye.
- Similarly, one end of other rod is formed into shape of a fork with an eye in each of the two fork legs.
- The knuckle pin passes through both, (i) eye hole and, (ii) fork holes.
- Knuckle pin can be secured by means of collar and taper pin or split pin.
- **Knuckle joint allows relative angular movement** of rods in the plane about axis of pin.

## 2.3.2 Applications of Knuckle Joint

1. Used to connect links and structural members.
2. Air brake arrangement on locomotive.
3. Connections of valve rods and eccentric rods.
4. Reversing gear mechanism in case of steam locomotives.
5. Remote controls of steam valves etc.

## 2.3.3 Advantages of Knuckle Joint

1. Simple to design.
2. Easy to manufacture.
3. Due to less number of parts, cost of knuckle joint is less.
4. Highly reliable.
5. Quick assembly as well as quick dismantling, whenever required.

## 2.3.4 Design Procedure of Kunckle Joint

- Knuckle joint is used to connect two circular rods, which are under the action of tensile load.
- Rods may have their ends either 'coincide' or 'intersect' and lie in one plane.
- Knuckle joint allows relative angular movement of rods in the plane about axis of pin.

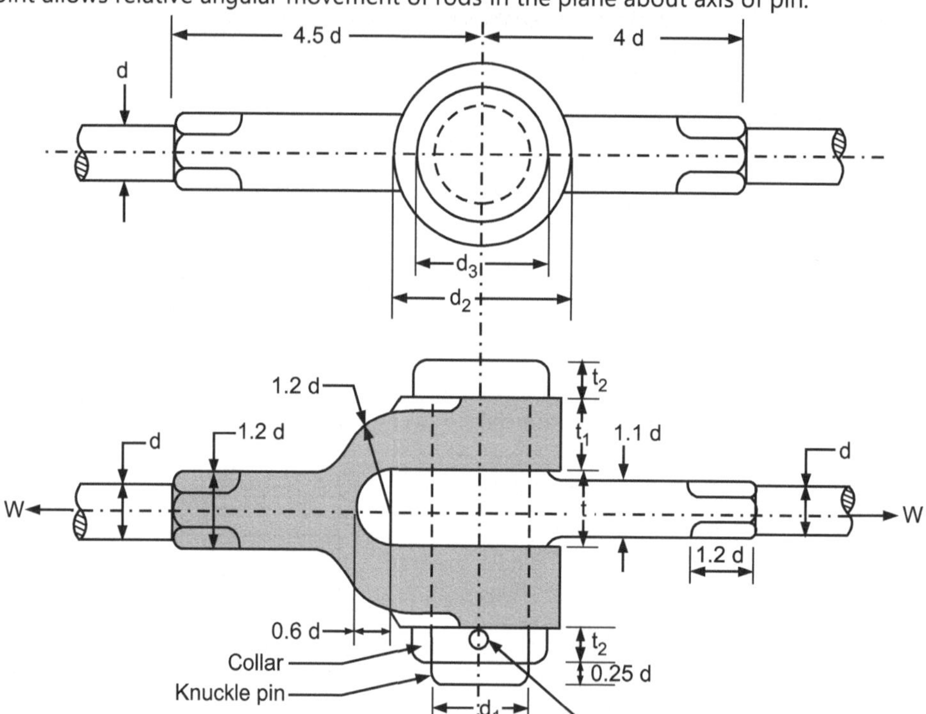


**Fig. 2.23 : Knuckle joint**

Let, W = Load carried by the rods, $\tau$ = Permissible shear stress

$\sigma_t$ = Permissible tensile stress, $\sigma_{ck}$ = Permissible crushing stress

**Step I : Failure of rod in tension :** The rod may fail in tension due to tensile load W.

We know that, Area resisting tearing $= \frac{\pi}{4} \times (d)^2$

where, d = Diameter of rods.

Considering tensile failure, $\sigma_t = \frac{W}{\frac{\pi}{4} \times d^2}$

Hence, value of 'd' can be determined.

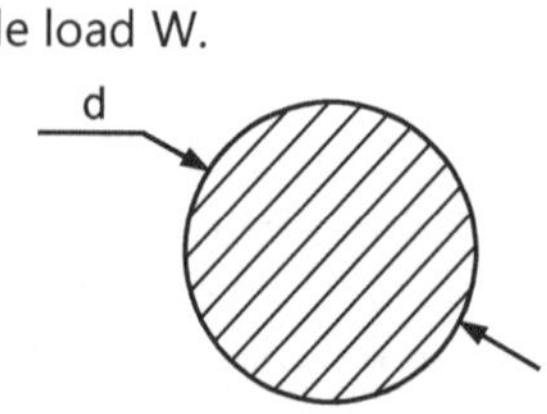


**Fig. 2.24 : Failure of rod in tension**

**Step II : Fix (i.e. calculate) dimensions of knuckle joint using empirical relations :**

| | | |
|---|---|---|
| (i) | Diameter of knuckle pin | $d_1 = d$ |
| (ii) | Outer diameter of single eye or double eye | $d_2 = 2d$ |
| (iii) | Diameter of knuckle pin head and collar | $d_3 = 1.5\ d$ |
| (iv) | Thickness of single eye | $t = 1.25\ d$ |
| (v) | Thickness of fork | $t_1 = 0.75\ d$ |
| (vi) | Thickness of collar | $t_2 = 0.5\ d$ |

After fixing (i.e. determining) the dimensions of knuckle joint using the above empirical relations, we will calculate values of various types of stresses induced and compare them with their corresponding permissible stresses, to find out, whether the dimensions obtained are safer or not?

**Step III : Checking the failure of single eye in tension across the slot (weakest section) :**

Area resisting tearing of single eye across the slot = $(d_2 - d_1) \times t$

Considering tensile failure,

Induced tensile stress, $\sigma_t = \dfrac{W}{(d_2 - d_1) \times t}$

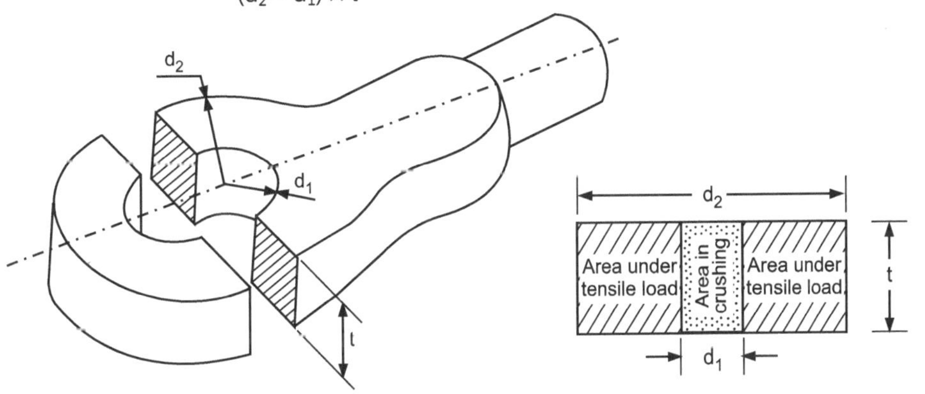


(a) (b)

**Fig. 2.25 : Area of single eye in tension**

**Step IV : Checking the failure of single eye in crushing:**

Area resisting crushing of single eye = $t \times d_1$

Induced crushing stress, $\sigma_{ck} = \dfrac{W}{t \times d_1}$

**Step V : Checking the failure of single eye in double shear :**

Area-resisting shear of single eye = $2 \times \left(\dfrac{d_2 - d_1}{2}\right) \cdot t = (d_2 - d_1) \times t$ $\left[\text{Term } \left(\dfrac{d_2 - d_1}{2}\right) \times t \text{ is multiplied by 2 due to failure of single eye in double shear}\right]$

Considering failure of single eye in double shear, we can write,

Induced shear stress, $\tau = \dfrac{W}{(d_2 - d_1) \cdot t}$

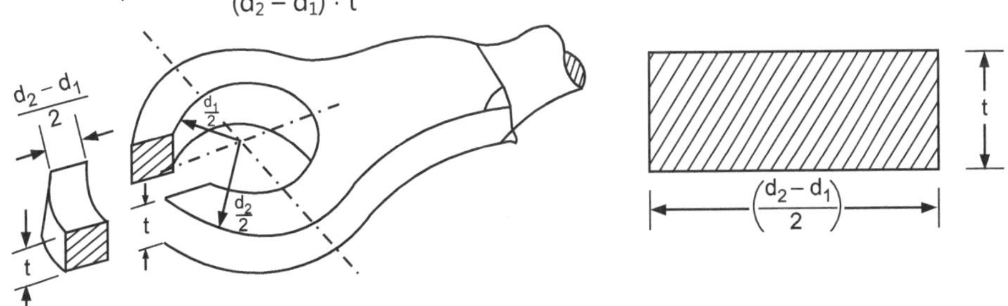


(a) (b)

**Fig. 2.26 : Area of single eye in double shear**

**Step VI : Checking the failure of fork end in tension across the slot :**

Area resisting tearing of double eye across the slot = $2 \times (d_2 - d_1) \cdot t_1$

Considering tensile failure,

$$\text{Induced tensile stress,} \quad \sigma_t = \frac{W}{2 \times (d_2 - d_1) \cdot t_1}$$

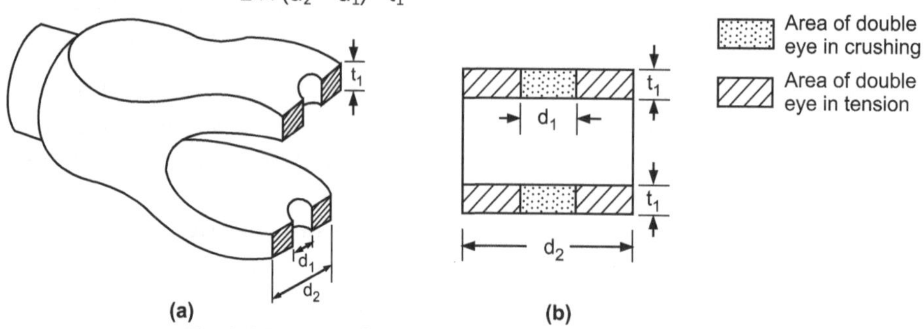


**Fig. 2.27 : Area of double eye in tension and crushing**

**Step VII : Checking failure of fork end in crushing :**

Area resisting crushing of fork end = $2 \times t_1 \cdot d_1$

Considering crushing failure,

$$\text{Induced crushing stress, } \sigma_{ck} = \frac{W}{2 \times t_1 \cdot d_1}$$

**Step VIII : Checking failure of fork end in shear :**

Area-resisting shear of fork end = $2 \times (d_2 - d_1) \times t_1$

Considering shear failure of fork end,

$$\text{Induced shear stress,} \quad \tau = \frac{W}{2 \times (d_2 - d_1) \times t_1}$$

**Step IX : Checking failure of knuckle pin in shear :** Knuckle pin may shear off at two places as shown in Fig. 2.28, i.e. double shear.

$$\therefore \quad \text{Area resisting shear} = 2 \times \frac{\pi}{4} \cdot d_1^2$$

$$\text{Induced shear stress,} \quad \tau = \frac{W}{2 \times \frac{\pi}{4} \cdot d_1^2}$$

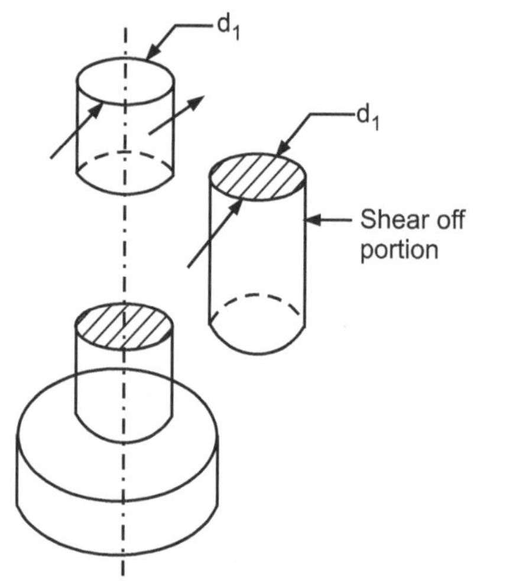


**Fig. 2.28 : Failure of knuckle pin in double shear**

- In this way, different types of induced stresses are calculated from step (III) to step (IX) and they are compared with corresponding permissible stress values, to find whether induced stresses are less than corresponding permissible stresses. If yes, then design is said to be safe. But, if no, then design is not safe. It means, dimensions of knuckle joint will have to be increased to withstand design load.

## 2.3.5 Modes of Failure in Knuckle Pin

**(1) Failure of knuckle pin in shear :** The knuckle pin is in double shear,

$$\therefore \quad \text{Cross-sectional area of pin under shearing} = 2 \times \frac{\pi}{4} \times (d_1)^2$$

$$\therefore \quad \text{Shear stress, } \tau = \frac{W}{2 \times \frac{\pi}{4} \times (d_1)^2} \qquad \dots (2.3)$$

Taking the value of permissible shear stress for pin material, diameter of knuckle pin ($d_1$) can be obtained from the above equation.

**(2) Failure of knuckle pin in bending :**

- In the above equation (2.3) of designing knuckle pin, considering its shear failure, it is assumed that, (i) There is no slack (loose) and (ii) There is no clearance between the pin and fork. Hence it is considered that, there will be no bending of pin.
- But, in actual practice, a small clearance is provided between pin and fork to permit angular movement.
- Therefore, pin is also subjected to bending, in addition to shearing.

  Let, $t$ = thickness of single eye

  $t_1$ = thickness of each fork of double eye
- If bending is taken into account, it is assumed that, the load on the knuckle pin is uniformly distributed in the middle portion (i.e. single eye) and varies uniformly over the forks, as shown by triangular distribution. Refer Fig. 2.29.
- Maximum bending moment will be induced at the centre of pin. It is given by,

$$\mathbf{M} = \frac{W}{2}\times\left(\frac{t_1}{3}+\frac{t}{2}\right)-\frac{W}{2}\times\left(\frac{t}{4}\right) = \frac{W}{2}\times\left[\frac{t_1}{3}+\frac{t}{2}-\frac{t}{4}\right] = \frac{\mathbf{W}}{\mathbf{2}}\left[\frac{\mathbf{t_1}}{\mathbf{3}}+\frac{\mathbf{t}}{\mathbf{4}}\right] \quad \text{... (2.4)}$$

- We know, Section modulus, $\mathbf{Z} = \frac{\pi}{32}\mathbf{d_1^3}$ ... (2.5)

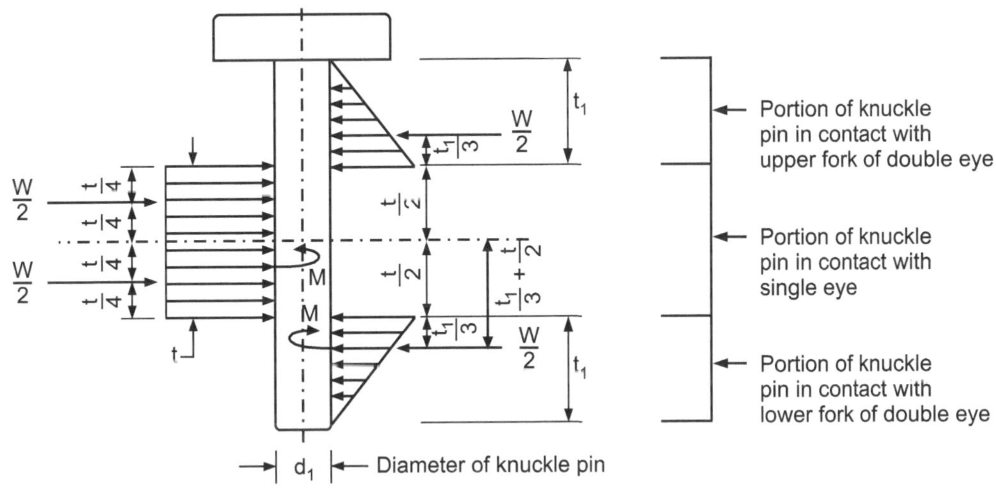


**Fig. 2.29 : Failure of knuckle pin in bending**

Maximum bending stress is given by,

$$\sigma_{b\,max} = \frac{M}{Z}$$

$$\therefore \quad \sigma_{b\,max} = \frac{\frac{W}{2}\left[\frac{t_1}{3}+\frac{t}{4}\right]}{\frac{\pi}{32}d_1^3} \quad \text{[From equations (2.4) and (2.5)]}$$

From this expression, value of $d_1$ may be obtained, i.e. diameter of knuckle pin can be designed considering bending failure.

## 2.3.6 Reasons for Designing Knuckle Pin with Bearing Consideration

- Pin is generally designed considering shear failure.
- But, knuckle joint is not a rigid joint. Being a flexible joint, it permits angular movement between the rods due to slackness (loose) of knuckle pin in forks.
- Slackness of knuckle pin also provides some turning movement (twisting moment or torque) of fork and eye about the pin.
- Referring to the Fig. 2.30, we can observe that, knuckle pin is pressed against the surfaces of eye as well as hole of fork. Therefore, friction occurs between the pin and the fork.
- Due to this friction, pin is subjected to bearing stress. Since knuckle joint permits relative movement, therefore, instead of crushing, the possibility of pin failure is considered under bearing pressure (also called as bearing stress).

- Therefore, knuckle pin is also designed by considering bearing pressure, in addition to shearing.

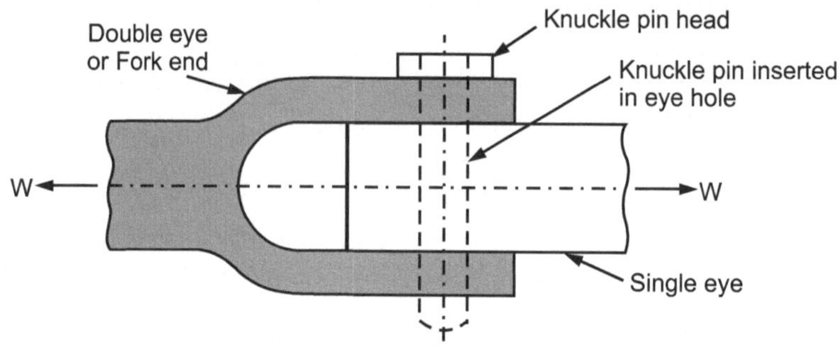


**Fig. 2.30 : Design of knuckle pin with bearing considerations**

## 2.3.7 Function of Split Pin in Knuckle Joint

- Split pin (also known as cotter pin), is a metal fastener having two tines, which are bent during installation.
- Here, 2 tines are like 2 thin pointed parts, which together form a fork, known as split pin.
- Split pin is made up of thick wire having semi-circular cross-section.
- In knuckle joint, after inserting the knuckle pin into eye, the split pin is pushed into the hole of knuckle pin provided at the bottom and the two tines are bent, so that, it works as a fastener or locking device.
- Its working is similar to a stapler or lever.

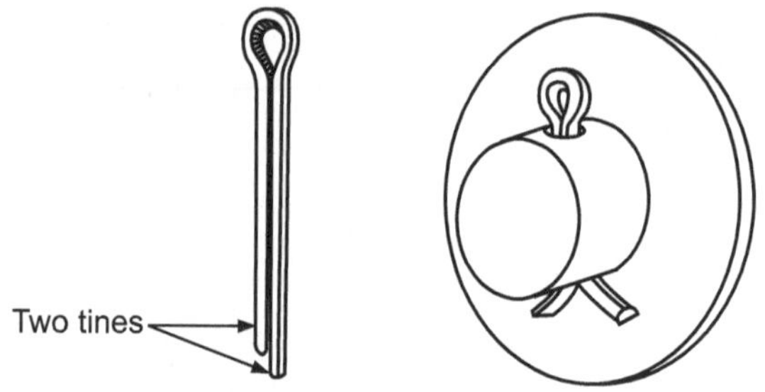


**(a) Before use** **(b) After use**

**Fig. 2.31 : Split pin**

## 2.3.8 Difference between Cotter Joint and Knuckle Joint

| Cotter Joint | Knuckle Joint |
|---|---|
| 1. Capable of taking both **tensile** as well as **compressive load**. | 1. Can take only **tensile load**. |
| 2. Cannot permit angular movement between rods. | 2. Permits limited angular movement between rods. |
| 3. Cotter is **not subjected to bearing failure**. | 3. Knuckle pin is **subjected to bearing failure**. |
| 4. It is known as **single cotter joint**. | 4. It is known as **forked pin joint**. |
| 5. **Taper and clearance** is provided to cotter. | 5. Theoretically, **no taper or clearance is provided to knuckle pin**. But, in actual practice, to permit angular movement between rods, a very small clearance exists. |
| 6. Cotter is **rectangular in cross-section**. | 6. Knuckle pin is **circular in cross-section**. |
| **Applications :**<br>(a) Cotter foundation bolt.<br>(b) Joining piston rod with cross-head.<br>(c) Joining two rods with a pipe.<br>(d) Big end of connecting rod of a steam engine. | **Applications :**<br>(a) Tie bar and roof truss.<br>(b) Links of suspension bridge.<br>(c) Valve mechanism.<br>(d) Fulcrum of the lever.<br>(e) Links of bicycle chain.<br>(f) Joint for rail shifting mechanism. |

## Numerical Type No. 2 : "Kunckle Joint"

**Problem 2.2 :** *A knuckle joint is to withstand a load of 150 kN. Design the joint, if permissible stresses are, $\sigma_t = 75\ N/mm^2$, $\tau = 60\ N/mm^2$ and $\sigma_c = 150\ N/mm^2$.*

**Solution : Given data : W** = load carried by the rods = 150 kN = $\mathbf{150 \times 10^3\ N}$

$\sigma_t = 75\ N/mm^2$, $\tau = 60\ N/mm^2$ and $\sigma_c = 150\ N/mm^2$.

**Procedure :** Value of permissible crushing stress is not given. **Let us assume, $\sigma_{ck} = 150\ N/mm^2$.**

**Step I : Failure of rod in tension :** The rod may fail in tension due to tensile load W.

We know that, Area resisting tearing $= \frac{\pi}{4} \times (d)^2$

Considering tensile or tearing failure of rod, we have,

$$\sigma_t = \frac{W}{\frac{\pi}{4} \times (d)^2}$$

$$\therefore \quad 75 = \frac{150 \times 10^3}{\frac{\pi}{4} \times (d)^2}$$

$$\therefore \quad d = 50.46\ mm \cong 52\ mm$$

Hence, value of d can be determined.

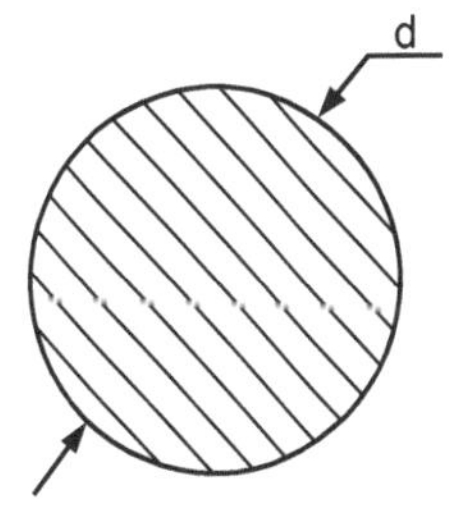


Fig. 2.32 : Failure of rod in tension

**Step II : Fix dimensions of knuckle joint using empirical relations :**

| | | |
|---|---|---|
| (i) | Diameter of knuckle pin | $d_1 = d = $ **52 mm** |
| (ii) | Outer diameter of single or double eye | $d_2 = 2 \times d = 2 \times 52 = $ **104 mm** |
| (iii) | Diameter of knuckle pin head and collar | $d_3 = 1.5 \times d = 1.5 \times 52 = $ **78 mm** |
| (iv) | Thickness of single eye | $t = 1.25 \times d = 1.25 \times 52 = $ **65 mm** |
| (v) | Thickness of fork | $t_1 = 0.75 \times d = 0.75 \times 52 = $ **39 mm** |
| (vi) | Thickness of collar | $t_2 = 0.5 \times d = 0.5 \times 52 = $ **26 mm** |

**Step III : Checking the failure of single eye in tension across the slot (weakest section) :**

Area resisting tearing of single eye across the slot $= (d_2 - d_1) \times t$

Induced tensile stress,

$$\sigma_t = \frac{W}{(d_2 - d_1) \times t} = \frac{150 \times 10^3}{(104 - 52) \times 65} = \mathbf{44.38\ N/mm^2 < 75\ N/mm^2\ (Permissible\ stress)}$$

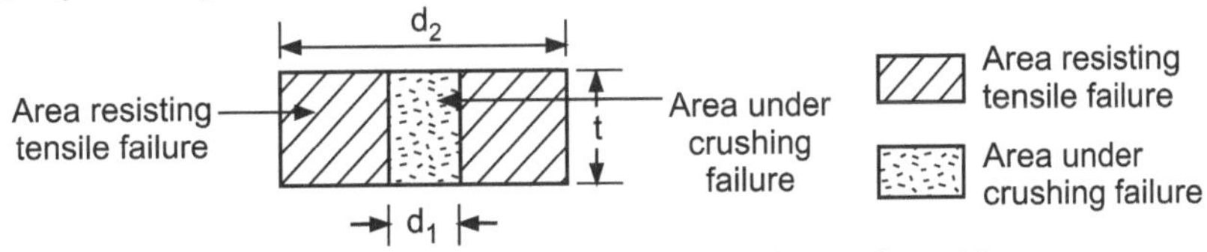


Fig. 2.33 : Failure of single eye in tension and crushing

**Step IV : Checking the failure of single eye in crushing :**

Area resisting crushing of single eye $= t \times d_1$

$\therefore$ Induced crushing stress,

$$\sigma_{ck} = \frac{W}{t \times d_1} = \frac{150 \times 10^3}{65 \times 52} = \mathbf{44.38\ N/mm^2 < 150\ N/mm^2\ (Permissible\ stress)}$$

**Step V : Checking the failure of single eye in shear :**

Area resisting shear of single eye $= 2 \times \left(\frac{d_2 - d_1}{2}\right) \cdot t = (d_2 - d_1) \times t$

$\therefore$ Induced shear stress,

$$\tau = \frac{W}{(d_2 - d_1) \times t} = \frac{150 \times 10^3}{(104 - 52) \times 65}$$

$= \mathbf{44.38\ N/mm^2 < 60\ N/mm^2}$ **(Permissible stress)**

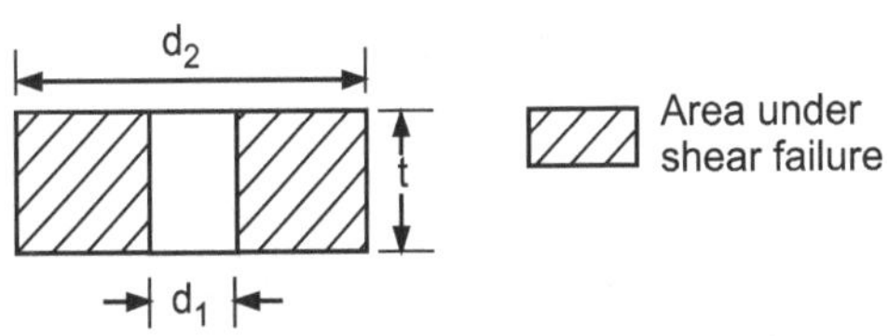


Fig. 2.34 : Failure of single eye in shear

**Step VI : Checking the failure of fork end in tension across the slot :**

Area resisting tearing of fork end across the slot $= 2 \times (d_2 - d_1) \times t_1$

∴ Induced tensile stress,

$$\sigma_t = \frac{W}{2 \times (d_2 - d_1) \times t_1} = \frac{150 \times 10^3}{2 \times (104 - 52) \times 39}$$

$$\sigma_t = \mathbf{36.98\ N/mm^2 < 75\ N/mm^2\ (Permissible\ stress)}$$

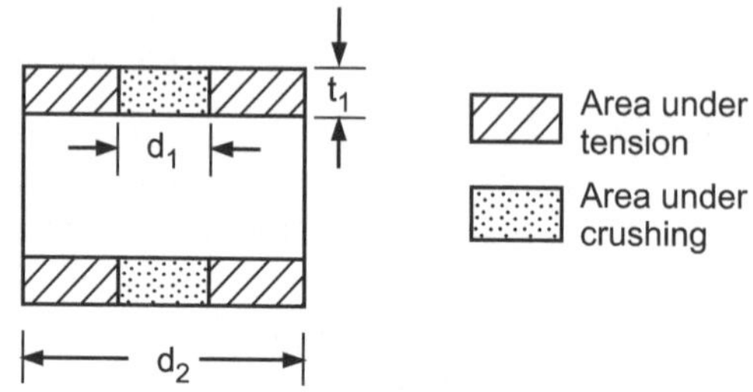


**Fig. 2.35 : Failure of fork end in tension and crushing**

**Step VII : Checking the failure of fork end in crushing :**

Area resisting crushing of fork end = $2 \times t_1 \times d_1$

∴ Induced crushing stress,

$$\sigma_{ck} = \frac{W}{2 \times t_1 \times d_1} = \frac{150 \times 10^3}{2 \times 39 \times 52}$$

∴ $\sigma_{ck}$ = **36.98 N/mm² < 150 N/mm² (Permissible stress)**

**Step VIII : Checking the failure of fork end in shearing :**

Area resisting shear of fork end = $2 \times (d_2 - d_1) \times t_1$

∴ Induced shear stress,

$$\tau = \frac{W}{2 \times (d_2 - d_1) \times t_1} = \frac{150 \times 10^3}{2 \times (104 - 52) \times 39} = \mathbf{36.98\ N/mm^2 < 60\ N/mm^2\ (Permissible\ stress)}$$

**Step IX : Checking the failure of knuckle pin in shear :** Knuckle pin is in double shear.

∴ Area resisting shear $= 2 \times \frac{\pi}{4} \cdot (d_1)^2$

∴ Induced shear stress, $\tau = \frac{W}{2 \times \frac{\pi}{4} \cdot (d_1)^2} = \frac{150 \times 10^3}{2 \times \frac{\pi}{4} \times (52)^2}$

∴ $\tau_{ind}$ = **35.31 N/mm² < 60 N/mm² (Permissible stress)**

As all the induced stresses (calculated in step III to IX) are less than their corresponding permissible limits, therefore, design of knuckle joint is safe.

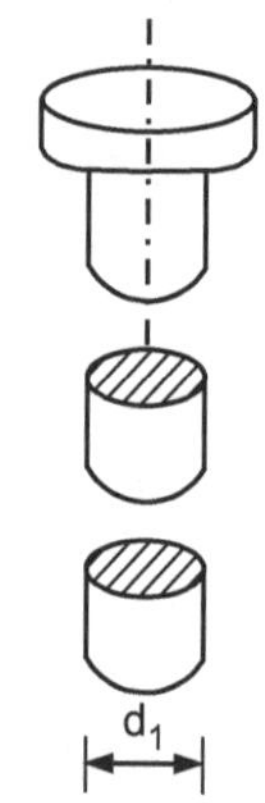


**Fig. 2.36 : Failure of knuckle pin in shear**

## 2.4 TURNBUCKLE

### 2.4.1 Introduction

- Turnbuckle is an **adjustable screwed joint** used to connect two circular rods.
- Turnbuckle is a form of coupling used to connect two circular rods (i.e. round tie rods). Both circular rods to be connected have threaded portion at their ends. One rod has right hand threads and the other rod has left hand threads. Both rods are screwed to a coupler, which has a threaded hole .
- Shape of coupler is hexagonal or rectangular in middle position across its length.
- Further, shape of coupler becomes round at its both ends to facilitate the rods to be loosen or tighten with the help of spanner (or rod), when needed.
- Two circular rods to be connected are under the action of tensile load.
- When the coupler is rotated by hand, then ends of rods either approach (move towards each other) or recede (move away from each other).
- Turnbuckle is made hollow in its middle portion to reduce its weight.

### 2.4.2 Purpose or Applications of Turnbuckle or Coupler

1. Turnbuckle is used to connect two tie rods. Therefore, it acts as a coupling.
2. Due to coupler nut, it is very easy task to tighten or loosen both rods with the help of spanner, whenever required.
3. Turnbuckle is used to couple two railway wagons.
4. Turnbuckle is used to connect links in the mechanisms to transfer motion.
5. Turnbuckle is used to connect structural members like roof trusses.

## 2.4.3 Design Procedure of Turnbuckle

**Design of Turnbuckle or Adjustable Screwed Joint for Round (i.e. Circular) Ends :**

Let, $W$ = Design load = 1.3 × Load carried by the rods in 'N'

$\tau$ = Permissible shear stress in 'N/mm²'

$\sigma_t$ = Permissible tensile stress in 'N/mm²'

$\sigma_{ck}$ = Permissible crushing stress in 'N/mm²'

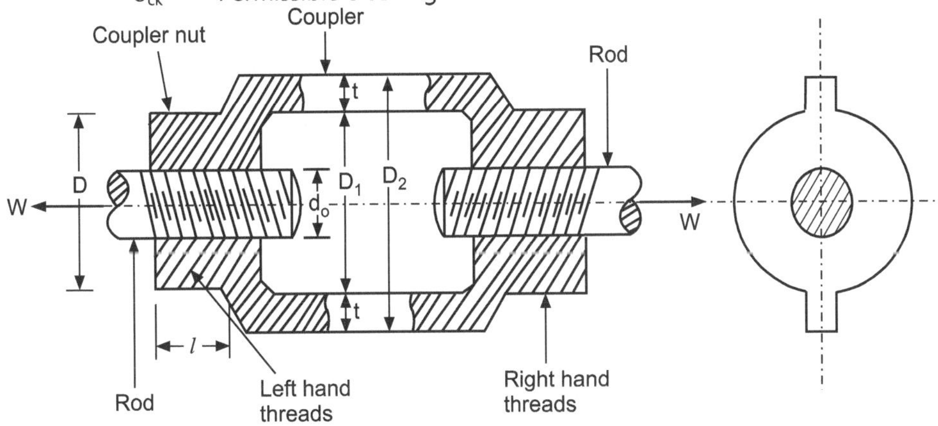


**Fig. 2.37 : Turnbuckle or Coupler**

**Step I : Failure of rod in tension :** The rod may fail in tension due to tensile load W.

We know that, Area resisting tearing = $\frac{\pi}{4} \cdot d_c^2$

Considering tensile stress, $\sigma_t = \frac{W}{\frac{\pi}{4} \cdot d_c^2}$

Hence, value of $d_c$ can be determined.

According to design principle, least area (weakest section) subjected to load should be taken. In case of threads, least area corresponding to least diameter i.e. core diameter ($d_c$) is taken for design purpose, which is equal to $\frac{\pi}{4} \cdot d_c^2$.

From the standard table of square threads (Refer Appendix 'A' attached at the end of textbook), values of nominal diameter ($d_o$) and pitch (p) can be determined by approximation.

**Step II : Considering the shear failure of threads at their roots :**

Area resisting shear $= \pi \cdot d_c \cdot l$

∴ Shear stress $= \tau = \frac{W}{\pi \cdot d_c \cdot l}$

Here, the value of '$l$', i.e. length of coupler nut can be determined.

However, for safe design, length of coupler nut ($l$) can be taken as [$d_o$ to $1.25d_o$ for steel] and [$1.5d_o$ to $2d_o$ for cast iron] for all practical purposes.

**Step III : Checking the crushing stress induced in threads :**

Area resisting crushing of threaded portion $= \frac{\pi}{4} \times (d_o^2 - d_c^2) \times n \times l$

where, $n$ = number of threads per mm of length $= \frac{1}{\text{pitch}}$

∴ Induced crushing stress,

$$\sigma_{ck} = \frac{W}{\frac{\pi}{4} \times (d_o^2 - d_c^2) \times n \times l}$$

Value of induced crushing stress should be less than permissible crushing stress for safe design.

**Step IV : Considering the tensile failure of coupler nut :**

Diameter (D) of coupler nut is found by considering tensile failure of coupler nut.

$$\text{Tensile stress} = \sigma_t = \frac{W}{\frac{\pi}{4}(D^2 - d_o^2)}$$

In general, diameter of coupler nut can be taken from $1.25\ d_o$ to $1.5\ d_o$.

**Step V : Considering tensile failure of coupler :**

Calculate inside diameter of coupler ($D_1$) using the formula, $D_1 = d_o + (6 \text{ to } 8 \text{ mm})$

Outer diameter ($D_2$) of coupler is found by considering tensile failure of coupler,

$$\text{Tensile stress} = \sigma_t = \frac{W}{\frac{\pi}{4}(D_2^2 - D_1^2)}$$

In general, outside diameter of coupler ($D_2$) can be taken as 1.5 $d_o$ to 1.7 $d_o$.

Other dimensions of coupler can be calculated as,

Length of coupler = 6 $d_o$ and Thickness of coupler = t = 0.75 $d_o$

**Numerical Type No. 3 : "Turnbuckle"**

**Problem 2.3 :** *A maximum pull of 100 kN acts on the round tie bar of the roof truss. If the safe tensile and shearing stress in the rod are 80 N/mm² and 30 N/mm² respectively, design the turnbuckle joint. Draw sketch.*

**Solution : Given data :** $W_1$ = Load carried by the rods = 100 kN = $\mathbf{100 \times 10^3}$ **N**

Let W = Design load = 1.3 × Load carried by the rods

∴ $\mathbf{W} = 1.3 \times W_1 = 1.3 \times 100 \times 10^3 = \mathbf{130 \times 10^3}$ **N**

$\sigma_t$ = Permissible tensile stress = 80 N/mm²

$\tau$ = Permissible shear stress = 30 N/mm²

**Procedure :** Value of permissible crushing stress is not given. **Let us assume, $\sigma_{ck}$ = 80 N/mm².**

**Step I : Failure of rod in tension :** The rod may fail in tension due to tensile load W.

We know that, Area resisting tearing $= \frac{\pi}{4} \cdot (d_c)^2$

Considering tensile failure of rod,

$$\sigma_t = \frac{W}{\frac{\pi}{4} \times (d_c)^2}$$

$$\therefore \quad 80 = \frac{130 \times 10^3}{\frac{\pi}{4} \times (d_c)^2}$$

$$\therefore \quad d_c = \mathbf{45.48 \text{ mm}}$$

From the standard table of square threads, value of nominal diameter ($d_o$) and pitch (p) can be taken.

$d_o$ = **58 mm** and pitch = p = **9 mm**

Therefore, $\mathbf{d_c} = d_o - p = 58 - 9 = \mathbf{49 \text{ mm}}$.

*[**Note :** Refer appendix 'A' of "Standard table for square threads" given at the end of text book.]*

**Step II : Considering shear failure of threads at their roots :**

Area resisting shear $= \pi \times d_c \times l$

$$\text{Considering shear failure, } \tau = \frac{W}{\pi \times d_c \times l}$$

$$\therefore \quad 30 = \frac{130 \times 10^3}{\pi \times 58 \times l}$$

$$\therefore \quad l = \mathbf{23.78 \text{ mm}}$$

But in actual practice, Length of coupler nut ($l$) can be taken in the range of $d_o$ to 1.25 $d_o$ for steel.

$$\therefore \quad \boldsymbol{l} = 1.25\, d_o = 1.25 \times 58 = \mathbf{72.5 \text{ mm}}$$

**Step III : Checking the crushing stress induced in threads :**

Area resisting crushing of fork end $= \frac{\pi}{4} \times (d_o^2 - d_c^2) \times n \times l$

Where, n = Number of threads per mm of length $= \frac{1}{\text{pitch}} = \frac{1}{9} = 0.111$

Considering crushing failure, we can find out crushing stress induced in threads,

$$\sigma_{ck} = \frac{W}{\frac{\pi}{4} \times (d_o^2 - d_c^2) \times n \times l} = \frac{130 \times 10^3}{\frac{\pi}{4} \times (58^2 - 49^2) \times 0.111 \times 72.5} = \mathbf{21.35 \text{ N/mm}^2}$$

As induced crushing stress (21.35 N/mm$^2$) is less than permissible crushing stress (80 N/mm$^2$), so we conclude that, design is safe.

**Step IV : Considering the tensile failure of coupler nut :**

Outside diameter (D) of coupler nut is found by considering tensile failure.

$$\sigma_t = \frac{W}{\frac{\pi}{4}(D^2 - d_o^2)}$$

$$\therefore \quad 80 = \frac{130 \times 10^3 \times 4}{\pi \times [(D)^2 - (58)^2]}$$

$$\therefore \quad \mathbf{D = 73.71\ mm \cong 74\ mm\ (say)}$$

**Step V : Considering tensile failure of coupler :**

Inside diameter of coupler = $\mathbf{D_1} = d_o + 6 = 58 + 6 =$ **64 mm**

Outer diameter ($D_2$) of coupler is found by considering tensile failure of coupler.

$$\sigma_t = \frac{W}{\frac{\pi}{4}(D_2^2 - D_1^2)}$$

$$\therefore \quad 80 = \frac{130 \times 10^3 \times 4}{\pi \times [(D_2)^2 - (64)^2]}$$

$$\therefore \quad \mathbf{D_2 = 78.51\ mm}$$

But, In actual practice, outside diameter of coupler is taken in the range of 1.5 $d_o$ to 1.7 $d_o$.

$\therefore$ Let us take, $\mathbf{D_2} = 1.7\ d_o = 1.7 \times 58 =$ **98.6 mm**

Also, Length of coupler $= 6\ d_o = 6 \times 58 =$ **348 mm** and

Thickness of coupler $= t = 0.75\ d_o = 0.75 \times 58 =$ **43.5 mm**

## 2.5 LEVERS

### 2.5.1 Introduction to Levers

- A lever is a rigid rod or bar capable of turning about a fixed point, called as **fulcrum**.
- **Function of lever:** To lift a large load by applying a small effort.
- 'Ratio of load lifted to effort applied' is called as **Mechanical Advantage (M.A.)**.
- A lever may be straight or curved. Lever may be used to apply force in any desired direction. This force may be horizontal or inclined. **Lever works on the principle of moments**.
- Consider a straight lever with parallel forces acting in the same plane as shown in Fig. 2.38. Points 'A' and 'B', through which, load and effort is applied, are known as **load point** and **effort point** respectively. 'F' is the fulcrum, about which, the lever is capable of turning.

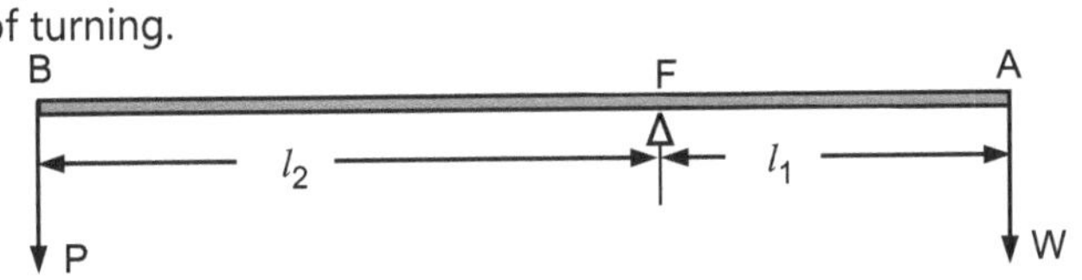


**Fig. 2.38**

- Perpendicular distance between the load point and fulcrum is known as **load arm ($l_1$)**, whereas, perpendicular distance between the effort point and fulcrum is known as **effort arm ($l_2$)**. Load arm and effort arm are also denoted by $l_w$ and $l_p$ respectively.
- Taking algebraic sum of moments of all forces about fulcrum and equating it to zero,

We get, $\quad \sum M_{F@_F} = -P \times l_2 + W \times l_1 = 0$

$$\therefore \quad W \times l_1 = P \times l_2$$

$$\therefore \quad \frac{W}{P} = \frac{l_2}{l_1} = \text{Mechanical advantage}$$

$$\therefore \quad \text{M.A.} = \frac{W}{P} = \frac{l_2}{l_1}$$

### 2.5.1.1 Different Values of Mechanical Advantage and Their Meaning

We know that,

$$\textbf{Mechanical advantage (M.A.)} = \frac{\text{Load lifted}}{\text{Effort applied}} = \frac{W}{P}$$

1. If M.A. = 1, then load lifted (W) and effort applied (P) will be same. Therefore, length of load arm and effort arm will also be same. In this case, lever is defined as, "a rigid bar or rod pivoted at a point, known as fulcrum and capable of burning about that fulcrum point".
2. If M.A. < 1, then it means that, $\frac{W}{P} < 1$. Therefore, W < P. In this case, Effort applied (P) will be more than the load lifted (W). This is not desirable. Such levers are rarely used. Since effort applied (P) is more, therefore, length of effort arm will be less than length of load arm.
3. If M.A. > 1, then it means that, $\frac{W}{P} > 1$. Therefore, W >P. In this case, effort applied (P) will be less than load lifted (W). This is desirable. Since effort applied (P) is less, therefore, length of effort arm will be more than length of load arm.

### 2.5.1.2 Various Applications of Lever in Engineering Practice

The various types of levers used are classified as :

1. **One arm lever :**

- One arm lever, as shown in Fig. 2.39, has only one arm and that is effort arm. This type of lever is used to apply external torque.
- **Examples :** Hand lever, Foot lever etc.

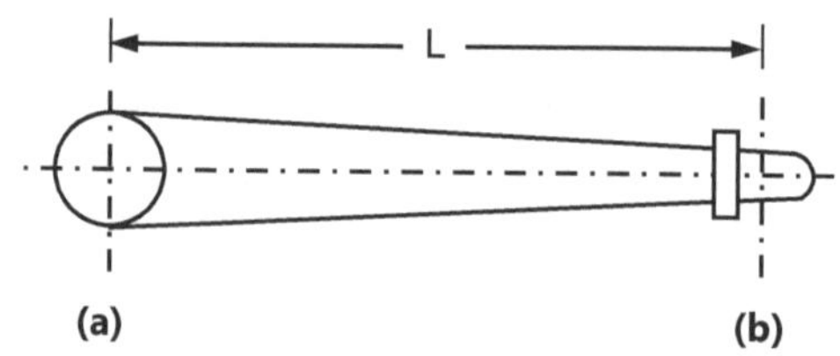


**Fig. 2.39 : One arm lever**

2. **Two arm levers :**

- Two arm levers have two arms : a load arm and an effort arm.
- The axes of load arm and effort arm are along the same line.
- Depending upon the relative positions of fulcrum, load and effort, two arm levers are of three types.

**(a) First type of lever :**

In first type of lever, the fulcrum is in between the load and effort. In this case, effort arm ($l_p$) is greater than load arm ($l_w$). Therefore, mechanical advantage is more than one.

**Examples :** Bell crank lever used in railway signaling arrangement, Rocker arm in I.C. engines, Handle of hand pump, Hand wheel of punching press, Beam of balance, Foot lever etc.

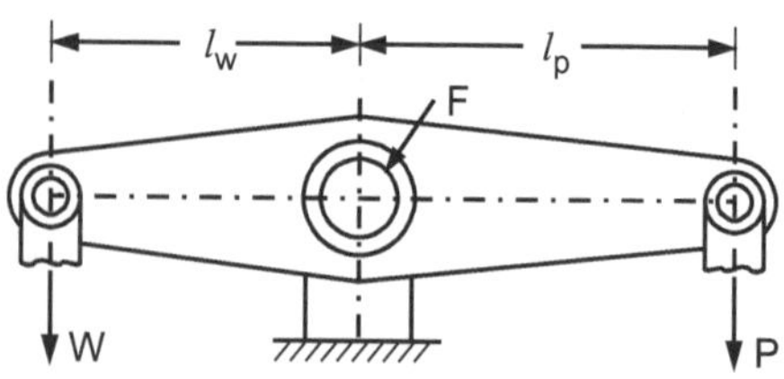


**Fig. 2.40**

**(b) Second type of lever :**

In second type of lever, the load is in between the fulcrum and effort. In this case, the effort arm is more than load arm. Therefore, the mechanical advantage is more than one.

**Examples :** Levers of loaded safety valve, Nut cutter etc.

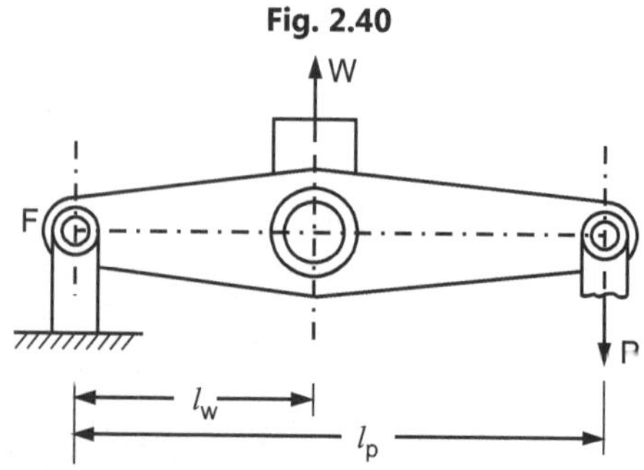


**Fig. 2.41**

**(c) Third type of lever :**

In third type of lever, the effort is in between the fulcrum and load. Effort arm is less than load arm, therefore, mechanical advantage is less than one. This type of lever is not recommended in engineering practice.

**Examples :** A pair of tongs, Treadle of a sewing machine etc.

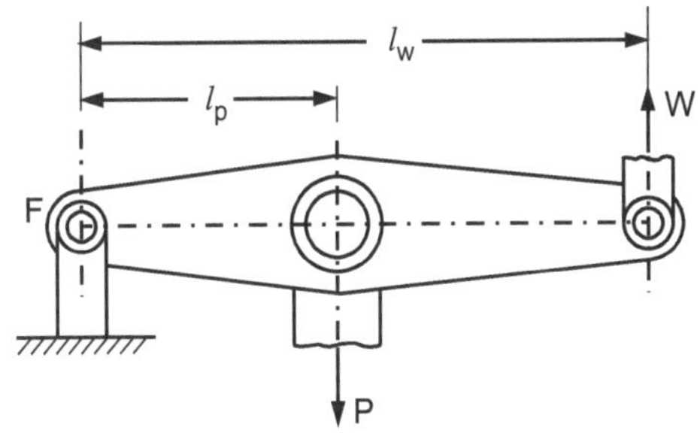


**Fig. 2.42**

**3. Angular lever :**

- In angular lever, as shown in Fig. 2.43, the load arm and effort arm are inclined at an angle other than 90° and 180°.
- Generally, the angle between the two arms is 135°. Such levers are used to facilitate the application of effort in the desired direction.
- **Examples :** Rocker arms in I.C. engines, Brake levers.

**4. Bell crank lever :**

- In bell crank lever, as shown in Fig. 2.44, the load arm and effort arm are at right angle to each other.
- **Examples :** Levers in Hartnell governor, Levers in railway signals.

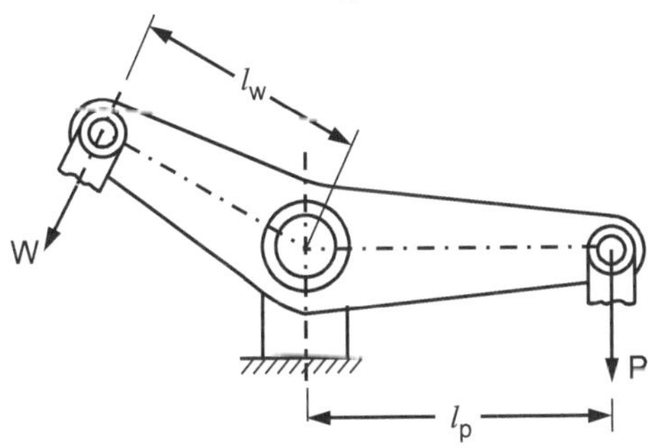


**Fig. 2.43 : Angular lever**

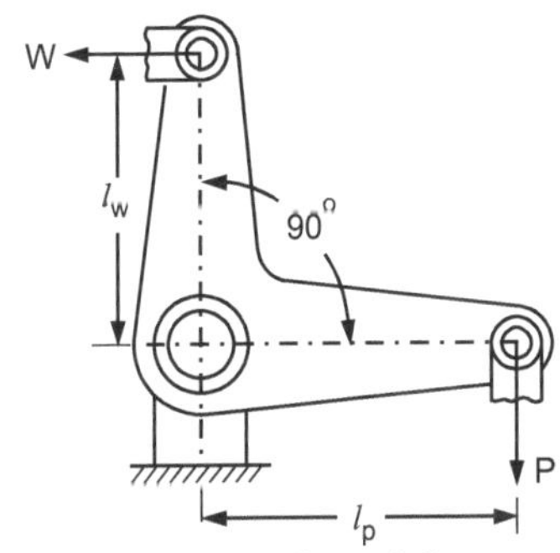


**Fig. 2.44 : Bell crank lever**

## 2.5.1.3 Leverage

- **Leverage** is defined as, *"the ratio of effort arm to load arm"*.
- Mathematically, leverage $= \frac{l_2}{l_1} = \frac{l_p}{l_w}$.
- In order to lift a large load by applying a small effort, the effort arm must be greater (longer) than the load arm, i.e. $l_2 > l_1$.

## 2.5.1.4 Why Levers are Tapered at the Ends ?

- Levers are subjected to eccentric load.
- It means that, load axis and lever (component) axis are not same.
- Load axis lies away from lever axis at a distance, called as eccentricity (e).
- Due to eccentric load, lever is subjected to bending moment.

  If W = Eccentric load applied in 'N'

  e = Eccentricity (distance between load axis and lever axis) in 'm'

  Then, Bending moment (M) = W × e in 'N·m'

  $\therefore \quad M \propto e.$
- It means that, bending moment is directly proportional to the distance of load from a point, about which, bending moment takes place.
- Bending moment is zero at the point of application of load. Therefore, resisting area will be minimum at the point of application of load.
- Conversely, resisting area will be maximum at a point, where bending takes place. Such conditions lead to stress concentration.
- This difficulty can be overcome by providing taper form gradually throughout the length of lever, with minimum area at its ends. Therefore, levers are tapered to avoid stress concentration.

## 2.5.2 Design Procedure of Hand Lever

Let,

| | | | |
|---|---|---|---|
| $P$ | = | Force applied at the end of handle/lever | in 'N' |
| $L$ | = | Effective length of lever | in 'mm' |
| $d$ | = | Diameter of shaft | in 'mm' |
| $d_1$ | = | Diameter of shaft at centre of bearing | in 'mm' |
| $d_2$ | = | Diameter of boss | in 'mm' |
| $l_2$ | = | Length of boss | in 'mm' |
| $l_1$ | = | Length of key | in 'mm' |
| $l$ | = | Overhang distance of bearing from the lever | in 'mm' |

**Step I :** Considering shaft under pure torsion, we have,

$$T = \frac{\pi}{16} \cdot \tau \cdot d^3 \quad \dots (2.6)$$

But, twisting moment on shaft,

$$T = P \times L \quad \dots (2.7)$$

∴ On equating (2.6) and (2.7), we have,

$$P \times L = \frac{\pi}{16} \cdot \tau \cdot d^3$$

∴ Diameter of shaft (d) may be obtained.

**Step II :** Using the empirical relations, fix the other dimensions as,

$$d_2 = 1.6d$$

$$t_2 = 0.3d$$

$$l_2 = 1.25\, d$$

$$l = 2 \times l_2$$

(Generally, the value of '$l$' is given. If not given, use this formula to calculate it)

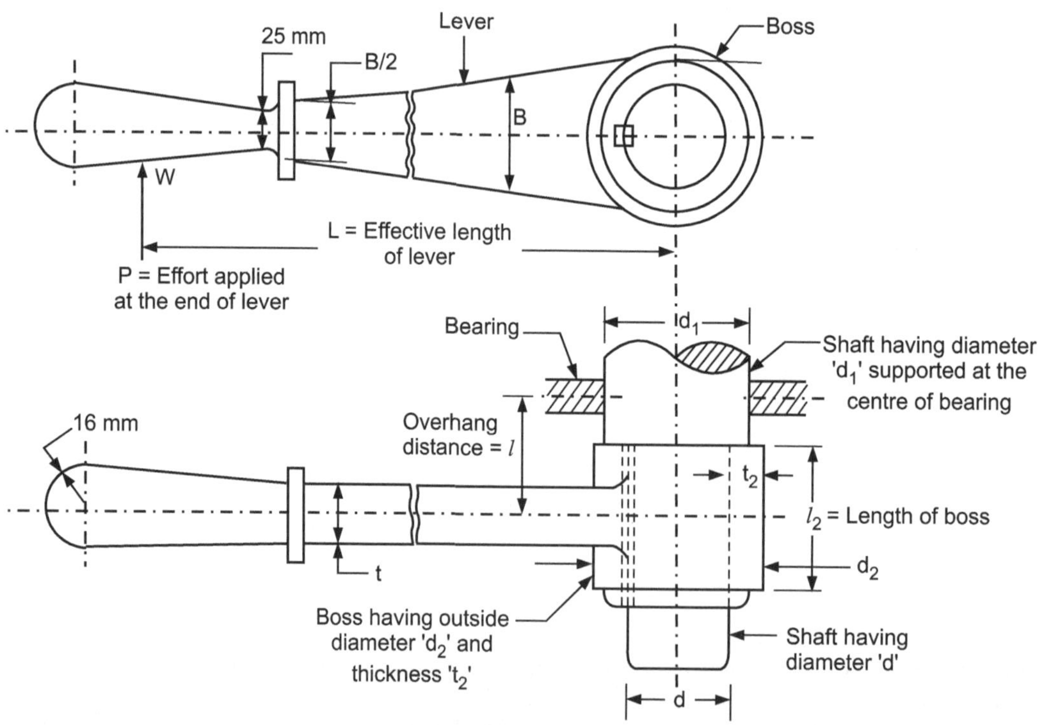


**Fig. 2.45 : Hand lever**

**Step III :** Considering the shaft supported at the centre of the bearing under combined twisting and bending moments, we have,

$$T = P \times L \quad \text{and} \quad M = P \times l$$

∴ Equivalent twisting moment,

$$T_e = \sqrt{M^2 + T^2} = \sqrt{(P \times l)^2 + (P \times L)^2} = P\sqrt{l^2 + L^2} \quad \dots (2.8)$$

Also, equivalent twisting moment,

$$T_e = \frac{\pi}{16} \cdot \tau_{max} \cdot d_1^3 \qquad \dots (2.9)$$

On equating equations (2.8) and (2.9), we get,

$$P\sqrt{l^2 + L^2} = \frac{\pi}{16} \cdot \tau_{max} \cdot d_1^3$$

Thus, value of $d_1$ = diameter of shaft supported at centre of bearing can be determined.

**Step IV : Design of key :** Let $t_1$ = thickness of key, w = width of key and $l_1$ = length of key.

After finding diameter of shaft (d), we can fix the dimensions for key using empirical relations as,

$$w = \frac{d}{4} \quad \text{and} \quad t_1 = \frac{d}{6}$$

Considering shear failure of key,

We have, $$T = (w \cdot l_1 \cdot \tau) \times \frac{d}{2}$$

$\therefore$ Length $l_1$ can be determined.

Since length of boss ($l_2$) can be taken equal to length of key ($l_1$), therefore, $l_2 = l_1$.

**Step V : Design of cross-section of lever :** Considering bending failure of lever, we can determine the cross-section of lever near the boss.

Let t = thickness of lever near the boss, and B = width or height of lever near the boss.

We have,

Bending moment on the lever = M = $P \times (L - r_b)$ and Section modulus = $Z = \frac{1}{6} \cdot t \cdot B^2$

where, $r_b$ = Radius of boss= $\frac{\text{Diameter of boss}}{2} = \frac{d_2}{2}$

To calculate dimensions of cross-section of lever, we have,

$$\text{Bending stress, } \sigma_b = \frac{M}{Z} = \frac{P \cdot (L - r_b)}{\frac{1}{6} \cdot t \cdot B^2}$$

Width of lever may be taken as, B = 2t to 5t. (If relation between 'B' and 't' is not given in numerical problem.)

From the above equation of bending stress, values of 't' and 'B' can be determined.

## Numerical Type No. 4 : "Hand Lever"

**Problem 2.5 :** *A hand lever is 800 mm long from the centre of gravity of spindle to the point of application of pull of 300 N. The effective overhang from the nearest bearing is 100 mm. If the permissible stresses in tension, shear and crushing are not to exceed 66 N/mm² each, design the spindle, key, lever and boss. Assume the arm of lever to be rectangular having width as twice of thickness.*

**Solution : Given data :** L = 800 mm, P = 300 N, $l$ = 100 mm, $\sigma_t = \sigma_c = \tau = 66$ N/mm², B = 2t

**Procedure :**

Let,
- P = Force or effort applied at the end of handle/lever.
- L = Effective length of lever.
- d = Diameter of shaft.
- $d_1$ = Diameter of shaft at centre of bearing.
- $d_2$ = Diameter of boss.
- $l_2$ = Length of boss.
- $l_1$ = Length of key.

**Step I :** Considering shaft under pure torsion, we have,

$$T = \frac{\pi}{16} \tau \cdot d^3 \qquad \dots (1)$$

But, twisting moment on shaft,

$$\mathbf{T} = P \times L = 300 \times 800 = \mathbf{240 \times 10^3\ N\text{-}mm}$$

∴ Equation (1) becomes,

$$240 \times 10^3 = \frac{\pi}{16} \times 66 \times d^3$$

∴ $\mathbf{d} = 26.45 \text{ mm} \cong \mathbf{28\ mm\ (say)}$

**Step II :** Using the empirical relations, fix the other dimensions as,

$$\mathbf{d_2} = 1.6d = 1.6 \times 28 = \mathbf{44.8\ mm}$$

$$\mathbf{t_2} = 0.3d = 0.3 \times 28 = \mathbf{8.4\ mm}$$

$$\boldsymbol{l_2} = 1.25d = 1.25 \times 28 = \mathbf{35\ mm}$$

**Step III :** We can observe that, shaft supported at the centre of bearing having diameter $d_1$ is under the action of combined twisting and bending moments. Therefore, diameter $d_1$ should be calculated according to maximum shear stress theory, which is given by,

Equivalent twisting moment,

$$T_e = \sqrt{M^2 + T^2} = \frac{\pi}{16} \tau_{max} \cdot d_1^3 \quad \dots (2)$$

We can find, $\mathbf{M} = P \times l = 300 \times 100 = \mathbf{30 \times 10^3\ N\text{-}mm}$

and $\mathbf{T} = P \times L = 300 \times 800 = \mathbf{240 \times 10^3\ N\text{-}mm}$

∴ Equation (2) gives,

$$\therefore \sqrt{(30 \times 10^3)^2 + (240 \times 10^3)^2} = \frac{\pi}{16} \times 66 \times d_1^3$$

∴ $d_1 = \mathbf{26.52\ mm}$

Modifying the diameter of shaft at centre of bearing = $\mathbf{d_1}$ = **32 mm (say).** This is done to make shaft at centre of bearing more stronger by keeping $d_1$ more as compared to diameter of shaft (i.e. d = 28 mm).

**Step IV : Design of key :** After finding diameter of shaft (d), we can fix the dimensions for key using empirical relations.

$$w = \frac{d}{4} = \frac{28}{4} = \mathbf{7\ mm} \text{ and } t_1 = \frac{d}{6} = \frac{28}{6} = \mathbf{4.67\ mm}$$

Considering shear failure of key, we have,

$$T = (w \cdot l_1 \cdot \tau) \times \frac{d}{2}$$

$$\therefore \quad 240 \times 10^3 = (7 \times l_1 \times 66) \times \frac{28}{2}$$

∴ $\boldsymbol{l_1}$ = length of key = $\mathbf{37.10\ mm} \cong \mathbf{38\ mm\ (say)}$

Since length of boss ($l_2$) may be taken as length of key ($l_1$), we can write, $\boldsymbol{l_2} = \boldsymbol{l_1} = \mathbf{38\ mm}$.

**Step V :** Considering bending failure of lever, we can find cross-section of lever.

Let t = thickness of lever near the boss.

and B = width or height of lever near the boss.

We have, bending moment on the lever,

$$\mathbf{M} = P \times (L - r_b) = P \times \left(L - \frac{d_2}{2}\right) = 300 \times \left(800 - \frac{44.8}{2}\right) = \mathbf{233.28 \times 10^3\ N\text{-}mm}$$

Section modulus, $\mathbf{Z} = \frac{1}{6} \cdot t \cdot B^2 = \frac{1}{6} \cdot t \cdot (2t)^2 = \mathbf{0.67 \times t^3}$ (∵ B = 2t)

We have, Bending stress, $\sigma_b = \frac{M}{Z}$

$$\therefore \quad 66 = \frac{233.28 \times 10^3}{0.67 \times t^3}$$

∴ $\mathbf{t} = 17.407 \text{ mm} \cong \mathbf{18\ mm\ (say)}$ and $\mathbf{B} = 2t = 2 \times 18 = \mathbf{36\ mm}$

## 2.5.3 Design Procedure of Foot Lever

Let,

| | | | |
|---|---|---|---|
| $P$ | = | Force of effort applied at the paddle | in 'N' |
| $L$ | = | Effective length | in 'mm' |
| $d$ | = | Diameter of shaft | in 'mm' |
| $d_1$ | = | Diameter of shaft at centre of bearing | in 'mm' |
| $d_2$ | = | Diameter of boss | in 'mm' |
| $l_2$ | = | Length of boss | in 'mm' |
| $l_1$ | = | Length of key | in 'mm' |

**Step I :** Considering shaft under pure torsion, we have,

$$T = \frac{\pi}{16} \cdot \tau d^3 \qquad \text{... (2.10)}$$

But, twisting moment on shaft can also be calculated by,

$$T = P \cdot L \qquad \text{... (2.11)}$$

∴ On equating (2.10) and (2.11), we get,

$$P \times L = \frac{\pi}{16} \cdot \tau d^3$$

From the above equation, diameter of shaft (d) may be obtained.

**Step II :** Using the empirical relations, fix the other dimensions, such as,

$$d_2 = 1.6d$$

$$t_2 = 0.3d$$

$$l_2 = 1.25d$$

$$l = 2 \cdot l_2$$

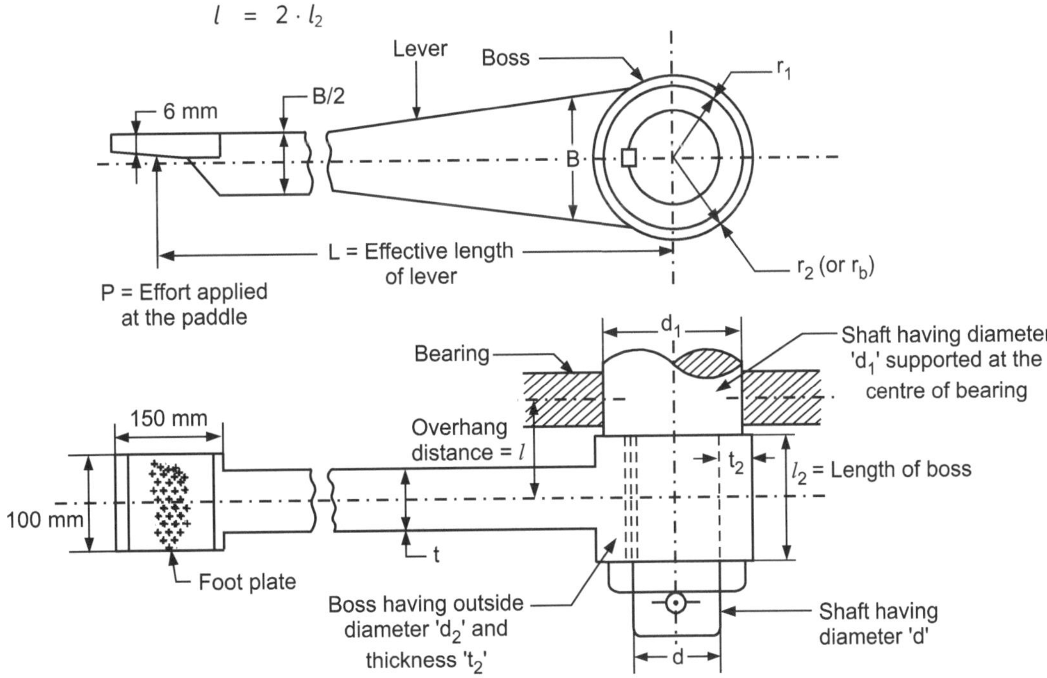


**Fig. 2.46 : Foot lever**

**Step III :** Since the shaft supported at the centre of the bearing having diameter $d_1$ is subjected to combined twisting and bending moments, value of $d_1$ must be designed on the basis of maximum shear stress theory, given as,

$$T_e = \sqrt{M^2 + T^2} = \frac{\pi}{16} \tau_{max} \cdot d_1^3 \qquad \text{... (2.12)}$$

We can find, $M = P \times l$

and $T = P \times L$

By putting these values of M and T in equation (2.12), we can find out the diameter of shaft ($d_1$) supported at the centre of bearing.

**Step IV : Design of key :** Let $t_1$ = thickness of key, w = width of key and $l_1$ = length of key.

After finding diameter of shaft (d), we can fix the dimensions for key using empirical relations.

$$w = \frac{d}{4} \text{ and } t_1 = \frac{d}{6}$$

Considering shear failure of key, we have,

$$T = (w \cdot l_1 \cdot \tau) \times \frac{d}{2}$$

∴ Length $l_1$ can be determined.

Since length of boss ($l_2$) can be taken equal to length of key ($l_1$), therefore, $l_2 = l_1$.

**Step V :** Considering bending failure of lever, we can find cross-section of lever near the boss.

Let, t = thickness of lever near the boss and B = width or height of lever near the boss.

We have, Bending moment acting on the lever = $\mathbf{M} = P \times \left(L - \frac{d_2}{2}\right) = \mathbf{P \cdot (L - r_b)}$ $\left[\because r_b = \frac{d_2}{2}\right]$

where, $r_b$ = Radius of boss

$d_2$ = Diameter of boss

and Section modulus = $\mathbf{Z = \frac{1}{6} \cdot t \cdot B^2}$

Considering the failure of lever in bending, we have,

$$\sigma_b = \frac{M}{Z} = \frac{P \cdot (L - r_b)}{\frac{1}{6} \cdot t \cdot B^2}$$

Here, width of lever may be taken as B = 2t to 5t to find out values of 't' and 'B'. (If relation between B and t is not given in the Numerical Problem).

**Numerical Type No. 5 : "Foot Lever"**

**Problem 2.6 :** *Design a foot brake lever from the following data :*

*(i) Length of lever from centre of gravity of spindle to point of application of load = 1 m.*

*(ii) Maximum load on the foot plate = 800 N*

*(iii) Overhang distance from nearest bearing = 100 mm*

*(iv) Permissible tensile and shear stress = 70 MPa*

**Solution : Given data : L** = 1 m = **1000 mm**, P = 800 N, $\sigma_t = 70$ N/mm², $\tau = 70$ N/mm².

**Procedure :** Value of permissible bending stress ($\sigma_b$) is not given in the numerical problem. Therefore, **let us assume $\sigma_b$ = 70 N/mm²**.

Let, P = Force applied at the paddle

L = Effective length

d = Diameter of shaft

$d_1$ = Diameter of shaft at centre of bearing

$d_2$ = Diameter of boss

$l_2$ = Length of boss

$l_1$ = Length of key

**Step I :** Considering shaft under pure tension, we have,

$$T = \frac{\pi}{16} \cdot \tau \cdot d^3 \quad \text{... (1)}$$

Twisting moment or torque acting on shaft can also be determined by,

$$\mathbf{T} = P \times L = 800 \times 1000 = \mathbf{800 \times 10^3 \text{ N-mm}}$$

By keeping the value of torque (T) in equation (1), we get,

$$\therefore \quad 800 \times 10^3 = \frac{\pi}{16} \times 70 \times d^3$$

$$\therefore \quad d = \mathbf{38.75 \text{ mm} \cong 40 \text{ mm (say)}}$$

**Step II :** Using the empirical relations, other dimensions can be fixed as,

$$d_2 = 1.6\,d = 1.6 \times 40 = \mathbf{64\ mm}$$
$$t_2 = 0.3\,d = 0.3 \times 40 = \mathbf{12\ mm}$$
$$l_2 = 1.25\,d = 1.25 \times 40 = \mathbf{50\ mm}$$

**Step III :** Since the shaft supported at centre of bearing having diameter '$d_1$' is under the action of combined twisting and bending moments, its diameter should be calculated by using maximum shear stress theory.

Bending moment and twisting moment acting on shaft supported at centre of bearing are calculated as,

$$T_e = \sqrt{M^2 + T^2} = \frac{\pi}{16} \cdot \tau_{max} \cdot d_1^3 \qquad \ldots (2)$$

We have, **M** = $P \times l = 800 \times 100 =$ **$80 \times 10^3$ N-mm** and **T** = $P \times L = 800 \times 1000 =$ **$800 \times 10^3$ N-mm**

Therefore, equation (2) becomes,

$$\sqrt{(80 \times 10^3)^2 + (800 \times 10^3)^2} = \frac{\pi}{16} \tau_{max} \cdot d_1^3$$

$$\therefore \quad 804 \times 10^3 = \frac{\pi}{16} \times 70 \times d_1^3$$

$$\therefore \quad d_1 = \mathbf{38.81\ mm}$$

To make shaft supported at the centre of bearing more stronger, its diameter should be increased. Also, value of $d_1$ must be more than d (40 mm).

$$\therefore \quad d_1 = \mathbf{44\ mm}$$

**Step IV : Design of key :** After finding diameter of shaft (d), we can fix the dimensions for key using empirical relations as,

$$\mathbf{w} = \frac{d}{4} = \frac{40}{4} = \mathbf{10\ mm} \text{ and } \mathbf{t_1} = \frac{d}{6} = \frac{40}{6} = \mathbf{6.67\ mm}$$

Considering shear failure of key,

We have, $$T = w \times l_1 \times \tau \times \left(\frac{d}{2}\right)$$

$$\therefore \quad 800 \times 10^3 = 10 \times l_1 \times 70 \times \left(\frac{40}{2}\right)$$

$$\therefore \quad l_1 = \text{length of key} = \mathbf{57.14\ mm \cong 58\ mm\ (say)}$$

In general, length of boss is taken as equal to length of key. Therefore, length of boss is increased from 50 mm to 58 mm.

$$\therefore \quad l_2 = l_1 = \mathbf{58\ mm}$$

**Step V :** Considering bending failure of lever, we can find out the cross-section of lever.

We have, $$r_b = \text{Radius of boss} = \frac{\text{Diameter of boss}}{2} = \frac{d_2}{2} = \frac{64}{2} = \mathbf{32\ mm}$$

Let, t = thickness of lever near the boss

and B = width or height of lever near the boss.

Bending moment acting on lever can be calculated by,

$$\mathbf{M} = P \times [L - r_b] = 800 \times (1000 - 32) = \mathbf{774.4 \times 10^3\ N.mm}$$

Also, $$\mathbf{Z} = \frac{1}{6} t B^2 = \frac{1}{6} \times t \times (3t)^2 = \mathbf{1.5 \times t^3} \qquad \textbf{[Assuming B = 3t]}$$

We have, $$\sigma_b = \frac{M}{Z}$$

$$\therefore \quad 70 = \frac{774.4 \times 10^3}{1.5\, t^3} \qquad (\text{Assuming } \sigma_b = 70 \text{ N/mm}^2)$$

$$\therefore \quad \mathbf{t} = 19.46 \text{ mm} \cong \mathbf{20\ mm} \text{ (say) and } \mathbf{B} = 3t = 3 \times 20 = \mathbf{60\ mm}$$

### 2.5.3.1 Applications of Hand Lever and Foot Lever

1. Handle of hand pump.
2. Handle of punching machine.
3. Lever of spring loaded safety valve.
4. Bell crank lever used in governors.
5. Paddle used in mechanical press machines.

## 2.5.4 Bell Crank Lever

### 2.5.4.1 Introduction

- In a bell crank lever, two arms (i.e. load arm and effort arm) of lever are perpendicular to each other.
- The arms of lever may be of rectangular, elliptical or I-section.

### 2.5.4.2 Applications of Bell Crank Lever

1. Governors.
2. Air extraction pump used in condenser.
3. Railway signal system.
4. Prime-movers (drive).

### 2.5.4.3 Design Procedure of Bell Crank Lever

**Step I : Calculate reaction ($R_F$) at fulcrum point 'F' by using following formula :**

$$R_F = \sqrt{W^2 + P^2} \quad \text{... (2.13)}$$

[To calculate W and P, (if one of them is given), take algebraic sum of moments of all forces about fulcrum and equate it to zero by applying the condition of equilibrium].

**Step II : Design of pin at fulcrum 'F' (i.e. fulcrum pin) :**

(a) Let, d = diameter of fulcrum pin and $l$ = length of fulcrum pin.

Load acting at fulcrum pin is $R_F$.

Considering the fulcrum pin subjected to bearing failure (under bearing pressure), we write,

$$\text{Bearing pressure, } P_b = \frac{\text{Load}}{\text{Bearing area}} = \frac{R_F}{l \times d} = \frac{R_F}{1.25d \times d} \quad \text{... [Assume } l = 1.25d]$$

Values of $l$ and d can be determined using the above equation.

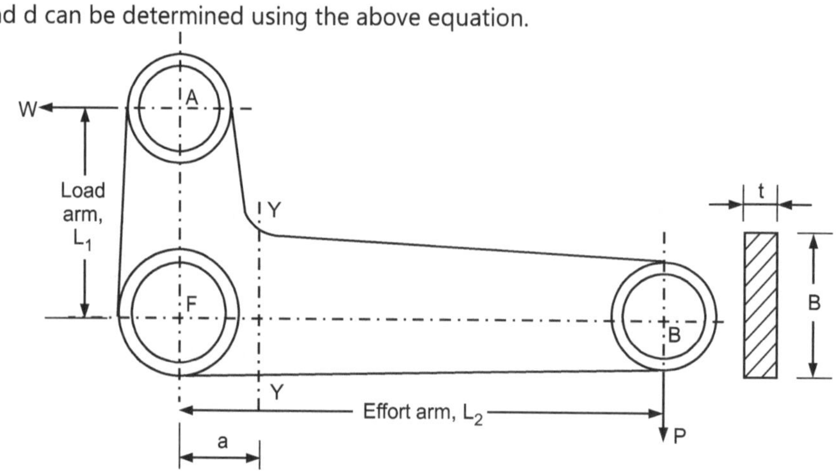


**Fig. 2.47 : Bell crank lever**

(b) To verify or confirm the calculated (designed) value of 'd', consider failure of fulcrum pin under shear failure. As the pin is in double shear, we have,

$$\tau_{ind} = \frac{R_F}{2 \times \left(\frac{\pi}{4} \cdot d^2\right)} \quad \text{... (2.14)}$$

Induced shear stress should be less than permissible stress value for safe design of fulcrum pin.

(c) A brass bush of 3 mm thickness is pressed into the boss of fulcrum to work as a bearing surface, to avoid friction between fulcrum pin and single or double eye (holes) of lever.

∴ Diameter of hole in the lever = d + (2 × 3) and Diameter of boss at fulcrum = 2d

**Step III : Design of pin at A :** Check whether, effort at 'A' is close to value of $R_F$ ? If yes, take same dimensions for pin at 'A' as that of fulcrum pin.

∴ $\mathbf{d_1}$ = diameter of pin at 'A' = **d**, And, $\boldsymbol{l_1}$ = length of pin at 'A' = $\mathbf{1.25\ d_1}$ or $\boldsymbol{l}$

**Step IV : Design of pin at B :**

Let, $d_2$ = diameter of pin at 'B', And, $l_2$ = length of pin at 'B'

Load acting at point B is effort (P) applied at the end of effort arm.

(a) Considering the pin at 'B' in bearing failure, we write,

$$P_b = \frac{\text{Load on pin located at point 'B'}}{\text{Bearing area}} = \frac{P}{d_2 \times l_2}$$

By assuming $l_2 = 1.25\ d_2$, we can find values of $d_2$ and $l_2$ using above equation.

(b) To verify or confirm the calculated (designed) value of $d_2$, we can consider the failure of pin at 'B' under shear.

We have, $$\tau_{ind} = \frac{P}{2 \times \left(\frac{\pi}{4} \cdot d_2^2\right)}$$ $\left[\begin{array}{l} \because \text{Pin is in double shear,} \\ \therefore A = 2 \times \frac{\pi}{4} d_2^2 \end{array}\right]$

Calculate the value of $\tau_{ind}$ by keeping values of P and $d_2$ in above equation.

If $\tau_{ind} < \tau_{permissible}$, then design of pin at 'B' is safe.

(c) **End B is forked end**. Calculate dimensions of forked end or double eye using the following empirical relations.

(i) Thickness of each eye = $t_1 = \frac{l_2}{2}$.

(ii) Inner diameter of each eye = $d_2 + (2 \times 3)$, where phosphor bronze bush of 3 mm thickness is provided in each eye.

(iii) Outer diameter of each eye = $D = 2d_2$.

**Step V : Design of cross-section of lever :**

Let, $t$ = Thickness of lever

$B$ = Width of lever

Now, let us consider a section, labelled as section Y-Y located on effort arm (horizontal arm) at a distance 'a' from centre of fulcrum point 'F'. We will find out thickness and width of lever at this section Y-Y. Distance 'a' can be arbitrarily chosen (i.e. approximated or assumed). While approximately choosing distance 'a', important tips given below may be helpful.

(i) Distance 'a' should be more than outer radius of boss of fulcrum point 'F'. Outer diameter of boss at fulcrum is equal to 2d. Therefore, outer radius of boss at fulcrum is equal to 'd', i.e. diameter of pin at fulcrum. In short, distance 'a' should be more than diameter 'd' of pin at fulcrum.

(ii) Section Y-Y, where dimensions of cross-section of lever are to be calculated, should not be located far away from fulcrum point 'F'. Therefore, 15 to 20 mm should be added to diameter of pin at fulcrum (d) to approximate the value of distance 'a'.

$\therefore$ Distance 'a' = d + 15 to 20 mm

**Maximum bending moment** acting at section Y-Y can be calculated as,

$$M = P \times (L_2 - a)$$

**Section modulus,** $$Z = \frac{1}{6} t B^2$$

Considering failure of lever due to bending, we can write,

**Bending stress,** $$\sigma_b = \frac{M}{Z} = \frac{P \times (L_2 - a)}{\frac{1}{6} t B^2}$$

If relation between B and t is unknown or not given in numerical problem, then width of lever may be assumed as, **B = 3t to 5t**.

Using the above equation of bending stress ($\sigma_b$), dimensions of cross-section of lever, i.e. B and t can be determined.

### 2.5.4.4 Necessity of Softer Material for Bushes Inserted in Eyes of Lever

- Refer step (IV) of article 2.5.4.3, where, it is mentioned to use hollow bushes of softer materials like gun metal, phosphor bronze etc.
- These hollow bushes are inserted in the eyes of lever, to avoid direct contact of pin to be inserted in the eyes of lever.
- Therefore, inside diameter of hollow bush should be equal to diameter of pin to be inserted.

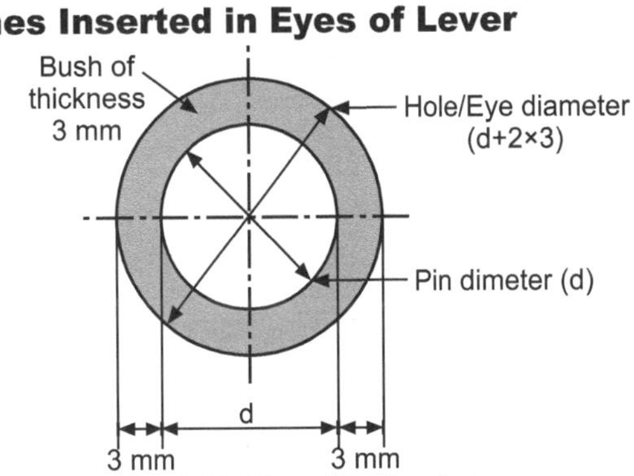


**Fig. 2.48 : Use of bushes in levers**

- Thickness of bush is generally taken as 3 mm. If d = diameter of pin, then diameter of hole/eye is equal to (d + 2 × 3) in milimetre. Outer diameter of bush is also equal to (d + 2 × 3) in milimetre.
- Refer the Fig. 2.48 to understand the following reasons for providing bushes of soft material.

**Advantages of inserting soft bush in between eye and pin :**

1. Friction is reduced to greater extent.
2. Wear rate of both materials, used for pin as well as pin hole is considerably reduced.
3. Life of lever increases.
4. If bushes are worn out, they can be easily replaced with new bushes.
5. Bushes are cheaper. Therefore, it is economically feasible to replace bushes, whenever preventive maintenance is carried out for levers.

**Numerical Type No. 6 : "Bell Crank Lever"**

**Problem 2.7 :** *A bell crank lever having one arm 500 mm and another arm 150 mm is used to lift a load of 5 kN. The permissible stresses for lever and pin materials in shear and tension are 60 N/mm² and 80 N/mm² respectively. The bearing pressure on the pin is to be limited to 10 N/mm². Determine,*

*(i) Diameter of pin, (ii) Diameter of boss, (iii) Dimensions of rectangular cross-section of lever.*

**Solution : Given :** W = 5000 N, $L_1$ = 150 mm, $L_2$ = 500 mm, $\tau$ = 60 N/mm², $\sigma_t$ = 80 N/mm², $P_b$ = 10 N/mm²

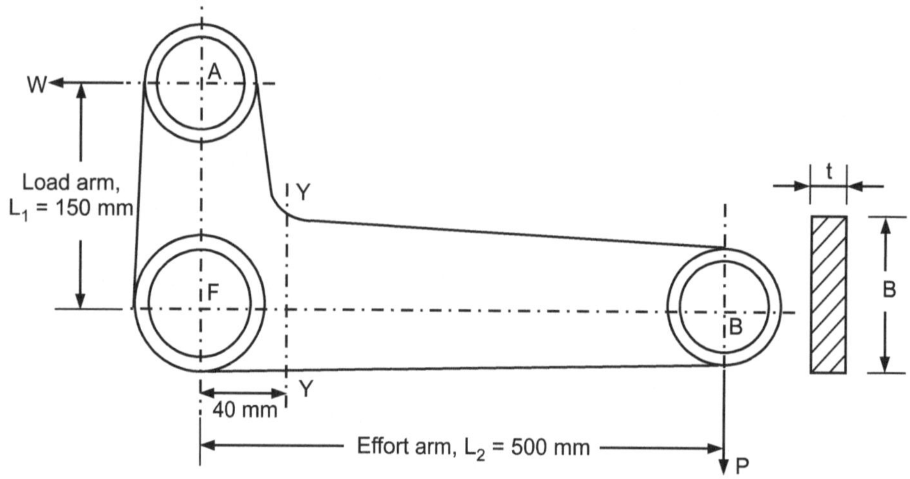


Fig. 2.49

**Procedure :** Value of permissible bending stress ($\sigma_b$) is not given in the Numerical Problem.

**Let us assume, $\sigma_b$ = 80 N/mm².**

**Step I : Calculate reaction ($R_F$) at fulcrum :**

By taking algebraic sum of moments of all forces about fulcrum 'F' and equating it to zero, we get,

$$P \times L_2 - W \times L_1 = 0$$

$$\therefore \quad P \times 500 = 5000 \times 150$$

$$\therefore \quad \mathbf{P = 1500\ N}$$

We have, $R_F = \sqrt{W^2 + P^2} = \sqrt{(5000)^2 + (1500)^2} = \mathbf{5220.15\ N}$

**Step II : Design of fulcrum pin :**

(a) Let, d = diameter of fulcrum pin

$l$ = length of fulcrum pin

We know that, the load on fulcrum pin is '$R_F$'.

Considering bearing failure of fulcrum pin,

$$\text{Bearing pressure } (P_b) = \frac{\text{Load}}{\text{Bearing area}} = \frac{R_F}{l \times d}$$

$$\therefore \quad 10 = \frac{R_F}{1.25d \times d} \qquad \textbf{[Assume } l = 1.25\ d\textbf{]}$$

$$\therefore \quad 10 = \frac{5220.15}{1.25d^2}$$

$$\therefore \quad \mathbf{d = 20.43\ mm \cong 21\ mm\ (say)}$$

$$\therefore \quad l = 1.25d = 1.25 \times 21 = \mathbf{26.25\ mm}$$

(b) To check the possibility of fulcrum pin under shear failure, we will find out the shear stress induced in the fulcrum pin, by keeping the values of $R_F$ and d in following design equation.

We have, $\tau_{ind} = \dfrac{R_F}{2 \times \left(\dfrac{\pi}{4} \cdot d^2\right)} = \dfrac{5220.15}{2 \times \left[\dfrac{\pi}{4} \times (21)^2\right]}$ [∵ Pin is in double shear]

∴ $\tau_{ind}$ = **7.53 N/mm² < 60 N/mm² (Permissible stress)**

Since the value of induced shear stress (7.53 N/mm²) is less than given permissible value of shear stress (60 N/mm²), therefore, design of pin at fulcrum is safe.

(c) A brass bush of 3 mm thickness is pressed into the boss of fulcrum as "bearing".

∴ Diameter of hole in the lever = d + (2 × 3) = 21 + 6 = **27 mm**

And Diameter of boss of fulcrum = 2 × d = 2 × 21 = **42 mm**

**Step III : Design of pin at 'A' :** As the load 'W' (5000 N) at 'A' is close to the value of $R_F$ (5220.15 N), we will take same dimensions for pin at 'A' as fulcrum pin (i.e. pin at fulcrum point 'F').

∴ $d_1$ = Diameter of pin at 'A' = d = **21 mm**

∴ $l_1$ = Length of pin at 'A' = $l$ = **26.25 mm**

**Step IV : Design of pin at 'B' :** Let, $d_2$ = diameter of pin at 'B', $l_2$ = length of pin at 'B'

(a) We know that, effort applied at point 'B' is equal to P, i.e. 1500 N.

Considering bearing failure of pin at 'B', we can write,

$$P_b = \frac{P}{d_2 \times l_2}$$

$$\therefore \quad 10 = \frac{1500}{d_2 \times 1.25 \times d_2} \qquad \textbf{(Assume } l_2 = 1.25\, d_2\textbf{)}$$

∴ $d_2$ = **10.95 mm ≅ 12 mm (say)**

and $l_2$ = 1.25 $d_2$ = 1.25 × 12 = **15 mm**

(b) To check the possibility of pin at point 'B', under shear failure, we will calculate shear stress induced in the pin by keeping values of P and $d_2$ in following design equation.

$$\text{Induced shear stress,} \quad \tau = \frac{P}{2 \times \left(\frac{\pi}{4} \times d_2^2\right)} = \frac{1500}{2 \times \left(\frac{\pi}{4} \times (12)^2\right)}$$

∴ $\tau_{ind}$ = **6.63 N/mm² < 60 N/mm² (Permissible stress)**

Since induced shear stress (6.63 N/mm²) is less than given permissible value of shear stress (60 N/mm²), design of pin at point 'B' is safe.

(c) As end 'B' is forked end (i.e. end having double eye), we will use empirical relations to find out dimensions of forked end.

Thickness of each eye = $t_1 = \dfrac{l_2}{2} = \dfrac{15}{2}$ = **7.5 mm**, and

Inner diameter of each eye = $d_2$ + (2 × 3) = 15 + 6 = **21 mm**

where, phosphor bronze bush of 3 mm thickness is provided in each eye to provide bearing surface.

Outer diameter of each eye = D = $2d_2$ = 2 × 15 = **30 mm**

**Step (V) : Design of lever :** Consider a section Y-Y at an arbitrarily or randomly chosen distance 'a' from centre of fulcrum.

Distance 'a' = d + 15 to 20 mm

∴ **a** = 21 + 19 = **40 mm** (Here, 19 mm is randomly chosen, which lies between 15 to 20 mm)

Let, t = Thickness of lever at section Y-Y and B = Width of lever at section Y-Y

Maximum bending moment at section Y-Y = P × ($L_2$ – a) = 1500 × (500 – 40)

∴ **M** = **6.9 × 10⁵ N.mm**

$$\text{Section modulus, } Z = \frac{1}{6} \times t \times B^2 = \frac{1}{6} \times t \times (3t)^2 \qquad \textbf{(Assume B = 3t)}$$

∴ **Z** = $\mathbf{1.5 \times t^3}$

We have, $\sigma_b = \dfrac{M}{Z}$

$$\therefore \quad 80 = \frac{6.9 \times 10^5}{1.5 \times t^3} \qquad \text{[Assumed value, } \sigma_b = 80 \text{ N/mm}^2\text{)}$$

∴ t = **17.91 mm ≅ 18 mm (say)**

∴ B = **3t = 3 × 18 = 54 mm**

## 2.7 ECCENTRIC LOADING

- An externally applied load on a machine part is said to be **eccentric**, if its line of action is parallel, but does not coincide with the centroidal axis of machine part.
- Distance between the line of action of eccentric load and centroidal axis of component is called as *eccentricity* and denoted by 'e'.
- **Examples :** 'C' clamp, Brackets, Offset links.
- Consider a short column of rectangular cross-section subjected to eccentric compressive load 'W' at a distance 'e' from the centroidal axis of column as shown in Fig. 2.50 (a).
- Let us introduce two equal and opposite forces of magnitude $W_1$ and $W_2$, without changing (altering) the equilibrium of column. Also, $W_1 = W_2 = W$. Refer Fig. 2.50 (b).
- Force $W_1$ will induce a direct compressive stress over the entire cross-section of column. Magnitude of this direct compressive stress is given by, $\sigma_o = \dfrac{W_1}{A} = \dfrac{W}{A}$

  where, A = cross-sectional area of column. Refer Fig. 2.50 (c).

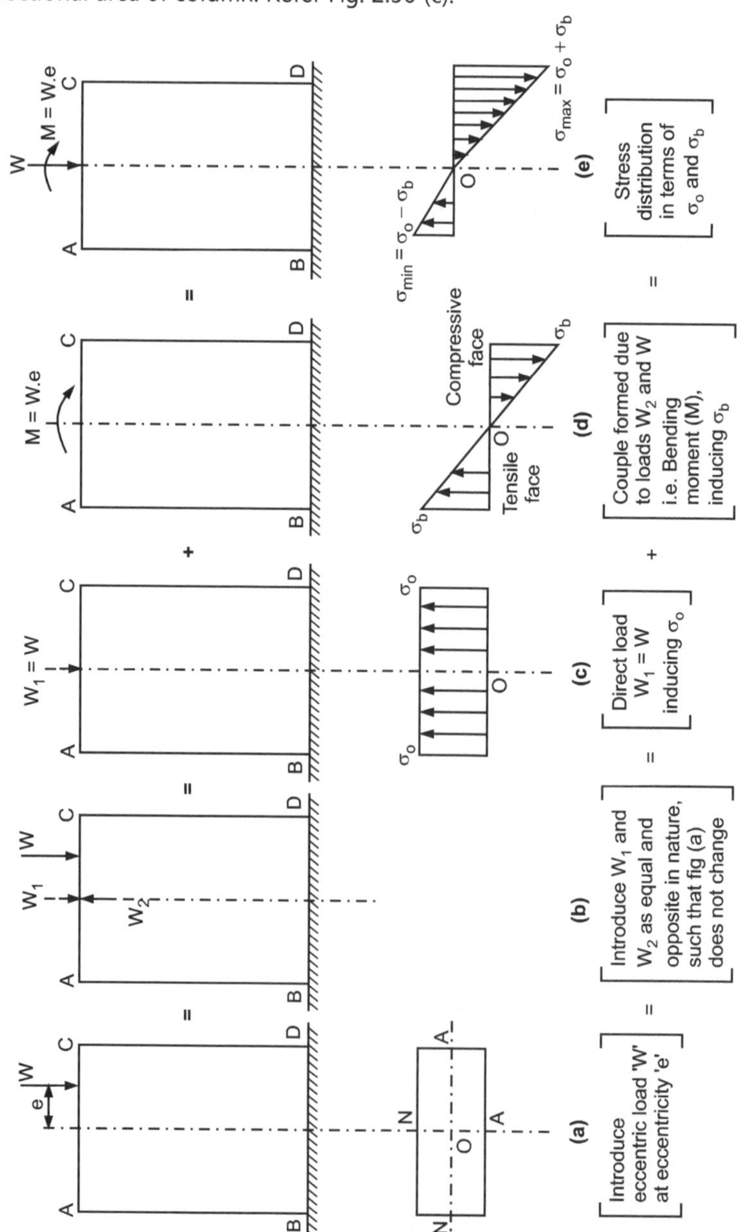


**Fig. 2.50 : Concept of eccentric loading**

- Forces $W_2$ and W will form a couple equal to '$W \times e$', which will induce bending stress.
- Bending stress induced on face 'AB' will be tensile in nature, whereas, bending stress induced on face 'CD' will be compressive in nature. Refer Fig. 2.50 (d).

∴ Maximum bending stress induced at faces 'AB' and 'CD' can be given by,

$$\sigma_b = \frac{M}{Z} = \frac{W \cdot e}{I/y} = \frac{W \cdot e \cdot y}{I}$$

where, $M$ = Bending moment = $W \cdot e$

$$Z = \text{Section modulus} = \frac{\text{Moment of inertia}}{\text{Distance of N.A. from outermost fibre}}$$

∴ $Z = \dfrac{I}{y_t}$ or $\dfrac{I}{y_c}$, for tensile and compressive stress respectively.

Here, $y_t$ and $y_c$ are the distances of neutral axis from the outermost fibres from tensile and compressive faces respectively.

∴ **Maximum resultant stress at CD = $\sigma_{max} = \sigma_o + \sigma_b$** $= \dfrac{W}{A} + \dfrac{M}{Z} = \dfrac{W}{A} + \dfrac{W \cdot e \cdot y}{I}$

For rectangular cross-section of width 'b' and thickness 't',

$$\sigma_{max} = \frac{W}{b \cdot t} + \frac{W \cdot e \cdot (b/2)}{\frac{1}{12} tb^3} \qquad \left[\because A = b \times t \text{ and } y = \frac{b}{2}\right]$$

∴ $$\sigma_{max} = \frac{W}{b \cdot t} + \frac{6W \cdot e}{t\,b^2}$$

**Note :** For understanding the induced stresses $\sigma_o$ (direct stress) and $\sigma_b$ (bending stress) please refer the stress distribution shown in Fig. 2.50 (c) and (d).

Also, **minimum resultant stress at AB = $\sigma_{min} = \sigma_o - \sigma_b$** $= \dfrac{W}{A} - \dfrac{M}{Z}$

∴ $$\sigma_{min} = \frac{W}{A} - \frac{W \cdot e \cdot y}{I}$$

∴ $$\sigma_{min} = \frac{W}{b \cdot t} - \frac{W \cdot e \cdot \left(\frac{b}{2}\right)}{\frac{1}{12} tb^3}$$

∴ $$\sigma_{min} = \frac{W}{b \cdot t} - \frac{6 \cdot W \cdot e}{t\,b^2}$$

The resultant stress distribution diagram (i.e. $\sigma_{max}$ and $\sigma_{min}$) is as shown in Fig. 2.50 (e).

## 2.7.1 Design Procedure of C-clamp

**Step I :** Considering load 'W' as direct tensile load, we find that, direct stress $\sigma_o$ will be induced in section.

$$\text{Direct stress, } \sigma_o = \frac{W}{A} = \frac{W}{(b \times t)}$$

∴ $$\sigma_o = \frac{W}{t \times 2t} = \frac{W}{2t^2} \quad \text{... (2.15) (Assuming b = 2t)}$$

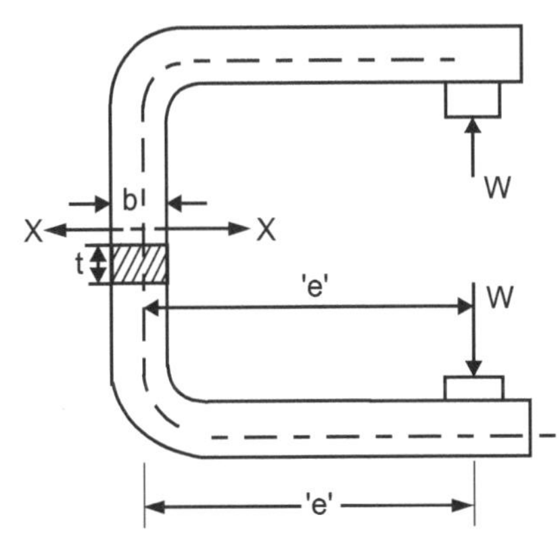


**Fig. 2.51 : C-clamp**

**Step II :** Considering load 'W' as eccentric load, we find that, since load axis does not pass through section of C-clamp and lies at an eccentricity 'e', section of C-clamp will be subjected to bending moment. Therefore, bending stress will be induced in the clamp.

Bending stress, $\sigma_b = \frac{M}{Z} = \frac{W \cdot e}{\frac{1}{6} t\, b^2} = \frac{6 \cdot W \cdot e}{t \times (2t)^2}$ $(\because b = 2t)$

$$\therefore \quad \sigma_b = \frac{6 \cdot W \cdot e}{4t^3} \quad \ldots (2.16)$$

**Step III :** Maximum resultant stress induced in C-clamp is given by,

$$\sigma_{tR} = \sigma_o + \sigma_b$$

$$\therefore \quad \sigma_{tR} = \frac{W}{2t^2} + \frac{6 \cdot W \cdot e}{4t^3} \quad \ldots (2.17)$$ [From equations (2.15) and (2.16)]

Equating the above equation (2.17) to permissible stress, we can find values of 't' and hence 'b'.

[**Note:** Ratio $\left(\frac{b}{t}\right)$ may be assumed in the range of 2 to 5, if not given in the numerical.]

## Numerical Type No. 7 : "C-clamp"

**Problem 2.9 :** *Design 'C' clamp frame for a total clamping force of 20 kN. Cross-section of frame is rectangular and width to thickness ratio is 2. Distance between load line and neutral axis of rectangular cross-section is 120 mm and gap between two faces is 180 mm. Frame is made of cast steel, for which, permissible tensile stress is 100 N/mm².*

**Solution : Given data : W** = 20 kN = **$20 \times 10^3$ N**, b = 2t,

Safe tensile stress for C-clamp = $\sigma_t$ = 100 N/mm²

**Procedure : Step I :** Considering 'W' as direct load, direct stress induced is given by,

$$\sigma_o = \frac{W}{A} = \frac{W}{(b \times t)} = \frac{20 \times 10^3}{(2t \times t)} = \frac{20 \times 10^3}{2t^2} = \frac{\mathbf{10 \times 10^3}}{\mathbf{t^2}} \quad \ldots (1)$$

**Step II :** Due to eccentricity in application of load, section will be subjected to bending moment (M). Therefore, bending stress will be induced in C-clamp, which is given by,

$$\sigma_b = \frac{M}{Z} = \frac{W \times e}{\frac{1}{6} t\, b^2} = \frac{6 \times W \times e}{tb^2} = \frac{6 \times 20 \times 10^3 \times 120}{t \times (2t)^2} = \frac{\mathbf{36 \times 10^5}}{\mathbf{t^3}} \quad \ldots (2)$$

where, M = W × e and Z = section modulus = $\frac{1}{6} t\, b^2$

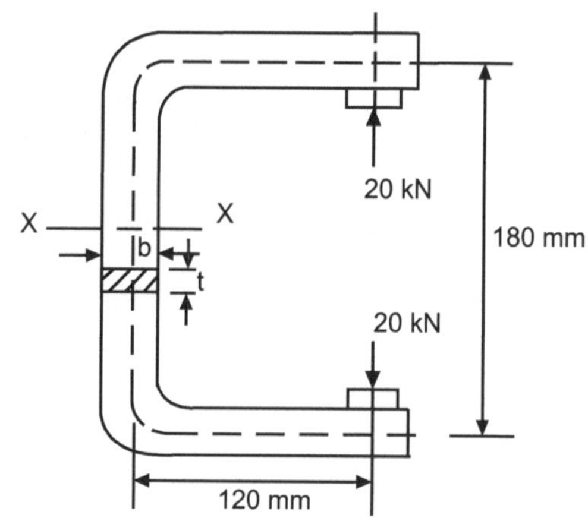

**Fig. 2.52**

**Step III :** Maximum resultant stress induced in C-clamp is given by,

$$\sigma_{tR} = \sigma_o + \sigma_b$$

On rearranging the above equation, we get

$$\sigma_o + \sigma_b = \sigma_{tR}$$

$$\therefore \quad \frac{10 \times 10^3}{t^2} + \frac{36 \times 10^5}{t^3} = 100$$ [From equations (1) and (2)]

$$\therefore \quad \frac{10 \times 10^3\, t + 36 \times 10^5}{t^3} = 100$$

$$\therefore \quad 100t^3 - 10 \times 10^3 t - 36 \times 10^5 = 0$$

$\therefore$ **t** = 34.02 mm ≅ **36 mm** and **b** = 2t = 2 × 36 = **72 mm**

**Problem 2.10:** *Spindle of a drilling machine as shown in Fig. 2.53 is subjected to a maximum axial load of 10 kN during operation. Determine the diameter of solid cast iron column, if the permissible tensile stress is 40 N/mm². Distance between axis of spindle and axis of column is 350 mm. Also, find direct stress and stress due to bending in column.*

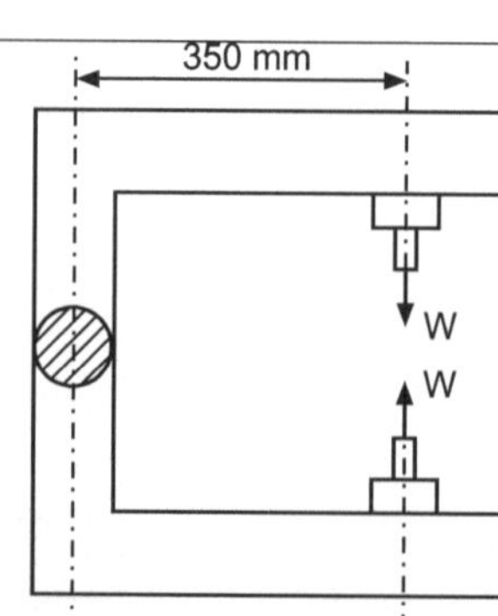

**Fig. 2.53**

**Solution : Given data : W** = 10 kN = **$10 \times 10^3$ N**, e = 350 mm, $\sigma_t$ = 40 N/mm$^2$

**Procedure :**

**Step I : Direct stress :** $\sigma_o = \frac{W}{A} = \frac{W}{\frac{\pi}{4}d^2} = \frac{10 \times 10^3 \times 4}{\pi \cdot d^2} = \frac{\mathbf{12732.39}}{\mathbf{d^2}}$ ... (1)

**Step II : Bending stress :** $\sigma_b = \frac{M}{Z} = \frac{W \times e}{\frac{\pi}{32}d^3} = \frac{10 \times 10^3 \times 350 \times 32}{\pi \times d^3} = \frac{\mathbf{35.65 \times 10^6}}{\mathbf{d^3}}$ ... (2)

**Step III : Maximum resultant stress :** $\sigma_{t_R} = \sigma_o + \sigma_b$

$\therefore \quad 40 = \frac{(12732.39)}{d^2} + \frac{(35.65 \times 10^6)}{d^3}$ [From equation (1) and (2)]

$\therefore \quad 40 = \frac{(12732.39\ d) + (35.65 \times 10^6)}{d^3}$

$\therefore \quad 40d^3 - 12732.39\ d - 35.65 \times 10^6 = 0$

$\mathbf{d = 97.34\ mm \cong 98\ mm\ (say)}$

From equation (1), **Direct stress,** $\sigma_o = \frac{12732.39}{d^2} = \frac{12732.39}{(98)^2} = \mathbf{1.32\ N/mm^2}$

From equation (2), **Bending stress,** $\sigma_b = \frac{35.65 \times 10^6}{d^3} = \frac{35.65 \times 10^6}{(98)^3} = \mathbf{37.87\ N/mm^2}$

---

**Problem 2.11 :** *Fig. 2.54 shows hacksaw. Blade is assembled with tension of W = 320 N. Frame of hacksaw is made of cold drawn steel, for which, yield point value in tension is 360 N/mm². Design a rectangular cross section of U-frame at sections 'A-A' and 'B-B'. Assume factor of safety = 4 and b = 2.5 t.*

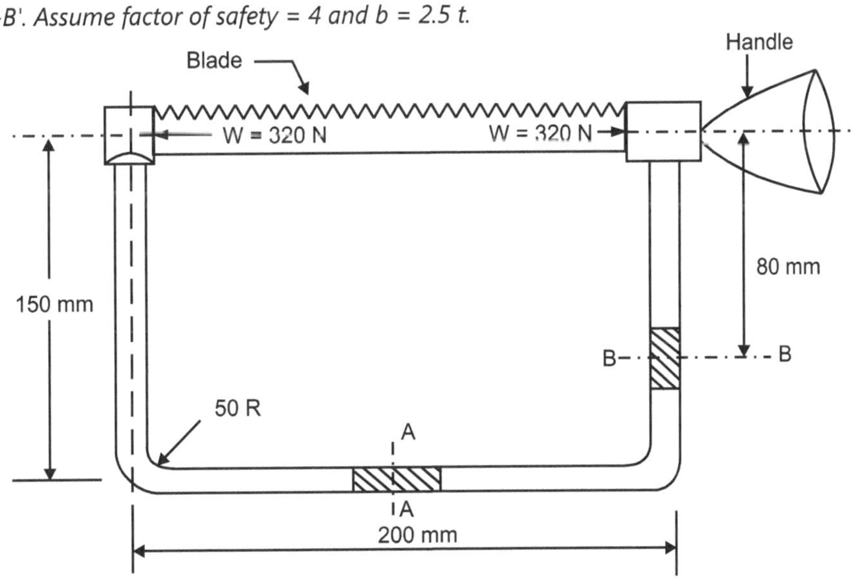


**Fig. 2.54**

**Solution : Data given :** W = 320 N, $\sigma_{yt}$ = 360 N/mm$^2$, F.O.S. = 4, b = 2.5 t

**Procedure :**

$\begin{bmatrix}\text{Allowable stress} \\ \text{in tension}\end{bmatrix} = \frac{\text{Yield point stress}}{\text{F.O.S.}} = \frac{360}{4} = \mathbf{90\ N/mm^2}$

**(1) At section A-A :**

**Step I : Direct stress :** $\sigma_o = \frac{W}{A} = \frac{320}{b \times t} = \frac{320}{(2.5t) \times t} = \frac{\mathbf{128}}{\mathbf{t^2}}$

**Step II : Bending stress :**

$\sigma_b = \frac{M}{Z} = \frac{W \times e}{(1/6)\ tb^2} = \frac{320 \times 150 \times 6}{t \times (2.5t)^2} = \frac{\mathbf{46080}}{\mathbf{t^3}}$ [$\because$ b = 2.5t]

**Step III : Maximum resultant stress :**

$$\sigma_{tR} = \sigma_o + \sigma_b$$

$$\therefore \quad 90 = \frac{128}{t^2} + \frac{46080}{t^3}$$

$$\therefore \quad 90 = \frac{128t + 46080}{t^3}$$

$$\therefore \quad 90t^3 - 128t - 46080 = 0$$

$$\therefore \quad \mathbf{t} = 8.06 \text{ mm} \cong \mathbf{9 \text{ mm (say)}}$$

$$\therefore \quad \mathbf{b} = 2.5\,t = 2.5 \times 9 = \mathbf{22.5 \text{ mm}}$$

**(2) At section B-B :**

- Frame is uniform in section throughout. Therefore, we must check the induced stresses considering section B-B. If the values of induced stresses are less than permissible values, then the above design is safe.
- Use b = 22.5 mm and t = 9 mm. Here, eccentricity is 80 mm. ∴ **e = 80 mm**.
- Bending stress induced at section B-B,

$$\sigma_{b\,ind} = \frac{M}{Z} = \frac{W \times e}{\frac{1}{6} tb^2}$$

$$\therefore \quad \sigma_{b\,ind} = \frac{320 \times 80 \times 6}{9 \times (22.5)^2} = \mathbf{33.71 \text{ N/mm}^2 < 90 \text{ N/mm}^2 \text{ (Permissible stress)}}$$

In addition to bending stress, the cross-section is also subjected to transverse shear stress. It is given by,

$$\tau_{ind} = 1.5\,\sigma_o$$

$$\therefore \quad \tau_{ind} = 1.5 \times \frac{W}{A} = 1.5 \times \frac{W}{(b \times t)} = 1.5 \times \frac{320}{(9 \times 22.5)} = \mathbf{2.37 \text{ N/mm}^2 < 90 \text{ N/mm}^2 \text{ (Permissible stress)}}$$

Since both induced stresses, $\sigma_b$ and $\tau$ are less than 90 N/mm$^2$, the design is said to be safe.

∴ **At section B-B :** We can take,

$$b = \mathbf{22.5 \text{ mm}}$$

$$t = \mathbf{9 \text{ mm}}$$

## 2.7.2 Design Procedure of Offset Links

Fig. 2.55 shows an offset link subjected to tensile load 'W'. Since, load axis and neutral axis (N.A.) of offset link are not same, therefore, distance between them is called as **eccentricity 'e'** and externally applied load 'W' is called as **eccentric load**. Centroidal Y-Y axis is nothing but, **neutral axis** of link.

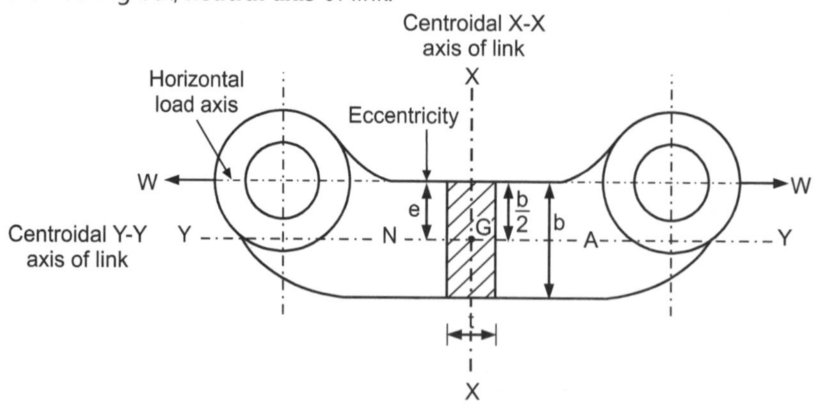


**Fig. 2.55 : Offset link**

Let,
- b = Width of rectangular section at X-X.
- t = Thickness of rectangular section at X-X.
- e = Eccentricity.
- W = Eccentric load on offset link [Tensile].
- $\sigma_o$ = Direct tensile stress
- $\sigma_b$ = Bending stress

As the load 'W' is eccentric load, cross-section of link will be simultaneously subjected to both types of stresses, direct stress as well as bending stress.

**Step I : Direct stress :** $\sigma_o = \frac{W}{A} = \frac{W}{b \times t}$

**Step II : Bending stress :** $\sigma_b = \frac{M}{Z_{yy}} = \frac{W \cdot e}{\frac{1}{6} \cdot tb^2} = \frac{6\,W \cdot e}{t\,b^2}$

**Step III :** Maximum resultant stress is obtained by,

$$\sigma_{t_R} = \sigma_o + \sigma_b$$

$$\therefore \quad \sigma_{t_R} = \frac{W}{b \times t} + \frac{6W \cdot e}{tb^2} \qquad \text{... (2.18)}$$

where, $e = \text{eccentricity} = \frac{b}{2}$ [Refer Fig. 2.55]

Now, in equation (2.18), put permissible stress for material used to made offset link as maximum resultant stress ($\sigma_{t_R}$), values of load (W) and eccentricity (e) to find out dimensions 'b' and 't' of the rectangular cross-section of offset link. If relation between width and thickness is not given in problem, then assume ratio $\left(\frac{b}{t}\right)$ in the range of 2 to 5.

## Numerical Type No. 8 : "Off-Set Link"

**Problem 2.12 :** *Design an off-set link for a load of 1000 N. Maximum permissible stress in tension for link material is limited to 60 N/mm². Assume b = 3t for rectangular cross-section of link.*

**Solution : Given Data :** $W = 1000$ N, $\sigma_t = 60$ N/mm², $b = 3t$

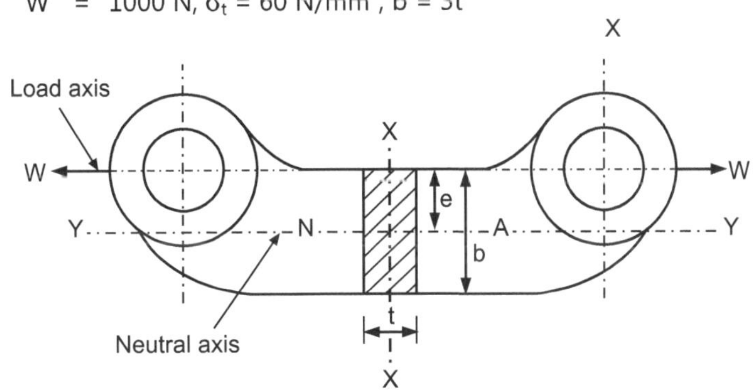


**Fig. 2.56**

**Procedure : Step I : Direct stress :**

$$\boldsymbol{\sigma_o} = \frac{W}{A} = \frac{1000}{b \times t} = \frac{1000}{3t \times t} = \mathbf{\frac{1000}{3t^2}} \qquad [\because b = 3t] \text{ ... (1)}$$

**Step II : Bending stress :** $\sigma_b = \frac{M}{Z_{yy}} = \frac{W \cdot e}{\frac{1}{6} \cdot t \cdot b^2}$

$$\therefore \quad \sigma_b = \frac{1000 \times \frac{b}{2}}{\frac{1}{6} \cdot t \cdot (3t)^2} = \frac{1000 \times 3t \times 6}{2 \times t \times 9t^2} \qquad \left[\because e = \frac{b}{2} \text{ and } b = 3t\right]$$

$$\therefore \quad \boldsymbol{\sigma_b} = \mathbf{\frac{1000}{t^2}} \qquad \text{... (2)}$$

**Step III : Maximum resultant stress will be,**

$$\sigma_{t_R} = \sigma_o + \sigma_b$$

$$\therefore \quad 60 = \frac{1000}{3t^2} + \frac{1000}{t^2} \qquad \text{[From equations (1) and (2)]}$$

$$\therefore \quad 60 = \frac{1000 + (1000 \times 3)}{3t^2}$$

$$\therefore \quad t^2 = \frac{4000}{60 \times 3}$$

$$\therefore \quad \mathbf{t} = 22.22 \text{ mm} \cong \mathbf{24\ mm\ (say)}$$

$$\text{and} \quad \mathbf{b} = 3t = 3 \times 24 = \mathbf{72\ mm}$$

**Problem 2.13 :** *Fig. 2.57 shows an offset link carrying load 'W' in N. The safe working stresses for link material are $\sigma_t$ = 20 MPa and $\sigma_c$ = 80 MPa. Find the magnitude of load 'W'.*

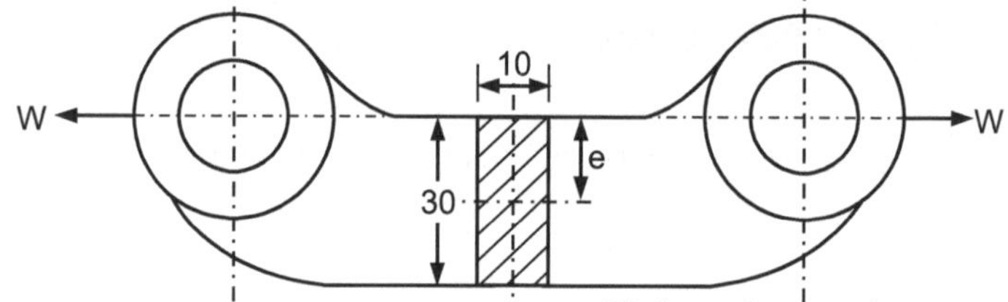


Fig. 2.57

**Solution : Given data :**

$$t = 10 \text{ mm, } b = 30 \text{ mm, } \therefore \mathbf{e} = \frac{b}{2} = \frac{30}{2} = \mathbf{15\ mm}$$

$$\sigma_t = 20 \text{ N/mm}^2, \sigma_c = 80 \text{ N/mm}^2$$

**Procedure :**

**Step I : Direct stress :**

$$\sigma_o = \frac{W}{A} = \frac{W}{b \times t} = \frac{W}{30 \times 10} = \mathbf{\frac{W}{300}} \qquad \text{... (1)}$$

**Step II : Bending stress : $\sigma_b$**

$$\sigma_b = \frac{M}{Z_{YY}} = \frac{W \times e}{\frac{1}{6} tb^2} = \frac{W \times 15}{\frac{1}{6} \times 10 \times (30)^2} = \mathbf{\frac{W}{100}} \qquad \text{... (2)}$$

**Step III :**

- When a component is subjected to direct stress as well as bending stress simultaneously, then maximum resultant stress will be equal to sum of direct stress and bending stress.
- From equations (1) and (2), it can be easily understood that, value of $\sigma_b$ will be more than $\sigma_o$. Therefore, minimum resultant stress, $\sigma_{min} = \sigma_o - \sigma_b$ will be negative and hence, it is of compressive nature, because offset link is subjected to tensile load. So, it is obvious that, maximum resultant stress will be of tensile nature.

$$\therefore \quad \sigma_{max} = \sigma_t = 20 \text{ N/mm}^2 \text{ and } \sigma_{min} = \sigma_c = 80 \text{ N/mm}^2$$

**Maximum resultant stress,**

$$\sigma_{max} = \sigma_t = \sigma_b + \sigma_o$$

$$\therefore \quad 20 = \frac{W}{100} + \frac{W}{300}$$

$$\therefore \quad W = \mathbf{1500\ N}$$

**Minimum resultant stress,**

$$\sigma_{min} = \sigma_c = \sigma_b - \sigma_o$$

$$\therefore \quad 80 = \left[\frac{W}{100} - \frac{W}{300}\right]$$

$$\therefore \quad W = \mathbf{12000\ N}$$

**Problem 2.14 :** *A mild steel link, as shown in Fig. 2.58, by full lines, transmits a pull of 80 kN. Find the dimensions b and t, if b = 3t. Assume the permissible tensile stress as 70 MPa. If an unsymmetrical one, as shown by dotted lines, replaces the original link having the same thickness t, find new depth $b_1$, using the same permissible stress as before.*

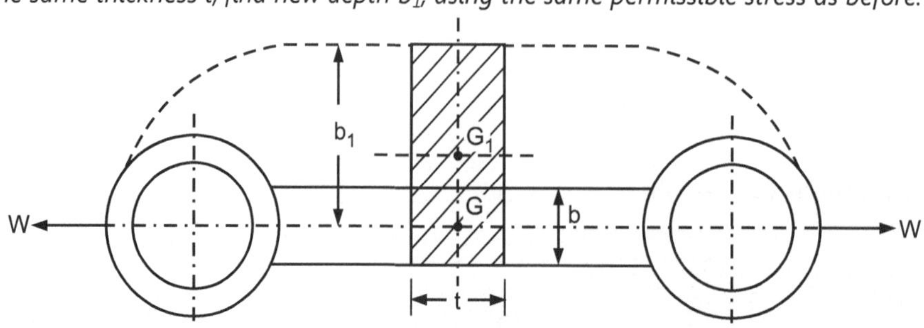


Fig. 2.58

**Solution : Given data : W** = 80 kN = **$80 \times 10^3$ N**, b = 3t, Safe tensile stress for link = $\sigma_t$ = 70 N/mm$^2$.

**Procedure : Case I :**

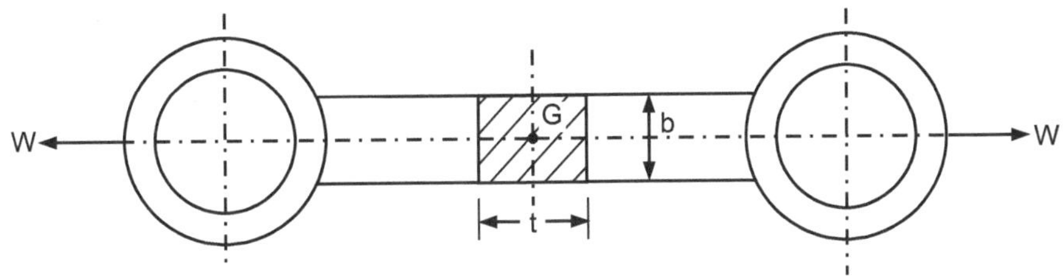


**Fig. 2.59 : Symmetrical link**

When the link is in the position shown by full lines (Refer Fig. 2.61), then it is subjected to direct tensile stress, which is given by,

$$\sigma_t = \frac{W}{A} = \frac{W}{(b \times t)} = \frac{W}{(3t \times t)} \qquad (\because b = 3t)$$

$$\therefore \quad 70 = \frac{80 \times 10^3}{3t^2}$$

$$\therefore \quad \mathbf{t = 19.51\ mm \cong 20\ mm\ (say)} \text{ and } \mathbf{b} = 3t = 3 \times 20 = \mathbf{60\ mm}$$

**Case II :**

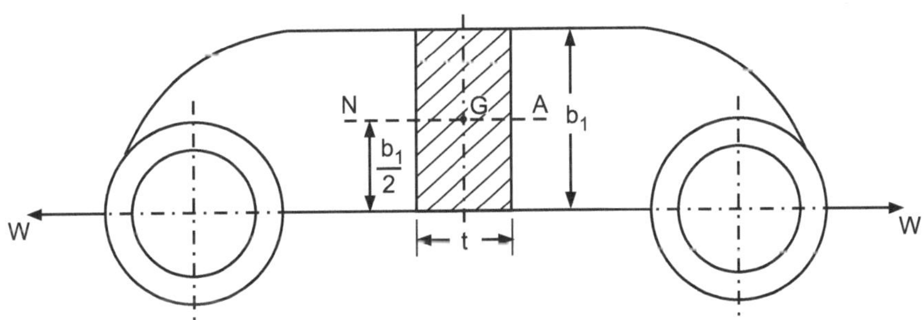


**Fig. 2.60 : Offset link**

When the link is in the position as shown in Fig. 2.60, it will be subjected to direct stress as well as bending stress. Here, the new depth will be $b_1$, but thickness will remain same, t = 20 mm.

$$\textbf{Direct stress, } \sigma_o = \frac{W}{A} = \frac{W}{(b_1 \times t)} = \frac{80 \times 10^3}{(b_1 \times 20)} = \frac{\mathbf{4 \times 10^3}}{\mathbf{b_1}} \qquad \ldots (1)$$

$$\textbf{Bending stress, } \sigma_b = \frac{M}{Z}$$

But, M = W × e and $\quad Z$ = Section modulus $= \frac{1}{6} t b_1^2$

$$\therefore \quad \sigma_b = \frac{W \times e}{\frac{1}{6} t\, b_1^2} = \frac{6 \times W \times e}{t\, b_1^2} = \frac{6 \times 80 \times 10^3 \times (b_1/2)}{20 \times b_1^2} = \frac{\mathbf{12 \times 10^3}}{\mathbf{b_1}} \qquad \ldots (2)$$

Maximum resultant stress induced in offset link will be,

$$\sigma_{max} = \sigma_o + \sigma_b$$

$$\therefore \quad \sigma_t = \frac{4 \times 10^3}{b_1} + \frac{12 \times 10^3}{b_1} \qquad \text{[From equations (1) and (2)]}$$

$$\therefore \quad 70 = \frac{16 \times 10^3}{b_1}$$

$$\therefore \quad \mathbf{b_1} = 228.57 \text{ mm} \cong \mathbf{230\ mm\ (say)}$$

## Numerical Type No. 9 : "Overhang Crank"

**Problem 2.15 :** *Shaft of an overhang crank subjected to a force (W) of 1 kN is shown in Fig. 2.61. The shaft is made of plain carbon steel 45C8 for which, tensile yield strength is 380 N/mm². The factor of safety is 2. Determine the diameter of the shaft using the maximum shear stress theory of failure.*

**Solution : Given data :** $\sigma_{yt}$ = 380 N/mm², **W** = 1 kN = **1000 N**, F.S. = 2.

**Procedure :** According the maximum shear stress theory,

$$\mathbf{S_{sy}} = 0.5\,\sigma_{yt} = 0.5 \times 380 = \mathbf{190\ N/mm^2}$$

∴ Permissible shear stress ($\tau$) will be,

$$\tau = \frac{S_{sy}}{F.S.} = \frac{190}{2} = \mathbf{95\ N/mm^2} \qquad \text{... (1)}$$

The stresses are critical at point 'A', which is subjected to a combined bending and twisting moment.

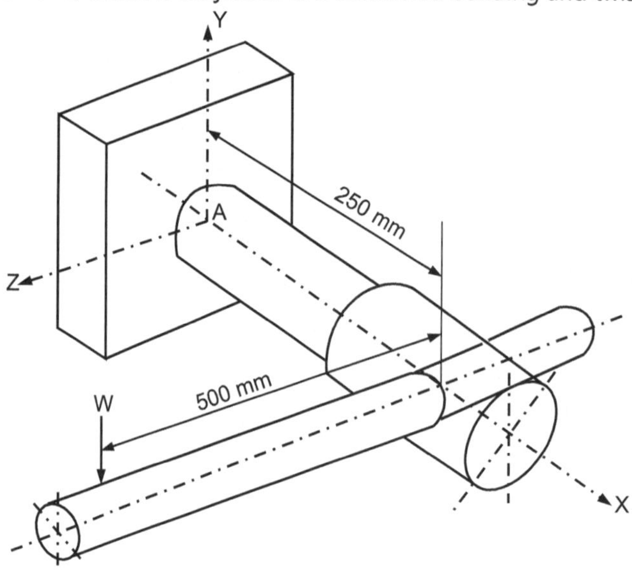


**Fig. 2.61**

$$\mathbf{M} = W \times 250 = 1000 \times 250 = \mathbf{250 \times 10^3\ N\text{-}mm}$$

$$\mathbf{T} = W \times 500 = 1000 \times 500 = \mathbf{500 \times 10^3\ N.mm}$$

We have,

$$\boldsymbol{\sigma_b} = \frac{M}{Z} = \frac{250 \times 10^3}{\frac{\pi}{32} \cdot d^3} = \mathbf{\frac{2546.48 \times 10^3}{d^3}}$$

and

$$\tau = \frac{16T}{\pi d^3} = \frac{16 \times 500 \times 10^3}{\pi \cdot d^3} = \mathbf{\frac{2546.48 \times 10^3}{d^3}}$$

**According to maximum shear stress theory,**

$$\tau_{max} = \frac{1}{2}\sqrt{\sigma_b^2 + 4\tau^2}$$

$$\therefore \quad 95 = \frac{1}{2}\sqrt{\left[\frac{2546.48 \times 10^3}{d^3}\right]^2 + 4\left[\frac{2546.48 \times 10^3}{d^3}\right]^2}$$

$$\therefore \quad d = \mathbf{31.06\ mm \cong 32\ mm\ (say)}$$

**Problem 2.16 :** *Find the maximum principal and shear stresses induced at the centre of a crankshaft bearing, if the overhang crank carries a tangential load of 20 kN at the centre of crank pin as shown in Fig. 2.62.*

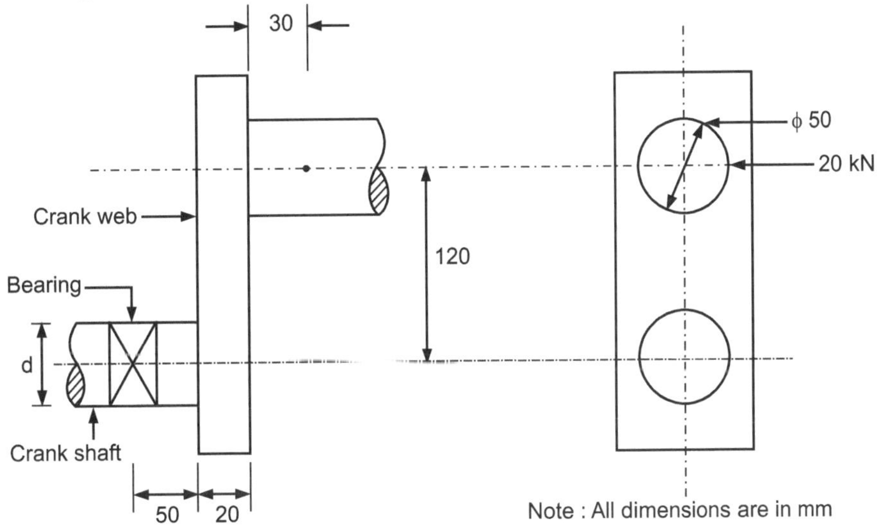


**Fig. 2.62**

**Solution : Given data : W** = 20 kN = **$20 \times 10^3$ N**, ***l*** = 50 + 20 + 30 = **100 mm**, L = 120 mm, d = 50 mm.

**Procedure :** Torque transmitted at the axis of shaft is given by,

$$\mathbf{T} = W \times L = 20 \times 10^3 \times 120 = \mathbf{2.4 \times 10^6\ N.mm}$$

Bending moment induced in the crank is given by,

$$\mathbf{M} = W \times l = 20 \times 10^3 \times 100 = \mathbf{2 \times 10^6\ N.mm}$$

Torsional shear stress induced in the crank,

$$\tau = \frac{16T}{\pi d^3} = \frac{16 \times 2.4 \times 10^6}{\pi \times (50)^3} = \mathbf{97.78\ N/mm^2}$$

Bending stress induced in the crank,

$$\sigma_b = \frac{M}{Z} = \frac{M}{\frac{\pi}{32} d^3} = \frac{2 \times 10^6}{\frac{\pi}{32} \times (50)^3} = \mathbf{162.97\ N/mm^2}$$

According to maximum principal stress theory, **maximum principal stress** is given by,

$$\sigma_{max} = \frac{1}{2}\left[\sigma_b + \sqrt{\sigma_b^2 + 4\tau^2}\right] = \frac{1}{2}[162.97 + \sqrt{(162.97)^2 + 4 \times (97.78)^2}] = \mathbf{208.76\ N/mm^2}$$

According to maximum shear stress theory, **maximum shear stress** is given by,

$$\tau_{max} = \frac{1}{2}\sqrt{\sigma_b^2 + 4\tau^2} = \frac{1}{2}\sqrt{(162.97)^2 + 4 \times (97.78)^2} = \mathbf{127.28\ N/mm^2}$$

## 2.7.3 Arm of Pulley (Arm of Flywheel)

Moment of inertia and section modulus for some frequently used sections (along X-X).

**(1) Circular section :**

$$I = \frac{\pi}{4} d^4,\ y = \frac{d}{2}$$

$$\therefore \quad Z = \frac{I}{y} = \frac{\pi}{32} d^3$$

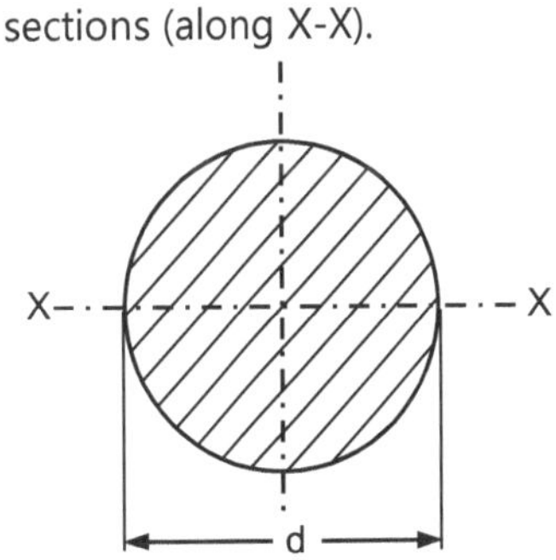


**Fig. 2.63**

**(2) Hollow circular section :**

$$I = \frac{\pi}{64}(d_o^4 - d_i^4), \quad y = \frac{d_o}{2}$$

$$Z = \frac{I}{y} = \frac{\pi}{32}\left(\frac{d_o^4 - d_i^4}{d_o}\right)$$

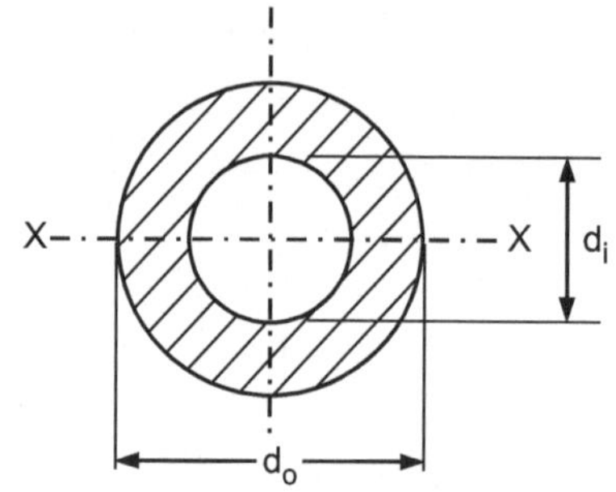


**Fig. 2.64**

**(3) Rectangular section :**

$$I = \frac{1}{12} tb^3, \quad y = \frac{b}{2}$$

$$\therefore \quad Z = \frac{I}{y} = \frac{1}{6} t\, b^2$$

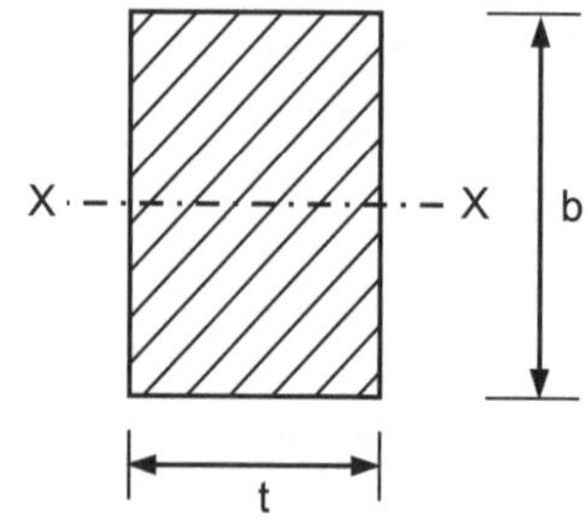


**Fig. 2.65**

**(4) Elliptical section :**

$$I = \frac{\pi}{64} b\, a^3, \quad y = \frac{a}{2}$$

$$Z = \frac{I}{y} = \frac{\pi}{32}\, ba^2$$

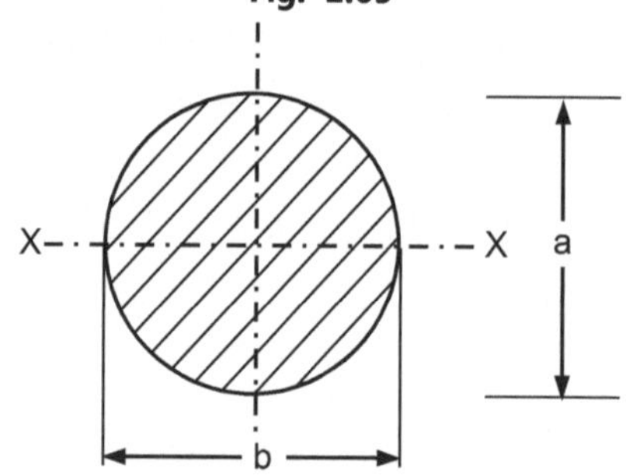


**Fig. 2.66**

**Numerical Type 10 : "Design of Flywheel Arm"**

**Problem 2.17 :** *A flywheel of mean diameter 3 m and having 6 arms has to transmit a maximum torque of $16 \times 10^6$ N.mm. The arms of the flywheel have elliptical section with major axis twice the minor axis. If the safe tensile stress for C.I., of which the flywheel is made is limited to 12.5 N/mm². Design the arms stating the assumptions made.*

**Solution : Data :** $d_f = 3 \text{ m} = 3000 \text{ mm}$

No. of arms = 6

Torque (T) = $16 \times 10^6$ N-mm

$\sigma_b = 12.5 \text{ N/mm}^2$

(Major axis) a = 2b (Minor axis)

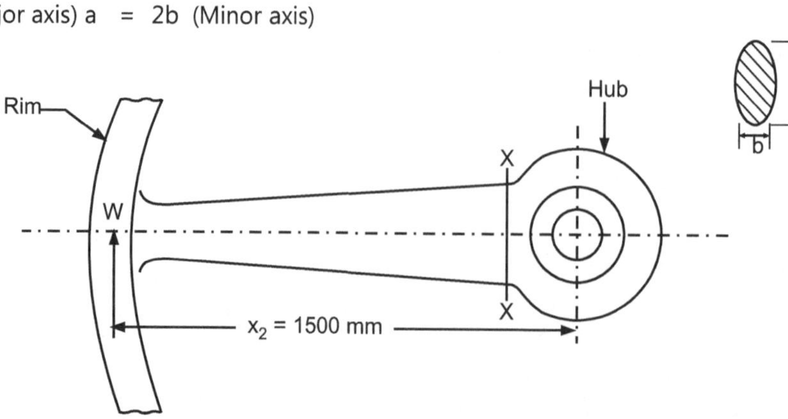


**Fig. 2.67**

Fig. 2.67 shows parts of the flywheel with one arm. For simplifying the problem – Make the following assumptions :

(1) Each arm transmits the torque equally.

(2) The arm extends from the centre of the rim to centre of the hub.

(3) Each arm is assumed to be cantilever fixed at the hub and loaded at the rim.

**Step I : Torque transmitted by each arm = $T_1$**

$$\therefore \quad T_1 = \frac{T}{\text{No. of arms}} = \frac{16 \times 10^6}{6} = 2.66 \times 10^6 \text{ N-mm}$$

**Step II :** If 'W' is the force on the arm at the rim, then

$$T_1 = W \times r$$

$$2.66 \times 10^6 = P \times 1500$$

$$\therefore \quad W = 1.77 \times 10^3 \text{ N}$$

**Step III : Maximum bending moment :**

$$M = W \times r$$

$$\therefore \quad M = 1.77 \times 10^3 \times 1500$$

$$\therefore \quad M = 2.66 \times 10^6 \text{ N.mm}$$

$$\therefore \quad M = T_1$$

**Step IV : For elliptical section, section modulus 'Z' :** Near the hub.

$$Z = \frac{\pi}{32} ba^2$$

**Given :**

$$a = 2b$$

$$\therefore \quad Z = \frac{\pi}{32} b \times (2b)^2$$

$$\therefore \quad Z = \frac{\pi}{32} \times 4b^3$$

$$\therefore \quad Z = \frac{\pi}{8} b^3$$

Using

$$\sigma_b = \frac{M}{Z}$$

$$\therefore \quad 12.5 = \frac{2.66 \times 10^6}{\frac{\pi}{8} b^3}$$

$$\therefore \quad 12.5 = \frac{8 \times 2.66 \times 10^6}{\pi b^3}$$

$$\therefore \quad 12.5 b^3 = \frac{8 \times 2.66 \times 10^6}{\pi}$$

$$\therefore \quad b = 81.52 \cong 82 \text{ mm}$$

$$\therefore \quad a = 2b = 2 \times 82 = 164 \text{ mm}$$

$\therefore$ Section near hub is $(a \times b) = 164 \text{ mm} \times 82 \text{ mm} = 13448 \text{ mm}$

$\therefore$ Section near rim is $2/3\ a \times b$

$$\therefore \quad \frac{2}{3} a = \frac{2}{3} \times 164 = 109.3 \text{ mm} \approx 110 \text{ mm}$$

$\therefore$ Section near rim is $(2/3\ a \times b) = 110 \text{ mm} \times 82 \text{ mm} = 9020 \text{ mm}.$

**Problem 2.18 :** *A cast iron pulley transmits 10 kW at 400 r.p.m. The diameter of pulley is 1.2 m and it has 4 straight arms of elliptical cross-section in which major axis is twice the minor axis. Determine the diameter of the arm if the available bending stress is 15 MPa.*

**Solution : Data :**

$$P = 10 \text{ kW} = 10 \times 10^3 \text{ W}$$
$$N = 400 \text{ r.p.m.}$$
$$d = 1.2 \text{ m} = 1200 \text{ mm}$$
$$\text{No. of arms} = 4$$
$$\sigma_b = 15 \text{ MPa}$$
$$a = 2b$$
$$P = \frac{2\pi NT}{60} \quad \therefore \quad T = \frac{P \times 60}{2\pi N} = \frac{10 \times 10^3 \times 60}{2 \times \pi \times 400}$$
$$T = 238.73 \text{ N.m}$$
$$\therefore \quad T = 238.73 \times 10^3 \text{ N.mm}$$

**Step I : Torque transmitted by each arm :**

$$\therefore \quad T_1 = \frac{238.73 \times 10^3}{4} = 59.68 \times 10^3 \text{ N.mm}$$

**Step II : If 'W' is the force acting on arm at rim :**

$$\therefore \quad T_1 = W \times \frac{D_f}{2}$$
$$\therefore \quad 59.68 \times 10^3 = W \times 600$$
$$\therefore \quad W = 99.46 \text{ N}$$

**Step III : Maximum bending moment on the arm at the hub :**

$$\therefore \quad M = W \times D_f/2$$
$$= 99.46 \times 600$$
$$\therefore \quad M = 59.68 \times 10^3 \text{ N.mm}$$
$$\therefore \quad M = T_1$$

**Step IV : For elliptical section, section modulus 'Z' near the hub :**

$$\therefore \quad Z = \frac{\pi}{32} ba^2$$

**Given :**

$$a = 2b$$
$$Z = \frac{\pi}{3} \times b \times (2b)^2$$
$$\therefore \quad Z = \frac{\pi}{8} b^3$$

Using

$$\sigma_b = \frac{M}{Z}$$
$$\therefore \quad 15 = \frac{59.68 \times 10^3}{\frac{\pi}{8} b^3}$$
$$\therefore \quad 15b^3 = \frac{8 \times 59.68 \times 10^3}{\pi}$$
$$\therefore \quad b = 21.63 \text{ mm}$$
$$\therefore \text{ Use} \quad b = 22 \text{ mm}$$
$$\therefore \quad a = 2b = 2 \times 22 = 44 \text{ mm}$$

## Theory Questions for Practice

1. What is a cotter joint ? Explain with a neat sketch.
2. State various applications of a cotter joint ?
3. Discuss the design procedure of a spigot and socket cotter joint with a neat sketch.
4. Draw a neat sketch of cotter joint and write the various strength equations for its complete design.
5. Why taper is provided on cotter ? What is its normal value ?
6. State the purpose of a turnbuckle. Describe its design procedure with a sketch.
7. Sketch an adjustable screwed joint for round rods and write the equations showing the strength of joint for most probable modes of failure.
8. Differentiate between 'direct stress' and 'bending stress'.

## Numberical Problems for Practice

1. Design a cotter joint for a maximum load of 6 kN. The cotter joint parts are made up of C20 steel. The safe stresses are as follows :

   Tensile stress = 60 N/mm$^2$, Compressive stress = 100 N/mm$^2$, Shear stress = 40 N/mm$^2$

   (**Ans.** d = 12 mm, $d_2$ = 14 mm, $d_1$ = 20 mm, modified $d_2$ = 16 mm, t = 4 mm, a = 5 mm, $d_3$ = 20 mm, $t_1$ = 3 mm, $d_1$ = 22 mm, $d_4$ = 32 mm, c = 6 mm, b = 20 mm)
2. A simple cotter joint is subjected to an axial load of 100 kN, which alternatively changes from tensile to compressive. Allowable stresses for the material are 80 N/mm$^2$, 35 N/mm$^2$ and 100 N/mm$^2$ in tension, shear and compression. The joint is subjected to 30% overload.

   (**Ans.** d = 46 mm, $d_2$ = 56 mm, modified $d_2$ = 74 mm, t = 18.5 mm, a = 26 mm, $d_3$ = 90 mm, $t_1$ = 16 mm, $d_1$ = 90 mm, $d_4$ = 146 mm, b = 102 mm, c = 26 mm, $l$ = 184 mm).
3. A knuckle joint is required to withstand a tensile load of 25 kN. Design the joint, if the permissible stresses are as :

   Tensile stress = 56 MPa, Shear stress = 46 MPa and compressive stress = 70 MPa.

   (**Ans.** d = 24 mm, $d_1$ = 24 mm, t = 30 mm, $d_2$ = 48 mm, $d_3$ = 36 mm, $t_1$ = 22.5 mm, $t_2$ = 15 mm)
4. Design a hand lever having a length 1100 mm from the center of shaft to the point of application 800 N load. Effective overhang from nearest bearing is 150 mm. Find : (a) Diameter of shaft. (b) Section of lever, if $\frac{t}{B}$ = 0.5.

   The allowable tensile stress is 73 MPa and allowable shear stress is 60 MPa.

   Also design the key by using the same stresses.

   (**Ans.** d = 43 mm, w = 11 mm, $t_1$ = 7.33 mm, t = 27 mm, b = 54 mm)
5. The effective length of hand lever is 1 m. The effective overhang from nearest bearing is 150 mm. The lever is made of alloy steel, for which, permissible tensile stress is 115 N/mm$^2$ and shear stress is 57.5 N/mm$^2$. Maximum force exerted on handle is 300 N. Design the lever and shaft. Assume B = 3t.

   (**Ans.** d = 30 mm, $d_2$ = 46 mm, $t_2$ = 9 mm, $l_2$ = 37.5 mm, w = 7.5 mm, $t_1$ = 5 mm, t = 12 mm, b = 36 mm)
6. A right angle bell crank lever is having one arm, 700 mm and other 400 mm long. A load of 175 kN is to be raised acting on a pin at the end of 700 mm arm and effort is applied at the end of 400 mm arm. The lever consist of a steel forgings, turning on a point at the fulcrum. The permissible stresses for the pin and lever are in tension and compression as 60 N/mm$^2$ and in shear as 80 N/mm$^2$ respectively. The bearing pressure on the pin is not to exceed 6 N/mm$^2$. Find :

   (a) Diameter and length of fulcrum pin.

   (b) Thickness (t) and depth (b) of rectangular cross-section of lever (assume B = 3t).

   (**Ans.** $d_F$ = 218 mm, $l_F$ = 272.5 mm, t = 124 mm, B = 372 mm)

7. A C-frame subjected to a force of 15 kN is shown in Fig. 2.68. It is made of cast iron FG300 with ultimate tensile strength of 300 MPa. The factor of safety is 2.5. Determine the dimensions of cross-section.

   (**Ans.** t = 15.81 mm ≅ 16 mm (say), b = 80 mm)

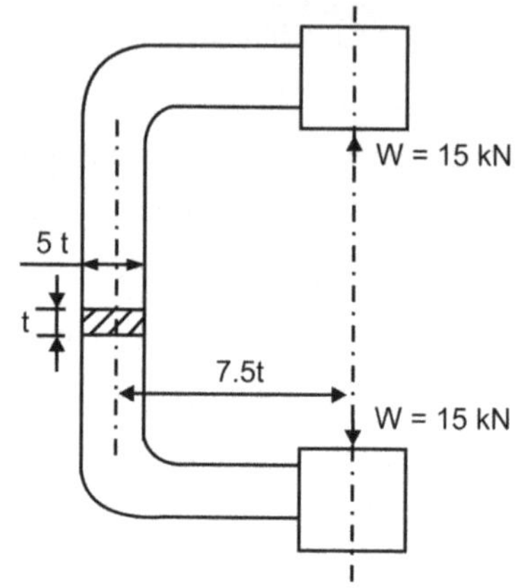


**Fig. 2.68**

# Chapter 3

# DESIGN OF SHAFTS, KEYS AND COUPLINGS

## 3.1 INTRODUCTION TO SHAFT

- Shaft is a rotating element, usually of circular cross-section.
- Shaft is used to transmit power, from one place to another.
- Power is delivered to the shaft by means of tangential force and resultant torque is set up within the shaft.
- Refer Fig. 3.1, which shows tangential force ($F_t$) applied at circumference of solid shaft having radius (r).
- Torque or twisting moment (T) causing to rotate the shaft is calculated by,

$$T = F_t \times r$$

- Due to the applied torque on shaft, torsional shear stress ($\tau$) is induced in shaft.

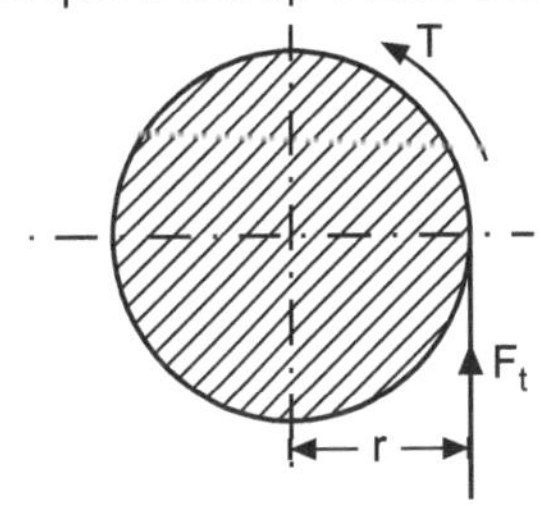


**Fig. 3.1 : Rotating shaft**

- To transfer the power from one place to another, various components such as pulleys, gears etc., are mounted on shaft, by means of keys and splines.
- Therefore, shaft is subjected to bending moment due to mounting of pulleys, gears, sprockets etc. Due to bending moment acting on the shaft, bending stress ($\sigma_b$) is induced in shaft.
- We observe, power transmitting shafts are subjected to bending moment and twisting moment (or torque) simultaneously. Hence, **bending and torsional shear stresses are induced in the shafts**.

### 3.1.1 Essential Characteristics for Shaft Material

**Essential characteristics for shaft material are :**

1. High strength.
2. Good heat treatment properties.
3. Good machinability.
4. Low notch sensitivity factor.
5. High wear resistant properties.
6. Ductility.
7. High resilience.
8. High fatigue strength.

### 3.1.2 Materials Used for Shaft

**Commonly used materials for shaft are :**

(a) For *ordinary use*, **carbon steels** are used. For example : 40C8, 45C8 and 50C4.

(b) For *high strength*, **alloy steels** such as nickel, nickel-chromium or chrome vanadium steel are used. For example : 40Cr1Mo28, 40Ni1Cr1Mo15 etc.

### 3.1.3 Manufacturing Processes used for Shaft

- Basic method of manufacturing shaft is hot rolling, followed by finishing up to required size by turning and grinding.
- Cold drawing produces a stronger shaft than hot rolling. However, cold drawn shafts have certain disadvantages. Cold drawing produces residual stresses at and near the surface of shaft.
- During machining operations like slotting and milling, required to make the key slot, the residual stresses are partially released causing distortion of shaft and producing twist.
- Therefore, most of the transmission shafts are turned and ground after hot rolling.
- Shafts of larger diameter are generally forged and turned to required size.
- In order to increase the wear resistance and toughness, heat treatment is done on the shaft material.

### 3.1.4 Types of Shafts

Shafts are classified into following types :

**(a) Transmission shaft :**

- Transmission shaft is used to transmit power from one point to another, i.e. from source of power to power absorbing (consuming) device. For example : Electric motor (source) and water pump (power absorbing device).
- It supports transmission elements like gears, pulleys, sprockets etc.
- **For example :** Line shafts, counter shafts and overhead shafts etc.
- Line shaft consists of number of shafts, which are connected in axial direction by means of couplings.
- Counter shaft is a secondary shaft, which is driven by main shaft and from which, power is supplied to a machine component.

**(b) Spindle :**

- A spindle is a short rotating shaft, which forms an integral part of any machine.
- It is used in most of the manufacturing machines.
  **For example :** Spindles of lathe, drilling, grinding etc. It imparts (gives) motion either to a cutting tool or workpiece.

**(c) Axle :**

- An axle is a non-rotating element.
- It supports a rotating element like wheel, hoisting drum or rope sheave, which is fitted to the housing by means of bearings.
- But, axle does not transmit any torque. Therefore, axle is subjected to only bending moment due to transverse loads like bearing or support reactions.
- **For example :** Rear axle of railway wagon.

**(d) Machine shaft :**

- It is an integral part of a machine itself.
- **For example :** Crank shaft.

### 3.1.5 Distinguish between Shaft and Axle

| Sr. No. | Shaft | Axle |
|---|---|---|
| 1. | Shaft is a rotating member. | Axle is a non-rotating member. |
| 2. | Purpose of a shaft is to transmit the torque and to support the transmission elements like gears, pulleys etc. | Purpose of an axle is only to support the transmission elements like wheels, pulleys etc. |
| 3. | Shaft is subjected to torque, bending moment, and/or axial force. | Axle is subjected to bending moment and/or axial force. |

### 3.1.6 Difference between Shaft, Axle and Spindle

| Shaft | Axle | Spindle |
|---|---|---|
| 1. It is a rotating element, transmitting power from one place to other. | 1. It is a stationary element, used to support a rotating body. | 1. It is a short shaft imparting motion either to a cutting tool or workpiece. |
| 2. It is subjected to torque, bending moment or axial force. | 2. It is subjected to only bending moment. | 2. It is subjected to torque, bending moment or axial force. |
| 3. Example : Propeller shaft. | 3. Example : Front and rear axles of motorcycle. | 3. Example : Drill spindle. |

### 3.1.7 Standard Sizes of Shafts

- Standard lengths of shafts are 5 m, 6 m and 7 m.
- The length of shafting should not be more than 7 m.
- Following are the common sizes of transmission shafts :
  (a) 25 to 60 mm by 5 mm increments.
  (b) 60 to 110 mm by 10 mm increments.
  (c) 110 to 140 mm by 15 mm increments.
  (d) 140 to 500 mm by 20 mm increments.

### 3.1.8 Design of Shaft on the Basis of Strength

- Shafts may be designed on the basis of (a) Strength, (b) Rigidity and (c) Stiffness.
- While designing a shaft on the basis of strength, following cases may be considered :
  (a) Shafts subjected to twisting moment only.
  (b) Shafts subjected to bending moment only.
  (c) Shafts subjected to combined twisting and bending moments.

**(a) Design of Shafts Subjected to Twisting Moment (Torque) Only :**

- Here, the diameter of shaft can be determined by using basic torsion equation.

We know that, $$\frac{T}{J} = \frac{\tau}{r} \quad \text{.... (3.1)}$$

where, T = Twisting moment (or torque) acting on the shaft in 'N.mm'

J = Polar moment of inertia (Polar M.I.) of shaft about axis of rotation in 'mm⁴'

$\tau$ = Torsional shear stress in 'N/mm²'

r = Distance of neutral axis to outermost fiber in 'mm'

$= \frac{d}{2}$ (for solid shaft) in 'mm'

$= \frac{d_o}{2}$ (for hollow shaft) in 'mm'

We have, $J = \frac{\pi}{32} \times d^4$ (for solid shaft) in 'mm⁴'

$= \frac{\pi}{32} \times (d_o^4 - d_i^4)$ (for hollow shaft) in 'mm⁴'

Therefore, for solid shaft, equation (3.1) becomes,

$$\frac{T}{\frac{\pi}{32} d^4} = \frac{\tau}{\frac{d}{2}}$$

i.e. $$\mathbf{T = \frac{\pi}{16} \cdot \tau \cdot d^3} \quad \text{... (3.2)}$$

From this equation, we can find diameter of shaft.

Similarly, for **hollow shaft**, equation (3.1) becomes,

$$\frac{T}{\frac{\pi}{32}[d_o^4 - d_i^4]} = \frac{\tau}{\frac{d_o}{2}}$$

$$\therefore \quad T \times \frac{d_o}{2} = \tau \times \frac{\pi}{32} \cdot [d_o^4 - d_i^4]$$

$$\therefore \quad T = \frac{\pi}{16} \cdot \tau \cdot \frac{(d_o^4 - d_i^4)}{d_o} = \frac{\pi}{16} \cdot \tau \cdot \frac{d_o^4}{d_o}\left[1 - \left(\frac{d_i}{d_o}\right)^4\right]$$

$$\therefore \quad \mathbf{T = \frac{\pi}{16} \cdot \tau \cdot d_o^3 (1 - k^4)} \quad \text{... (3.3)}$$

where, $k = \frac{d_i}{d_o}$ = Ratio of inside diameter to outside diameter

From the above equation (3.3), diameters of hollow shaft ($d_o$ and $d_i$) can be determined,.

**(b) Design of Shafts Subjected to Bending Moment Only :**

- Here, the diameter of shaft can be determined by using bending equation (flexural formula).

$$\frac{M}{I} = \frac{\sigma_b}{y} \quad \text{... (3.4)}$$

where, M = Maximum bending moment on the shaft in 'N.mm'

I = Moment of inertia of cross-section of shaft about axis of rotation in 'mm⁴'

$\sigma_b$ = Bending stress in 'N/mm²'

y = Distance from neutral axis to outermost fibre in 'mm'

**For solid shaft,** $I = \frac{\pi}{64} d^4$

and $y = \frac{d}{2}$

∴ Equation (3.4) becomes,

$$\frac{M}{\frac{\pi}{64} d^4} = \frac{\sigma_b}{\frac{d}{2}}$$

∴ $$\mathbf{M = \frac{\pi}{32} \cdot \sigma_b \cdot d^3} \quad \text{... (3.5)}$$

From this equation, diameter of solid shaft can be determined.

**For hollow shaft,** $I = \frac{\pi}{64} [(d_o)^4 - (d_i)^4]$

∴ $I = \frac{\pi}{64} \cdot d_o^4 [1 - k^4]$ $\left[\because k = \frac{d_i}{d_o}\right]$

and $y = \frac{d_o}{2}$

∴ Substituting values of I and y, equation (3.4) becomes,

$$\frac{M}{\frac{\pi}{64} \cdot d_o^4 (1 - k^4)} = \frac{\sigma_b}{\frac{d_o}{2}}$$

∴ $$\mathbf{M = \frac{\pi}{32} \cdot \sigma_b \cdot d_o^3 (1 - k^4)} \quad \text{... (3.6)}$$

From this equation, diameters of hollow shaft ($d_o$ and $d_i$) can be determined.

**(c) Design of Shafts Subjected to Combined Twisting and Bending Moment :**

When shaft is subjected to combined twisting and bending moment, then shaft must be designed on the basis of two moments (i.e. bending moment as well as twisting moment) simultaneously.

**(1) According to maximum shear stress theory :**

We have, $\tau_{max}$ = Maximum shear stress = $\frac{1}{2}\sqrt{\sigma_b^2 + 4(\tau)^2}$ ... (3.7)

We know that, $T = \frac{\pi}{16} \cdot \tau \cdot d^3$ [From equation (3.2)]

∴ $\tau = \frac{16T}{\pi d^3}$

Also, $M = \frac{\pi}{32} \cdot \sigma_b \cdot d^3$ [From equation (3.3)]

∴ $\sigma_b = \frac{32\,M}{\pi \cdot d^3}$

Put the values of $\tau$ and $\sigma_b$ in equation (3.7),

$$\tau_{max} = \frac{1}{2}\sqrt{\left(\frac{32\,M}{\pi d^3}\right)^2 + 4\left(\frac{16T}{\pi d^3}\right)^2}$$

∴ $$\tau_{max} = \frac{1}{2} \cdot \left(\frac{32}{\pi d^3}\right)\sqrt{M^2 + T^2}$$

∴ $$\sqrt{M^2 + T^2} = \frac{\pi}{16} \tau_{max} \cdot d^3$$

Expression $[\sqrt{M^2 + T^2}]$ is called as ***equivalent twisting moment*** and it is denoted by $T_e$.

By limiting the maximum shear stress ($\tau_{max}$) equal to allowable shear stress ($\tau$), we have,

$$\mathbf{T_e = \sqrt{M^2 + T^2} = \frac{\pi}{16} \tau_{max} \cdot d^3} \quad \text{... (3.8)}$$

**(2) According to maximum normal stress theory :**

We have, $\sigma_{b\,max}$ = Maximum normal stress

$$\therefore \quad \sigma_{b\,max} = \frac{1}{2}\sigma_b + \frac{1}{2}\sqrt{\sigma_b^2 + 4(\tau)^2}$$

After putting values of $\tau$ and $\sigma_b$ in above equation, we get,

$$\sigma_{b\,max} = \frac{1}{2}\cdot\left(\frac{32M}{\pi d^3}\right) + \frac{1}{2}\sqrt{\left(\frac{32M}{\pi d^3}\right)^2 + 4\cdot\left(\frac{16T}{\pi d^3}\right)^2} \quad \text{(For solid shaft)}$$

$$\therefore \quad \sigma_{b\,max} = \frac{1}{2}\left(\frac{32M}{\pi d^3}\right) + \frac{1}{2}\cdot\left(\frac{32}{\pi d^3}\right)\cdot\sqrt{M^2 + T^2}$$

$$\therefore \quad \sigma_{b\,max} = \frac{32}{\pi d^3}\left[\frac{1}{2}\cdot M + \frac{1}{2}\cdot\sqrt{M^2 + T^2}\right]$$

$$\therefore \quad \sigma_{b\,max} = \frac{32}{\pi d^3}\left[\frac{1}{2}(M + \sqrt{M^2 + T^2})\right]$$

$$\therefore \quad \frac{1}{2}(M + \sqrt{M^2 + T^2}) = \frac{\pi}{32}\sigma_{b\,max}\, d^3$$

Expression $\frac{1}{2}\{M + \sqrt{M^2 + T^2}\}$ is called as ***equivalent bending moment*** and it is denoted by $M_e$. By limiting the maximum normal stress ($\sigma_{b\,max}$) equal to the allowable bending stress $\sigma_b$, we have,

$$\mathbf{M_e = \frac{1}{2}[M + \sqrt{M^2 + T^2}] = \frac{\pi}{32}\cdot \sigma_{b\,max}\cdot d^3} \qquad \text{... (3.9)}$$

Thus, we can find diameter of **solid shaft** using above expression.

Similarly, **for hollow shaft**, we can derive,

$$\mathbf{T_e = \frac{\pi}{16}\cdot \tau_{max}\cdot d_o^3\cdot (1 - k^4) = \sqrt{M^2 + T^2}} \qquad \text{... (3.10)}$$

and

$$\mathbf{M_e = \frac{\pi}{32}\cdot \sigma_{b\,max}\cdot d_o^3\,(1 - k^4) = \frac{1}{2}(M + \sqrt{M^2 + T^2})} \qquad \text{... (3.11)}$$

## 3.1.9 Design of Shaft on the Basis of Rigidity

- In some applications, torsional deformation of shaft (torsional rigidity) and lateral deflection (lateral rigidity) may not be tolerated.
- For such cases, rigidity or stiffness of shaft is very important.

### 3.1.9.1 Torsional Rigidity

- Torsional rigidity means the **'resistance offered against angular deflection of a shaft'**.
- Torsional stiffness or Rigidity of a shaft is given by,

$$S = \frac{\text{Torque}}{\text{Torsional deflection or Angle of twist}} = \frac{T}{\theta} \qquad \text{... (3.12)}$$

- Torsional deflection may be obtained by using basic torsion equation.

$$\frac{T}{J} = \frac{G\theta}{L}$$

$$\therefore \quad \theta = \frac{T\cdot L}{G\cdot J} \qquad \text{... (3.13)}$$

where,

$\theta$ = Torsional deflection or Angle of twist — in 'radians'

T = Twisting moment or Torque — in 'N.mm'

J = Polar M.I.

$= \frac{\pi}{32}\cdot d^4$ (for solid shaft) — in 'mm$^4$'

$= \frac{\pi}{32}\cdot [(d_o)^4 - (d_i)^4]$ (for hollow shaft) — in 'mm$^4$'

L = Length of shaft — in 'mm'

G = Modulus of rigidity — in 'N/mm$^2$'

- Therefore, equation (3.12) becomes

Torsional stiffness or Rigidity of shaft, $S = \frac{T}{\theta} = \frac{T}{\frac{T \cdot L}{G \cdot J}} = \frac{GJ}{L}$. [Refer equation 3.13]

$$\therefore \qquad \mathbf{S = \frac{T}{\theta} = \frac{GJ}{L}} \qquad \ldots (3.14)$$

- Allowable deflection for shafts is limited to 1° for a length of 20 times the diameter of shaft.
- **Precautions to reduce the deflection of shaft are :**
  1. Parts such as pulleys, gears mounted on shaft should be arranged very close to the bearings (which are support reactions).
  2. Pulleys, gears and other parts mounted should be lighter in weight.

### 3.1.9.2 Lateral Rigidity

- Lateral rigidity refers to **'lateral deflection of shaft'**.
- In case, if shaft is of uniform cross-section, then lateral deflection of a shaft may be obtained by using deflection formulae as studied in strength of materials.
- But, when the shaft is of variable cross-section, then lateral deflection may be determined from fundamental equation of elastic curve of a beam.

$$\frac{d^2y}{dx^2} = \frac{M}{E \cdot I}$$

- Lateral deflection should not exceed *0.85 mm per metre length of shaft.*
- Maximum permissible lateral deflection for transmission shafts is taken as,

$$\delta = 0.001 \text{ L to } 0.003 \text{ L} \qquad \text{in 'mm'}$$

where, $L$ = Length of shaft in 'mm'

- The product 'E · I' represents **lateral rigidity**, which is given by,

Lateral rigidity of shaft $= \frac{\text{Force}}{\text{Maximum deflection}} = \frac{F}{\delta}$

### 3.1.9.3 Necessity of Designing Shaft on the Basis of Rigidity

- While designing a shaft in high speed applications, its torsional rigidity should be high.
- It means that, the angle of twist should not be more than 0.25° per metre length of shafts, in case of I.C. engine, where timing of valves would be affected.
- In case of shaft running at high speed, where a small deflection would cause huge out of balance of forces, lateral rigidity is very important.
- Insufficient rigidity or stiffness can result in poor performance due to various elements such as bearings, clutches, gears etc. mounted on shaft.

### 3.1.9.4 Applications, where Shaft is Designed on the Basis of Rigidity

(1) Propeller shaft of an automobile.
(2) Shaft of an electric motor.
(3) Shaft of a water pump.
(4) Shaft of marine engine.

## 3.1.10 ASME Code for Design of Shaft Subjected to Fluctuating Load

- Transmission shafts or line shafts or overhead shafts are meant for transmitting power from one end to another end.
- A prime mover, such as, a constant speed electric motor is used to supply necessary power to the shaft. The power is taken out at various points from the shaft through various machine elements like belts and pulleys, chains and sprocket, gear pairs etc., which are directly mounted on it.
- Various shaft lengths are coupled together by using rigid coupling, thus, forming required length of line shafting. These shafts are supported on sleeve bearings or pedestal bearings.
- These shafts are thus subjected to combined bending and twisting moments. Since, loads on most of the shafts in connected machinery are not constant, therefore, it is necessary to make proper allowance for the harmful effects of load fluctuations.

- According to ASME (American Society of Mechanical Engineering) code, this harmful effect is compensated by introducing two factors '$k_m$' and '$k_t$' in the usual design equations used for design of transmission shafts on the basis of maximum principal shear stress theory (Guest's theory) as follows.

**Maximum principal shear stress,** $\tau_{max} = \frac{16}{\pi d^3}\sqrt{(k_m \cdot M)^2 + (k_t \cdot T)^2}$

where, $k_m = k_b$ = Combined shock and fatigue factor in bending

$k_t$ = Combined shock and fatigue factor in twisting

$\therefore \quad \sqrt{(k_m \cdot M)^2 + (k_t \cdot T)^2} = \frac{\pi}{16}\tau_{max} \cdot d^3$

- The recommended values of $k_m$ and $k_t$ are given in the Table 3.1.

**Table 3.1 : Recommended (ASME Code) values for $k_m$ and $k_t$**

| Types of load | $k_m$ or $k_b$ | $k_t$ |
|---|---|---|
| 1. Stationary shaft : | | |
| (a) Gradually applied load | 1.0 | 1.0 |
| (b) Suddenly applied load | 1.5 to 2.0 | 1.5 to 2.0 |
| 2. Rotating shafts : | | |
| (a) Gradually applied or steady load | 1.5 | 1.0 |
| (b) Suddenly applied load with minor shocks | 1.5 to 2.0 | 1.5 to 2.0 |
| (c) Suddenly applied load with heavy shocks | 2.0 to 3.0 | 1.5 to 3.0 |

- According to ASME code, permissible shear stress for shaft **without keyway** is taken as **30% of yield strength** or **18% of ultimate tensile strength**, whichever is minimum.

$\therefore \quad \tau_{max} = 0.30\ \tau_{yt}$

or $\quad \tau_{max} = 0.18\ \tau_{ut}$

- But, if keyways are present, then these values are to be reduced by 25 %.

## 3.1.11 Advantages of Hollow Shaft

**Advantages of Hollow Shafts Over Solid Shafts :**

(1) Hollow shaft has higher torque transmitting capacity than the solid shaft of same weight. i.e. hollow shaft is lighter in weight than the solid shaft for same torque transmission capacity.

(2) Hollow shaft has higher torsional as well as lateral rigidity than the solid shaft of same weight.

## 3.1.12 Applications of Hollow Shaft

- Hollow shafts are used in : (1) Machine tools. (2) Marine applications.

## 3.1.13 Comparison of Solid Shaft and Hollow Shaft

| Solid shaft | Hollow shaft |
|---|---|
| 1. Design equations :<br>(a) $T = \frac{\pi}{16} \cdot \tau \cdot d^3$<br>(b) $M = \frac{\pi}{32} \cdot \sigma_b \cdot d^3$ | 1. Design equations :<br>(a) $T = \frac{\pi}{16} \cdot \tau \cdot d_o^3 \cdot (1 - k^4)$<br>(b) $M = \frac{\pi}{32} \cdot \sigma_b \cdot d_o^3\ (1 - k^4)$ |
| 2. Weight of solid shaft is more than hollow shaft for same torque transmitted. | 2. Weight of hollow shaft is less. So material cost is less as compared to solid shaft. |

## Additional Formulae

(1) Power, $\quad P = \frac{2\pi NT}{60}$ in 'Watts'

where, $T$ = Torque or twisting moment in 'N.m,'

$N$ = Speed of shaft in 'r.p.m.'

(2) Torque transmitted by a pulley,

$T = (T_1 - T_2) \times R$

where, $T_1$ = Tension in tight side of belt in 'N'

$T_2$ = Tension in slack side of belt in 'N'

$R$ = Radius of pulley in 'mm'

**Numerical Type No. 1 : "Shafts Subjected to Twisting Moment Only"**

**Problem 3.1 :** *A solid shaft is transmitting 1 MW at 240 r.p.m. Determine the diameter of shaft, if the maximum torque transmitted exceeds the mean torque by 20%. Take the maximum allowable shear stress as 60 N/mm².*

**Solution : Given data : P** = 1 MW = $\mathbf{1 \times 10^6}$ **W**, N = 240 r.p.m., $\tau$ = 60 N/mm².

**Procedure :**

We have,

$$\text{Power transmitted} = P = \frac{2\pi N \cdot T_{mean}}{60}$$

$$\text{Torque transmitted, } \mathbf{T_{mean}} = \frac{P \times 60}{2\pi N} = \frac{1 \times 10^6 \times 60}{2 \times \pi \times 240} = \mathbf{39788.73\ N\text{-}m}$$

It is given that : $T_{max}$ exceeds more than $T_{mean}$ by 20%.

$$\therefore \quad \mathbf{T_{max}} = 1.2 \times T_{mean} = 1.2 \times 39788.73 = \mathbf{47746.47\ N\text{-}m = 47746.47 \times 10^3\ N\text{-}mm}$$

Shaft is designed on the basis of strength from the torsion equation given as,

$$T_{max} = \pi/16 \times \tau \times d^3$$

$$\therefore \quad 47746.47 \times 10^3 = \pi/16 \times 60 \times d^3$$

$$\therefore \quad d = 159.43\ \text{mm} \cong \mathbf{160\ mm\ (say)}$$

---

**Problem 3.2 :** *A hollow shaft of a rotary compressor is to be designed to transmit a maximum torque of 4750 N-m. The shear stress in the shaft is limited to 50 N/mm². Determine the inside and outside diameters of the shafts, if the ratio of inside to outside diameter is 0.4.*

**Solution : Given data : T** = 4750 N-m = $\mathbf{4.75 \times 10^6}$ **N-mm**, $\tau$ = 50 N/mm².

$$(d_i/d_o) = \mathbf{k = 0.4}$$

**Procedure :** Hollow shaft is designed on the basis of strength from the derived torsion equation,

$$T = \frac{\pi}{16} \times \tau \times d_o^3 (1 - k^4)$$

$$\therefore \quad 4.75 \times 10^6 = \frac{\pi}{16} \times 50 \times d_o^3 \{1 - (0.4)^4\}$$

Thus, $\mathbf{d_o}$ = 79.18 mm ≈ **80 mm (say)** and $\mathbf{d_i}$ = 0.4 × $d_o$ = 0.4 × 80 = **32 mm**

---

**Problem 3.3 :** *A hollow shaft is required to transmit 50 kW power at 600 r.p.m. Calculate its inside and outside diameters, if ratio of diameters is 0.8. Consider yield strength of material as 380 N/mm² and factor of safety as 4.*

**Solution : Given data : P** = 50 kW = $\mathbf{50 \times 10^3}$ **W**, N = 600 r.p.m.

$$\mathbf{k} = \frac{d_i}{d_o} = \mathbf{0.8}, \quad \sigma_{yt} = 380\ \text{N/mm}^2 \text{ and F.O.S.} = 4$$

**Procedure :** Here, value of ultimate shear stress is not given, which can be divided by F.O.S. to get permissible value of shear stress. Instead of it, value of yield strength of shaft material is given, $\sigma_{yt}$ = 380 N/mm².

**According to maximum shear stress theory (Guest's theory) of elastic failures**, "Shear stress at yield point is equal to half of yield point tensile stress".

$$\therefore \quad \boldsymbol{\tau_y} = \frac{1}{2} \times \sigma_{yt} = \frac{1}{2} \times 380 = \mathbf{190\ N/mm^2}$$

∴ Permissible shear stress,

$$\tau = \frac{\tau_y}{\text{F.O.S.}} = \frac{190}{4} = \mathbf{47.5\ N/mm^2}$$

We have,

$$\text{Power transmitted, } P = \frac{2\pi NT}{60}$$

$$\therefore \quad 50 \times 10^3 = \frac{2\pi \times 600 \times T}{60}$$

$$\therefore \quad \mathbf{T} = 795.77\ \text{N.m} = \mathbf{795.77 \times 10^3\ N.mm}$$

For Hollow shaft,

Torque transmitted, $T = \frac{\pi}{16} \tau \cdot d_o^3 \cdot (1 - k^4)$

$$\therefore \quad 795.77 \times 10^3 = \frac{\pi}{16} \times 47.5 \times d_o^3 \times [1 - (0.8)^4]$$

$$\therefore \quad d_o^3 = 144516.59$$

$$\therefore \quad d_o = \mathbf{52.477\ mm \cong 54\ mm\ (say)}$$

We know, $k = \frac{d_i}{d_o} = 0.8$

$$\therefore \quad \mathbf{d_i} = 0.8 \times d_o = 0.8 \times 54 = \mathbf{43.2\ mm}$$

---

**Problem 3.4 :** *A hollow shaft is to be designed to transmit 600 kW at 110 r.p.m. The maximum torque being 20% greater than the mean. The shear stress is not to exceed 63 MPa and angle of twist in a length of 3 metres not to exceed 1.4°. Find external diameter of the shaft, if the internal diameter to external diameter is 3/8. Take modulus of rigidity 84 GPa.*

**Solution : Given data : P** = 600 kW = **600 × 10³ W**, N = 110 r.p.m., $T_{max} = 1.2\ T_{mean}$

$$\tau_{max} = 63\ N/mm^2,\ \theta = 1.4° = 1.4 \times \frac{\pi}{180} = \mathbf{0.0244^c},\ \mathbf{L} = 3\ m = \mathbf{3000\ mm},$$

$$\mathbf{k} = \frac{d_i}{d_o} = \frac{3}{8} = \mathbf{0.375},\ \mathbf{G} = 84\ GPa = \mathbf{84 \times 10^3\ MPa}$$

**Procedure : (1) Design of shaft on the basis of 'Strength' :**

We have, $P = \frac{2\pi N\ T_{mean}}{60}$

$$\therefore \quad \mathbf{T_{mean}} = \frac{600 \times 10^3 \times 60}{2\pi \times 110} = 52.087 \times 10^3\ N.m = \mathbf{52.087 \times 10^6\ N.mm}$$

But, $\mathbf{T_{max}} = 1.2 \times T_{mean} = 1.2 \times 52.087 \times 10^6 = \mathbf{62.504 \times 10^6\ N.mm}$

For hollow shaft, we have, $T = \frac{\pi}{16} \tau\, d_o^3 \times (1 - k^4)$

$$\therefore \quad 62.504 \times 10^6 = \frac{\pi}{16} \times 63 \times [1 - (0.375)^4] \times d_o^3$$

$$\mathbf{d_o} = 172.74\ mm \cong \mathbf{174\ mm\ (say)}$$

$$\therefore \quad \mathbf{d_i} = 0.375 \times d_o = 0.375 \times 174 = \mathbf{65.25\ mm}$$

**(2) Design of shaft on the basis of 'Rigidity' :**

$$\frac{T}{J} = \frac{G\theta}{L}$$

$$\therefore \quad T \times L = J \times G \times \theta$$

$$\therefore \quad 62.504 \times 10^6 \times 3000 = \frac{\pi}{32} \times (d_o^4 - d_i^4) \times 84 \times 10^3 \times 0.0244$$

$$\therefore \quad 931.88 \times 10^6 = d_o^4 \times \left[1 - \left(\frac{d_i}{d_o}\right)^4\right]$$

$$\therefore \quad 931.88 \times 10^6 = d_o^4 \times [1 - (0.375)^4]$$

$$\therefore \quad \mathbf{d_o} = 175.59\ mm \cong \mathbf{176\ mm\ (say)}$$

$$\therefore \quad \mathbf{d_i} = 0.375 \times d_o = 0.375 \times 176 = \mathbf{66\ mm}$$

- **For safe design**, we need to take **higher values** for dimensions of any product.

  Therefore, External diameter of hollow shaft, $\mathbf{d_o}$ = **176 mm**.

  Internal diameter of hollow shaft, $\mathbf{d_i}$ = **66 mm**.

---

**Problem 3.5 :** *Find the diameter of solid shaft to transmit 20 kW at 200 r.p.m. The ultimate shear stress for steel may be taken as 360 MPa and F.S. = 8. If a hollow shaft is used in place of solid shaft, find inside and outside diameter. Ratio of outside to inside diameter for hollow shaft is 2.*

**Solution : Given data : P** = 20 kW = **20 × 10³ W**, N = 200 r.p.m.,

$$\tau_{ut} = 360\ MPa,\ F.O.S. = 8,\ \frac{d_o}{d_i} = 2;\ \therefore\ \frac{d_i}{d_o} = \frac{1}{2} = 0.5 = k$$

**Procedure :** We have, $\tau = \frac{\text{Permissible shear stress}}{\text{(Working stress)}} = \frac{\text{Ultimate shear stress}}{\text{Factor of safety}} = \frac{360}{8}$ = **45 N/mm²**

Given that, $\frac{d_o}{d_i} = 2$

$$\therefore \quad \mathbf{k} = \frac{d_i}{d_o} = \frac{1}{2} = \mathbf{0.5}$$

We know that, $P = \frac{2\pi NT}{60}$

$$\therefore \quad \mathbf{T} = \frac{P \times 60}{2\pi N} = \frac{20 \times 10^3 \times 60}{2 \times \pi \times 200} = 954.92 \text{ N.m} = \mathbf{954.92 \times 10^3 \text{ N.mm}}$$

**(1) For solid shaft :** Torque transmitted,

$$T = \frac{\pi}{16} \times \tau \times d^3$$

$$\therefore \quad 954.92 \times 10^3 = \frac{\pi}{16} \times 45 \times d^3$$

$$\therefore \quad d = 47.63 \text{ mm} \cong \mathbf{48 \text{ mm (say)}}$$

**(2) For hollow shaft :** Torque transmitted, $T = \frac{\pi}{16} \times \tau \times d_o^3 \times [1 - k^4]$

$$\therefore \quad 954.92 \times 10^3 = \frac{\pi}{16} \times 45 \times d_o^3 \times [1 - 0.5^4] \qquad [k = 0.5]$$

$$\therefore \quad \mathbf{d_o} = 48.66 \text{ mm} \cong \mathbf{50 \text{ mm (say)}}$$

$$\therefore \quad \mathbf{d_i} = 0.5 \times d_o = 0.5 \times 50 = \mathbf{25 \text{ mm}} \qquad \left[\because k = \frac{d_i}{d_o} = 0.5; \ \because d_i = 0.5\, d_o\right]$$

---

**Problem 3.6 :** *The shaft running at 125 r.p.m. transmits 440 kW. Find the diameter of shaft (d), if allowable shear stress in shaft material is 55 N/mm² and the angle of twist must not be more than 1° on a length of 16 (d). The modulus of rigidity is, G = 0.80 × 10⁵ N/mm².*

**Solution : Given data : P** = 440 kW = **440 × 10³ W**, N = 125 r.p.m., θ = 1°, L = 16 × d, G = 0.8 × 10⁵ N/mm².

**Procedure :** We have,

$$\text{Power transmitted, } P = \frac{2\pi N \cdot T_{mean}}{60}$$

$$\therefore \quad \text{Torque, } T_{mean} = \frac{P \times 60}{2\pi N} = \frac{440 \times 10^3 \times 60}{2 \times \pi \times 125}$$

$$\therefore \quad T_{mean} = 33.613 \times 10^3 \text{ N-m} = \mathbf{33.613 \times 10^6 \text{ N-mm}}$$

Also, angle of twist, $\theta = 1° = 1 \times \frac{\pi}{180} = 0.0174$ radian, i.e. $\boldsymbol{\theta = 0.0174^c}$

**(1) If shaft is designed on the basis of torsional rigidity**, then

We have, $\frac{T_{mean}}{J} = \frac{G\theta}{L}$

$$\therefore \quad \frac{33.613 \times 10^6}{\pi/32 \times d^4} = \frac{0.8 \times 10^5 \times 0.0174}{16 \times d}$$

$$\therefore \quad d = \mathbf{157.88 \text{ mm}} \qquad \dots (1)$$

**(2) If shaft is designed on the basis of strength**, then

We have, $T_{mean} = \frac{\pi}{16} \times \tau \times d^3$

$$\therefore \quad 33.613 \times 10^6 = \frac{\pi}{16} \times 55 \times d^3$$

$$\therefore \quad d = \mathbf{146.007 \text{ mm}} \qquad \dots (2)$$

- For safe design, we must take higher value of diameter after comparing equations (1) and (2).

∴ Safe diameter of shaft = 157.88 mm ≅ **160 mm (say)**

**Problem 3.7 :** *A shaft is required to transmit 1 MW power at 240 r.p.m. The shaft must not twist more than 1° on a length of one metre. If the modulus of rigidity for the material of the shaft is 80 kN/mm², find diameter of shaft and shear stress induced in it.*

**Solution : Given data :** $\mathbf{P}$ = 1 MW = $\mathbf{1 \times 10^6}$ **W**, N = 240 r.p.m., $\theta = 1° = 1 \times \frac{\pi}{180} = \mathbf{0.0174^c}$

$\mathbf{L}$ = 1 m = **1000 mm**, $\mathbf{G}$ = 80 kN/mm² = $\mathbf{80 \times 10^3}$ **N/mm²**.

**Procedure :** We have, Power transmitted, $P = \frac{2\pi N \cdot T_{mean}}{60}$

$$\text{Torque, } \mathbf{T_{mean}} = \frac{P \times 60}{2\pi N} = \frac{1 \times 10^6 \times 60}{2 \times \pi \times 240} = 39.788 \times 10^3 \text{ N-m} = \mathbf{39.788 \times 10^6 \text{ N-mm}}$$

**(1) Diameter of shaft :** If the shaft is designed on the basis of torsional rigidity, then we use,

$$\frac{T_{mean}}{J} = \frac{G\theta}{L}$$

$$\therefore \quad \frac{39.788 \times 10^6}{\pi/32 \times d^4} = \frac{80 \times 10^3 \times 0.0174}{1000}$$

$$\therefore \quad d = 130.62 \text{ mm} \cong \mathbf{132 \text{ mm (say)}}$$

**(2) Induced shear stress :**

$$T_{mean} = \frac{\pi}{16} \times \tau \times d^3$$

$$\therefore \quad 39.788 \times 10^6 = \frac{\pi}{16} \times \tau \times (132)^3$$

$$\therefore \quad \tau = \mathbf{88.104 \text{ N/mm}^2}$$

**Numerical Type No. 2 : "Shaft Subjeced to Bending Moment Only"**

**Problem 3.8 :** *A pair of wheels of a railway wagon carries a load of 50 kN on each axle box acting at a distance of 100 mm outside the wheel base. The gauge of the rails is 1.4 m. Find the diameter of axle between the wheels, if the stress is not to exceed 100 MPa.*

**Solution : Given data :**

$\mathbf{W}$ = 50 kN = $\mathbf{50 \times 10^3}$ **N**,

L = 100 mm,

$\mathbf{L_1}$ = 1.4 m = **1400 mm**,

$\boldsymbol{\sigma_b}$ = 100 MPa = **100 N/mm²**

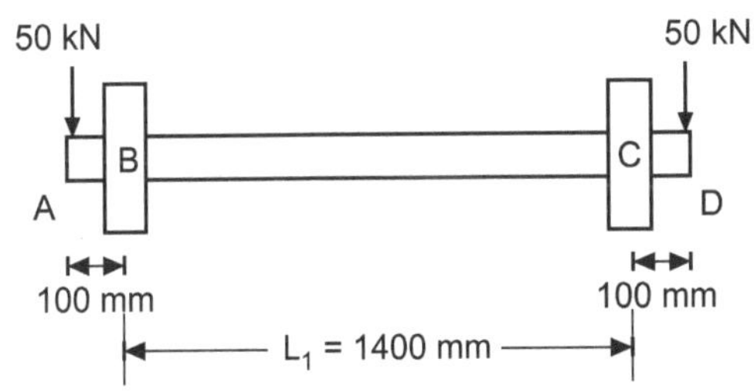

**Fig. 3.2**

**Procedure :** Considering the axle as an overhanging beam, we know that, bending moments at end points are zero.

$$\therefore \quad M_A = 0 \text{ and } M_D = 0$$

**Finding maximum B.M. :**

B.M. at B, $\mathbf{M_B} = W \times L = 50 \times 10^3 \times 100 = \mathbf{50 \times 10^5 \text{ N-mm}}$

B.M. at C, $\mathbf{M_C} = W \times L = 50 \times 10^3 \times 100 = \mathbf{50 \times 10^5 \text{ N-mm}}$

$\therefore$ **M** = Maximum B.M. acting at B and C = $\mathbf{50 \times 10^5}$ **N-mm**

We have, $$M = \frac{\pi}{32} \cdot \sigma_b \cdot d^3$$

$$\therefore \quad 50 \times 10^5 = \frac{\pi}{32} \times 100 \times d^3$$

$$\therefore \quad d = \mathbf{79.85 \text{ mm} \cong 80 \text{ mm (say)}}$$

**Problem 3.9 :** *An axle 1 m long is supported in bearings at its ends and carries a flywheel weighing 30 kN at centre. If the bending stress is not to exceed 60 MPa, find diameter of axle and sketch arrangement.*

**Solution : Given data :** $\mathbf{W_B}$ = 30 kN = $\mathbf{30 \times 10^3}$ **N**, $\boldsymbol{\sigma_b}$ = 60 MPa = **60 N/mm²**

**Procedure :**

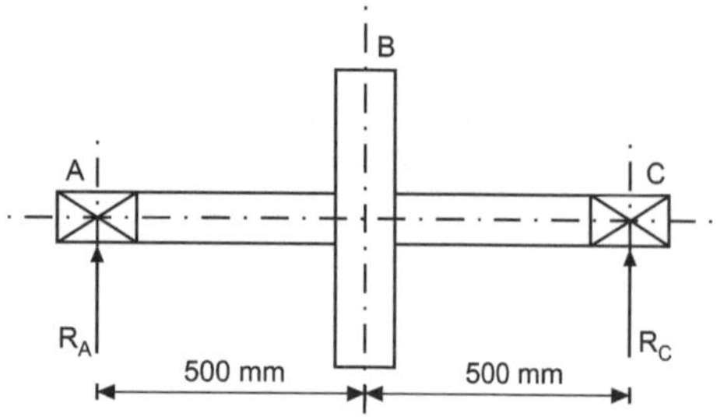


**(a) Arrangement**

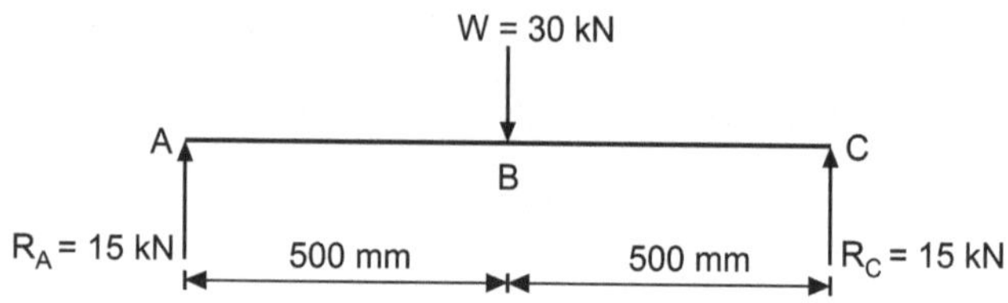


**(b) Vertical loading diagram**

**Fig. 3.3**

Considering the vertical load diagram, $R_A + R_C = 30 \times 10^3$ N. ... (1)

To find $R_C$, taking moment about point A, $\Sigma M_A = 0$.

$$\therefore \quad 30 \times 10^3 \times 500 - R_C \times 1000 = 0$$

$$\therefore \quad \mathbf{R_C = 15 \times 10^3\ N}$$

Putting the value of $R_C$ in equation (1), we get,

$$R_A + 15 \times 10^3 = 30 \times 10^3$$

$$\therefore \quad \mathbf{R_A = 15 \times 10^3\ N}$$

**Finding maximum bending moment :** Considering the axle as simply supported beam, we know that, bending moments at end points is zero. $\therefore$ $M_A = 0$ and $M_C = 0$.

Bending moment at point B = $M_B = 15 \times 10^3 \times 500 = \mathbf{7500 \times 10^3\ N\text{-}mm}$

$\therefore$ Maximum bending moment occurs at point B.

$$\therefore \quad \mathbf{M} = \text{Maximum bending moment} = M_B = \mathbf{7500 \times 10^3\ N.mm}$$

Axles are subjected to bending moment only.

$$\therefore \text{ We take,} \quad M = \frac{\pi}{32} \cdot \sigma_b \cdot d^3$$

$$\therefore \quad 7500 \times 10^3 = \frac{\pi}{32} \times 60 \times (d)^3$$

$$\therefore \quad d = \mathbf{108.38\ mm \cong 110\ mm\ (say)}$$

---

**Numerical Type No. 3 : "Shaft Subjeced to Both Twisting and Bending Moments"**

**Problem 3.10 :** *A mild steel shaft transmits 20 kW at 200 r.p.m. It carries a central load of 900 N and is simply supported on bearings 2.5 metres apart. Determine the size of the shaft, if the allowable shear stress is 42 MPa and maximum tensile or compressive stress is not to exceed 56 MPa.*

**Solution : Given data : P** = 20 kW = $\mathbf{20 \times 10^3}$ **W**, N = 200 r.p.m., $\tau$ = 42 MPa, **L** = 2.5 m = **2500 mm**

$\sigma_t = \sigma_c$ = 56 MPa, Central load = 900 N. **Assume, $\sigma_b$ = 56 MPa,**

**Procedure :**

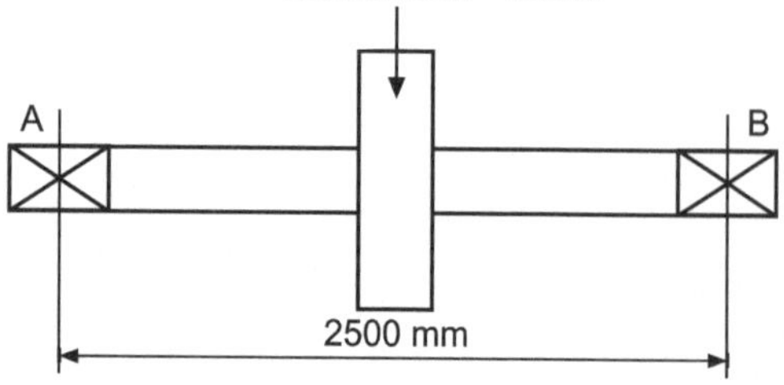


**(a)**

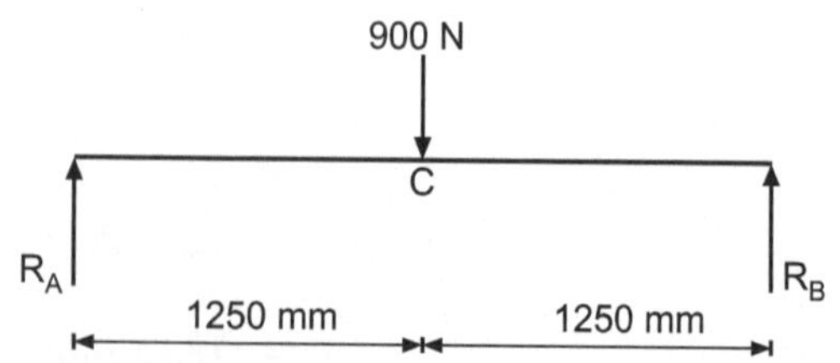


**(b) Free body diagram or Vertical loading diagram**

**Fig. 3.4**

From the figure, $\quad R_A = R_B = \dfrac{900}{2} = \mathbf{450\ N}$ [Since 900 N is central load]

**Step (1) : To calculate torque transmitted by shaft (T) :**

We have, $P = \frac{2\pi NT}{60}$

$\therefore$ $$\mathbf{T} = \frac{P \times 60}{2\pi N} = \frac{20 \times 10^3 \times 60}{2\pi \times 200} = 954.92 \text{ N.m} = \mathbf{954.92 \times 10^3 \text{ N.mm}}$$

**Step (2) : To calculate maximum bending moment acting on shaft (M) :**

Considering the shaft as simply supported beam, we know that, bending moments at end points is zero.

$\therefore$ $M_A = 0$ and $M_B = 0$

Bending moment at point 'C' = $M_C = R_A \times 1250 = 450 \times 1250 = 562.5 \times 10^3$ N.mm

$\therefore$ Maximum bending moment occurs at point C.

$\therefore$ $$M = M_{max} = M_C = \mathbf{562.5 \times 10^3 \text{ N.mm}}$$

**Step (3) : To calculate diameter of shaft (d) :**

As the shaft is subjected to both twisting moment and bending moment, then

**(a) According to maximum shear stress theory,**

$$\text{Equivalent twisting moment, } T_e = \frac{\pi}{16} \tau_{max} \cdot d^3$$

$$\therefore \quad \sqrt{M^2 + T^2} = \frac{\pi}{16} \tau_{max} \cdot d^3$$

$$\therefore \quad \sqrt{(562.5 \times 10^3)^2 + (954.92 \times 10^3)^2} = \frac{\pi}{16} \times 42 \times d^3$$

$$\therefore \quad d = \mathbf{51.22 \text{ mm}} \qquad \ldots(1)$$

**(b) According to maximum normal stress theory,**

$$\text{Equivalent bending moment, } M_e = \frac{\pi}{32} \cdot \sigma_{b\,max} \cdot d^3$$

$$\therefore \quad \frac{1}{2}\left\{M + \sqrt{M^2 + T^2}\right\} = \frac{\pi}{32} \cdot \sigma_b \max \cdot d^3$$

$$\therefore \quad \frac{1}{2}\left\{562.5 \times 10^3 + \sqrt{(562.5 \times 10^3)^2 + (954.92 \times 10^3)^2}\right\} = \frac{\pi}{32} \times 56 \times d^3$$

$$\therefore \quad d = \mathbf{53.36 \text{ mm}} \qquad \ldots(2)$$

For safe design, we will take larger value of diameter, by comparing equations (1) and (2).

$$\therefore \quad d = 53.36 \text{ mm} \cong \mathbf{54 \text{ mm (say)}}$$

---

**Problem 3.11 :** *A solid circular shaft is subjected to a B.M. of 3000 N-m and T = 10,000 N-m. Shaft is made of 46C8 steel having yield tensile stress of 700 MPa and ultimate shear stress of 500 MPa. Assuming F.S. = 6, determine diameter of shaft.*

**Solution : Given data :**

$\mathbf{M} = 3000 \text{ N.m} = 3000 \times 10^3 \text{ N.mm} = \mathbf{3 \times 10^6 \text{ N.mm}}$

$\mathbf{T} = 10000 \text{ N.m} = 10000 \times 10^3 \text{ N-mm} = \mathbf{10 \times 10^6 \text{ N-mm}}$

$\sigma_{yt} = 700$ MPa, $\tau_{ut} = 500$ MPa, F.O.S. = 6

**Procedure :**

$$\text{Working bending stress, } \sigma_b = \frac{\sigma_{yt}}{\text{F.O.S.}} = \frac{700}{6} = \mathbf{116.66 \text{ N/mm}^2} \qquad \left[\begin{array}{l}\textbf{Assumed} \\ \boldsymbol{\sigma_b = \sigma_t}\end{array}\right]$$

$$\text{Working shear stress, } \tau = \frac{\tau_{ut}}{\text{F.O.S.}} = \frac{500}{6} = \mathbf{83.33 \text{ N/mm}^2}$$

**Equivalent Twisting Moment :** $T_e = \sqrt{M^2 + T^2} = \sqrt{(3 \times 10^6)^2 + (10 \times 10^6)^2} = \mathbf{10.44 \times 10^6 \text{ N-mm}}$

$$\text{We know that,} \quad T_e = \frac{\pi}{16} \cdot \tau \cdot d^3$$

$$\therefore \quad 10.44 \times 10^6 = \frac{\pi}{16} \times 83.33 \times d^3$$

$$\therefore \quad d = \mathbf{86.09 \text{ mm}} \qquad \ldots(1)$$

**Equivalent Bending Moment :**

$$M_e = \frac{1}{2}\left[M + \sqrt{M^2 + T^2}\right] = \frac{1}{2}[M + T_e]$$

$$\therefore \quad M_e = \frac{1}{2}[3 \times 10^6 + 10.44 \times 10^6] = \mathbf{6.72 \times 10^6\ N\text{-}mm}$$

We know that, $$M_e = \frac{\pi}{32} \times \sigma_b \times d^3$$

$$\therefore \quad 6.72 \times 10^6 = \frac{\pi}{32} \times 116.66 \times d^3$$

$$\therefore \quad d = \mathbf{83.71\ mm} \qquad \ldots (2)$$

Comparing equations (1) and (2), for safe design, we will take larger value,

$$d = \mathbf{86.09\ mm \cong 88\ mm\ (say)}$$

---

**Problem 3.12 :** *A shaft supported at the ends in the ball bearings carries a straight tooth spur gear and is required to transmit 7.5 kW at 300 r.p.m. P.C.D. of gear is 150 mm. Distances between centre line of bearings and gear are 100 mm each. If shaft is made of steel and allowable shear stress is 45 MPa, determine the diameter of shaft. Show the arrangement of gear mounted on the shaft and bearings with a neat sketch. Pressure angle of gear is 20°.*

**Solution : Given data :** $P = 7.5\ kW = \mathbf{7.5 \times 10^3\ W}$, N = 300 r.p.m.,

Diameter of gear, D = 150 mm, $\tau$ = 45 MPa, L = 200 mm, $\alpha = 20°$

**Procedure :**

**Step (1) : To calculate torque transmitted by shaft (T) :**

Torque transmitted by shaft,

$$T = \frac{P \times 60}{2\pi N} = \frac{7.5 \times 10^3 \times 60}{2 \times \pi \times 300} = 238.73\ N\text{-}m$$

$$\therefore \quad \mathbf{T = 238.73 \times 10^3\ N\text{-}mm}$$

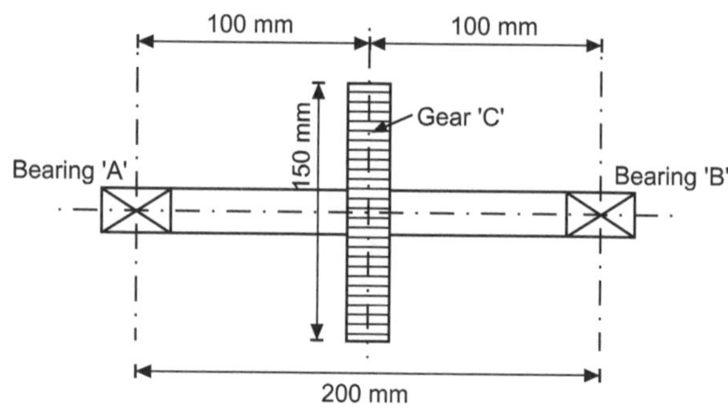


**Fig. 3.5 : Arrangement of gear mounted on shaft**

**Step (2) : To calculate maximum bending moment acting on the shaft (M) :**

We know that, Torque = Tangential force × Radius

$$\therefore \quad T = F_t \times R$$

$$\therefore \quad T = F_t \times \frac{D}{2}$$

$\therefore$ Tangential force acting on gear,

$$\mathbf{F_t} = \frac{T}{\frac{D}{2}} = \frac{238.73 \times 10^3}{\frac{150}{2}} = \mathbf{3.18 \times 10^3\ N}$$

Normal load acting on tooth of gear,

$$\therefore \quad \mathbf{W} = \frac{F_t}{\cos \alpha} = \frac{3.18 \times 10^3}{\cos (20°)} = \mathbf{3.38 \times 10^3\ N}$$

Considering shaft as simply supported beam having central load 'W', maximum bending moment is given by,

$$\therefore \quad \mathbf{M} = \frac{WL}{4} = \frac{3.38 \times 10^3 \times 200}{4} = \mathbf{169 \times 10^3\ N\text{-}mm}$$

**Step (3) : To calculate diameter of shaft (d) :**

**Equivalent Twisting Moment :** $$T_e = \sqrt{M^2 + T^2} = \sqrt{(169 \times 10^3)^2 + (238.73 \times 10^3)^2} = \mathbf{292.49 \times 10^3\ N\text{-}mm}$$

We have, $$T_e = \frac{\pi}{16} \tau d^3$$

$$\therefore \quad 292.49 \times 10^3 = \frac{\pi}{16} \times 45 \times d^3$$

$$\therefore \quad d = \mathbf{32.10\ mm \cong 34\ mm\ (say)}$$

**Numerical Type No. 4 : "Shafts with One Pulley Subjected to Both Twisting and Bending Moments"**

**Problem 3.13 :** *A mild steel shaft is supported in two bearings 1 metre apart and transmits 15 kW at 300 r.p.m. to a pulley of 200 mm diameter located at a distance of 300 mm from left hand bearing. Belt passing over the pulley is vertical and ratio of belt tensions is 2 : 1. Pulley weighs 500 N. Design the diameter of shaft. Take $\sigma_t$ = 70 N/mm$^2$ and $\tau$ = 56 N/mm$^2$.*

**Solution : Given data : P** = 15 kW = **$15 \times 10^3$ W**, N = 300 r.p.m., **L** = 1 m = **1000 mm**

D = 200 mm; ∴ **R** = $\frac{200}{2}$ = **100 mm**, $\sigma_t$ = 70 N/mm$^2$, **$\sigma_b$ = 70 N/mm$^2$ (Assumed)**,

$\tau$ = 56 N/mm$^2$, $W_p$ = 500 N, $\frac{T_1}{T_2} = 2$.

**Procedure :**

**Step (1) : Calculation of torque transmitted by shaft (T) :**

We know that, power transmitted,

$$P = \frac{2\pi NT}{60}$$

$$\therefore \quad 15 \times 10^3 = \frac{2\pi \times 300 \times T}{60}$$

$$\therefore \quad T = 477.46 \text{ N-m} = \mathbf{477.46 \times 10^3 \text{ N-mm}}$$

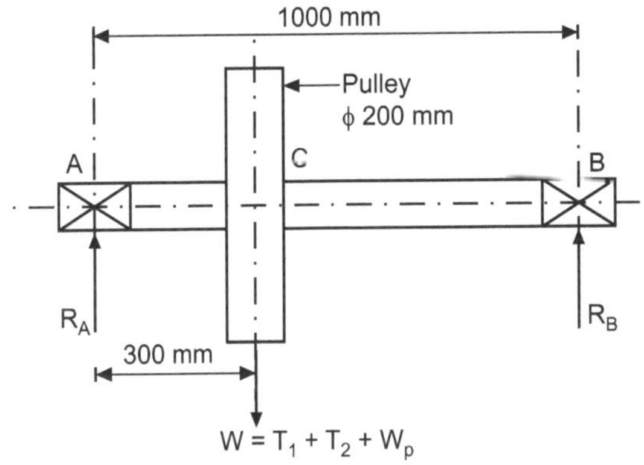


**Fig. 3.6 (a) : Arrangement**

**Step (2) : Calculation of maximum bending moment (M) :**

Tension ratio for belt passing over pulley is given as,

$$\frac{T_1}{T_2} = 2; \therefore \mathbf{T_1 = 2T_2}$$

We know that, Torque transmitted by pulley is given by,

$$T = (T_1 - T_2) \cdot R$$

$$\therefore \quad 477.46 \times 10^3 = (2\,T_2 - T_2) \times 100 \qquad [\because T_1 = 2T_2]$$

$$\therefore \quad \mathbf{T_2 = 4774.6 \text{ N}}$$

$$\therefore \quad T_1 = 2 \times T_2 = 2 \times 4774.6$$

$$\therefore \quad \mathbf{T_1 = 9549.2 \text{ N}}$$

∴ **Total load acting at point 'C'** = $T_1 + T_2 + W_p$

= 9549.2 + 4774.6 + 500

= **14823.8 N**

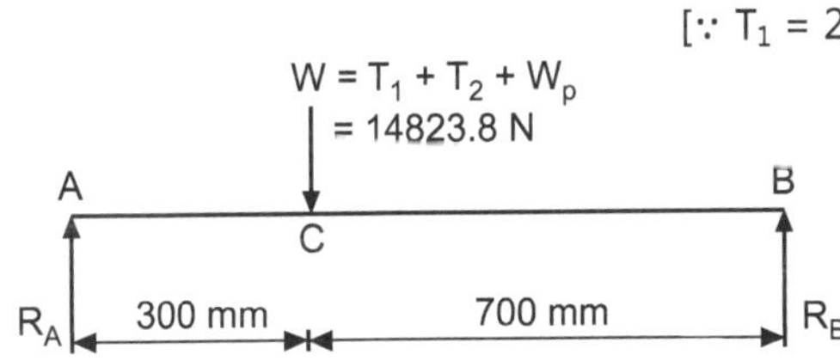


**Fig. 3.6 (b) : Vertical load diagram**

Taking algebraic sum of moments of all forces about 'A' and equating it with zero, we get,

$$\Sigma M_{@A} = 14823.8 \times 300 - R_B \times 1000 = 0$$

$$\therefore \quad \mathbf{R_B = 4447.14 \text{ N}}$$

Also, $\Sigma F_y = 0$

$$R_A + R_B = 14823.8 \text{ N}$$

$$\therefore \quad R_A + 4447.14 = 14823.8 \text{ N} \qquad (\because R_B = 4447.14 \text{ N})$$

$$\therefore \quad \mathbf{R_A = 10376.66 \text{ N}}$$

Considering shaft as simply supported beam, we know that, bending moments at end points are zero.

$$\therefore \quad M_A = 0 \text{ and } M_B = 0$$

Bending moment acting at point 'C' = $M_C = R_A \times 300 = 10376.66 \times 300 = 3.113 \times 10^6$ N.mm

∴ Maximum bending moment occurs at point C.

$$\therefore \quad \mathbf{M} = M_{max} = M_C = \mathbf{3.113 \times 10^6 \text{ N.mm}}$$

**Step (3) : Calculation of diameter of shaft (d) :**

**(a) Considering maximum shear stress theory,**

$$T_e = \sqrt{M^2 + T^2} = \frac{\pi}{16} \cdot \tau_{max} \cdot d^3$$

$$\therefore \sqrt{(3.113 \times 10^6)^2 + (477.46 \times 10^3)^2} = \frac{\pi}{16} \times 56 \times d^3$$

$$\therefore \quad d = \mathbf{65.91 \text{ mm}} \qquad \dots (1)$$

**(b) Considering maximum principal stress theory,**

$$M_e = \frac{1}{2}\{M + \sqrt{M^2 + T^2}\} = \frac{\pi}{32} \cdot \sigma_{b\,max} \cdot d^3$$

$$\therefore \quad \frac{1}{2} \times [3.113 \times 10^6 + \sqrt{(3.113 \times 10^6)^2 + (477.46 \times 10^3)^2}] = \frac{\pi}{32} \times 70 \times d^3$$

$$\therefore \quad d = \mathbf{76.94\ mm} \qquad \ldots (2)$$

On comparing equations (1) and (2), for safe design, we take larger value of diameter.

$$\therefore \quad d = \mathbf{76.94\ mm \approx 78\ mm\ (say)}$$

**Problem 3.14 :** *A shaft 1000 mm long is supported between two bearings. A pulley of 250 mm diameter is keyed at 400 mm distance away from left hand bearing. The power transmitted by shaft is 10 kW at 800 r.p.m. The pulley gives power to another pulley vertically below it having an angle of contact between pulley and belt as 180°. The weight of pulley is 300 N. The coefficient of friction between belt and pulley is 0.15. Take shear stress for shaft material as 60 MPa. Find the diameter of the shaft.*

**Solution : Given data :**

$$D_p = 250\text{ mm};\ \therefore\ \mathbf{R_p} = \frac{D_P}{2} = \frac{250}{2} = \mathbf{125\ mm}$$

$$W_p = 300\text{ N},\ L = 100\text{ mm}$$

$$\mathbf{P} = 10\text{ kW} = \mathbf{10 \times 10^3\ W},\ N = 800\text{ r.p.m.}$$

$$\boldsymbol{\theta} = 180° = \pi^c,\ \mu = 0.15,\ \tau = 60\text{ MPa}$$

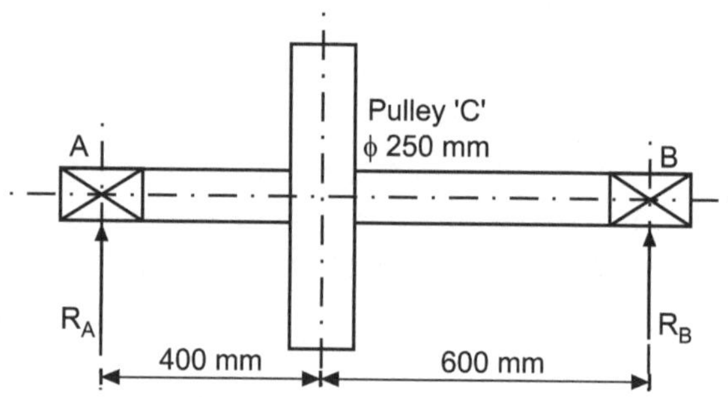


**Fig. 3.7 (a) : Arrangement**

**Procedure :**

**Step 1 : Calculation of torque (T) :**

We have, $$P = \frac{2\pi NT}{60}$$

$$\therefore \quad T = \frac{P \times 60}{2\pi N} = \frac{10 \times 10^3 \times 60}{2 \times \pi \times 800}$$

$$\therefore \quad \mathbf{T} = 119.36\text{ N.m} = \mathbf{119.36 \times 10^3\ N.mm}$$

**Step 2 : Calculation of maximum bending moment (M) :**

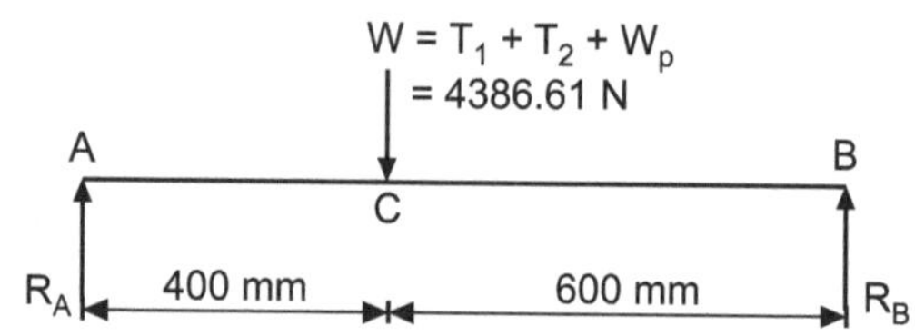


**Fig. 3.7 (b) : Vertical load diagram**

We have, belt tension ratio, $$\frac{T_1}{T_2} = e^{(\mu\theta)}$$

$$\therefore \quad \frac{T_1}{T_2} = e^{(0.15 \times \pi)}$$

$$\therefore \quad \mathbf{T_1 = 1.61\ T_2}$$

Torque transmitted by pulley, $T = (T_1 - T_2) \times R_p$

$$\therefore \quad 119.36 \times 10^3 = (1.61\ T_2 - T_2) \times 125$$

$$\therefore \quad \mathbf{T_2 = 1565.37\ N}$$

$$\therefore \quad \mathbf{T_1} = 1.61\ T_2 = 1.61 \times 1565.37 = \mathbf{2520.24\ N}$$

Total load acting at point 'C' $= T_1 + T_2 + W_p$

$\therefore$ **Total load acting at point 'C'** $= 2520.24 + 1565.37 + 300 = \mathbf{4386.61\ N}$

Taking moment @ point A, $\Sigma M_{@A} = 0$

$$\therefore \quad 4386.61 \times 400 - R_B \times 1000 = 0$$

$$\therefore \quad \mathbf{R_B = 1754.64\ N}$$

We have, $\Sigma F_y = 0$

$$\therefore \quad R_A + R_B = 4386.61$$

$$\therefore \quad \mathbf{R_A} = 4386.61 - 1754.64 = \mathbf{2631.966\ N} \qquad [R_B = 1754.64\text{ N}]$$

Considering the shaft as simply supported beam, we know that, bending moments acting at end points are zero.

$$\therefore \quad M_A = 0 \text{ and } M_B = 0$$

Bending moment at point 'C' $= M_C = R_A \times 400 = 2631.966 \times 400 = 1052786.4$ N.mm

$\therefore$ Maximum bending moment occurs at point C,

$$\mathbf{M} = M_C = \mathbf{1052786.4\ N.mm}$$

**Step 3 : Calculation of diameter of shaft (d) :**

According to maximum shear stress theory,

$$T_e = \sqrt{M^2 + T^2} = \frac{\pi}{16} \tau_{max} \cdot d^3$$

$$\therefore \sqrt{(1052786.4)^2 + (119.36 \times 10^3)^2} = \frac{\pi}{16} \times 60 \times d^3$$

$$\therefore \quad d = \mathbf{44.8\ mm \cong 46\ mm\ (say)}$$

**Problem 3.15 :** *Design a shaft transmitting power of 60 kW at 600 r.p.m. as shown in Fig. 3.8. The shaft is hollow having ratio of inner to outer diameter 0.5. The shaft carries pulley 'C' as shown. The angle of contact between belt and pulley is 180°. The belts are vertical as shown. The diameter and weight of pulley are 250 mm and 700 N respectively and the ratio of tight-side to slack side tensions is 3. Take $K_m$ = 2, $K_t$ = 1.5. Design shaft, if shear stress in material is not to exceed 52 N/mm².*

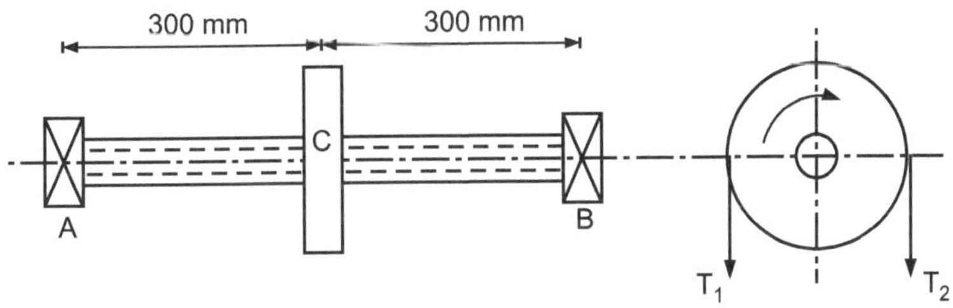


**Fig. 3.8**

**Solution : Given data : P** = 60 kW = **60 × 10³ W**, N = 600 r.p.m., $\mathbf{k} = \frac{d_i}{d_o} = \mathbf{0.5}$, $\theta = 180° = \pi^c$, $W_c$ = 700 N,

$D_c = 250$ mm; $\therefore \mathbf{R_c} = \frac{D_c}{2} = \frac{250}{2} = \mathbf{125\ mm}$, $\frac{T_1}{T_2} = 3$; $\therefore \mathbf{T_1 = 3T_2}$, $k_m = 2$, $k_t = 1.5$, $\tau_{max} = 52$ N/mm²

**Procedure :**

**Step (1) : To calculate torque (T) :**

We have, $$P = \frac{2\pi NT}{60}$$

$$\therefore \quad \mathbf{T} = \frac{P \times 60}{2\pi N} = \frac{60 \times 10^3 \times 60}{2\pi \times 600} = 954.93 \text{ N.m} = \mathbf{954.93 \times 10^3\ N.mm}$$

**Step (2) : To calculate maximum bending moment (M) :**

Torque transmitted by pulley 'C' $= (T_1 - T_2) \times R_c$

$$\therefore \quad 954.93 \times 10^3 = (3T_2 - T_2) \times 125$$

$$\therefore \quad \mathbf{T_2 = 3819.72\ N}$$

$$\therefore \quad \mathbf{T_1} = 3T_2 = 3 \times 3819.72 = \mathbf{11459.16\ N}$$

Total load acting at point 'C' = $T_1 + T_2 + W_c$ = 11459.16 + 3819.72 + 700 = **15978.88 N**

Since point 'C' is midpoint of line AB, therefore, total load acting at point 'C' is **central load**. Reactions $R_A$ and $R_B$ can be calculated as,

$$\mathbf{R_A = R_B} = \frac{15978.88}{2} = \mathbf{7989.44\ N}$$

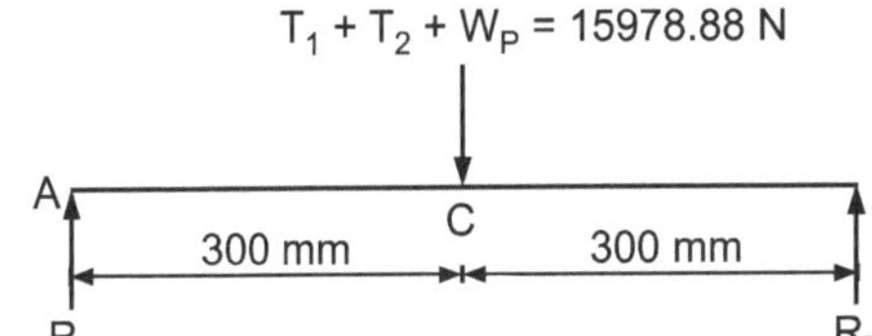


**Fig. 3.9 : Vertical loading diagram**

Considering shaft AB as simply supported beam, bending moment at each point can be calculated as,

$M_A = 0$, $M_B = 0$

$M_C = R_A \times 300 = 7989.44 \times 300$

$$\therefore \quad M_C = \mathbf{2396.832 \times 10^3\ N.mm}$$

$\therefore$ Maximum bending moment occurs at point 'C'.

$$\mathbf{M} = M_C = \mathbf{2396.832 \times 10^3\ N.mm}$$

**Step (3) : To calculate diameters ($d_o$ and $d_i$) :**

Since shock and fatigue factors ($k_m$ and $k_t$) are given, the design equation for hollow shaft is,

$$\sqrt{(k_m \times M)^2 + (k_t \times T)^2} = \frac{\pi}{16} \tau_{max}\ d_o^3 \times (1 - K^4)$$

$\therefore \quad \sqrt{(2 \times 2396.832 \times 10^3)^2 + (1.5 \times 954.93 \times 10^3)^2} = \frac{\pi}{16} \times 52 \times d_o^3 \times [1 - (0.5)^4]$

$\therefore \quad \mathbf{d_o} = 80.55 \cong \mathbf{82\ mm\ (say)}$

$\therefore \quad \mathbf{d_i} = 0.5\ d_o = 0.5 \times 82 = \mathbf{41\ mm}$ $\left[\because k = \frac{d_i}{d_o} = 0.5;\ \therefore d_i = 0.5 d_o\right]$

## Numerical Type No. 5 : "Shaft with an Overhanging Pulley"

**Problem 3.16 :** *Determine the diameter of a hollow shaft having inside diameter 0.6 times outside diameter. The maximum allowable shear stress for shaft is 85 MPa. Shaft is driven by a 900 mm diameter overhang pulley placed vertically below it. The weight of pulley is 600 N. Tensions in tight and slack side of belt are 2900 N and 1000 N respectively. The overhang is 250 mm. Assume angle of lap of belt on the pulley to be 180°.*

**Solution : Given data :** $d_i = 0.6\ d_o$, $\therefore$ $\mathbf{k} = d_i/d_o = \mathbf{0.6}$; $\therefore$ $d_i = 0.6\ d_o$, $D_p$ = Diameter of pulley = 900 mm,

$\therefore \quad \mathbf{R_p} = \frac{D_p}{2} = \frac{900}{2} = \mathbf{450\ mm}$, $\tau = 85\ N/mm^2$, $W_p = 600$ N, $T_1 = 2900$ N, $T_2 = 1000$ N.

**Procedure :**

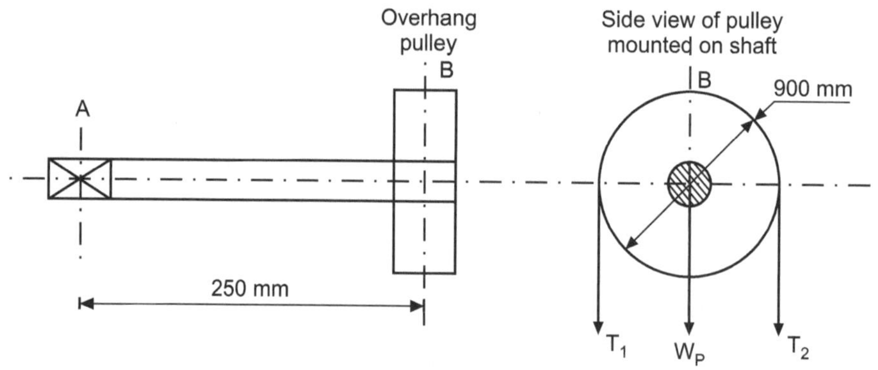


**Fig. 3.10 (a) : Arrangement**

**Step (1) : Calculation of torque transmitted by shaft (T) :**

**Torque transmitted by pulley,** $T = (T_1 - T_2) \times R_p = (2900 - 1000) \times 450 = \mathbf{855 \times 10^3\ N.mm}$

Assuming no losses, torque transmitted by pulley can be taken as equal to torque transmitted by shaft.

$\therefore$ Torque transmitted by shaft, $T = \mathbf{855 \times 10^3\ N.mm}$

**Step (2) : Calculation of maximum bending moment on shaft (M) :**

**Total load acting at point B** $= T_1 + T_2 + W_P = 2900 + 1000 + 600 = \mathbf{4500\ N}$

Considering the shaft as a cantilever beam, we know that,

Bending moment at free end is zero. $\therefore$ $M_B = 0$.

$\therefore$ Maximum bending moment at point 'A'

$= M = W \times 250 = 4500 \times 250 = \mathbf{1.125 \times 10^6\ N\text{-}mm.}$

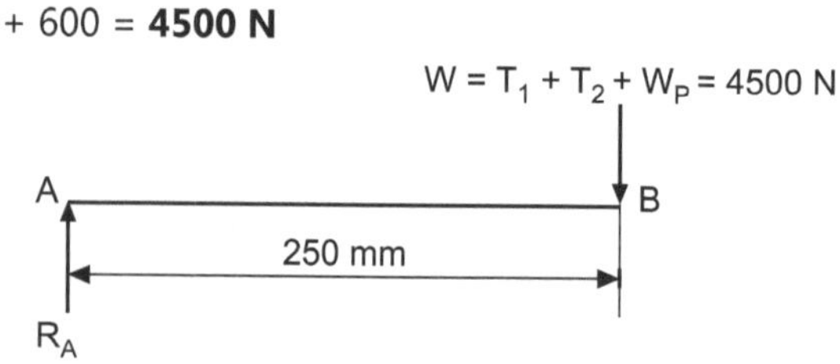


**Fig. 3.10 (b) : Vertical load diagram**

**Step (3) : Calculation of diameters of hollow shaft ($d_o$ and $d_i$) :**

As the shaft is subjected to both twisting and bending moment, then for hollow shaft, according to maximum shear stress theory, equivalent twisting moment is given by,

$$T_e = \sqrt{M^2 + T^2} = \frac{\pi}{16} \times \tau \times d_o^3 (1 - k^4)$$

$\therefore \quad \sqrt{(1.125 \times 10^6)^2 + (855 \times 10^3)^2} = \frac{\pi}{16} \times 85 \times d_o^3 \{1 - (0.6)^4\}$

$\therefore \quad \mathbf{d_o} = 45.989$ mm $\cong$ **46 mm (say)** and $\mathbf{d_i} = 0.6 \times d_o = 0.6 \times 46 = \mathbf{27.6\ mm}$

**Problem 3.17 :** *Design a shaft to transmit power from an electric motor to a lathe headstock through a pulley by means of a belt drive. The pulley weighs 200 N and is located at 300 mm from the centre-line of bearing. Diameter of the pulley is 200 mm and maximum power transmitted is 1 kW at 120 r.p.m. Angle of lap of the belt is 180° and coefficient of friction between the belt and pulley is 0.3. Allowable shear stress in the shaft may be taken as 35 N/mm$^2$.*

**Solution : Given data :** $W_p$ = 200 N, $D_p$ = 200 mm; $\therefore$ $\mathbf{R_p} = \frac{200}{2}$ = **100 mm**, **P** = 1 kW = **1000 W**, N = 120 r.p.m.,

$\theta = 180° = \pi^c$, $\mu = 0.3$, $\tau = 35$ N/mm$^2$.

**Procedure :**

**Step (1) : Calculation of torque transmitted by shaft (T) :**

We have, $$P = \frac{2\pi NT}{60}$$

$$\therefore \quad 1000 = \frac{2\pi \times 120 \times T}{60}$$

$$\therefore \quad \mathbf{T} = 79.58 \text{ N.m} = \mathbf{79.58 \times 10^3 \text{ N.mm}}$$

**Step (2) : Calculation of maximum bending moment acting on shaft (M) :**

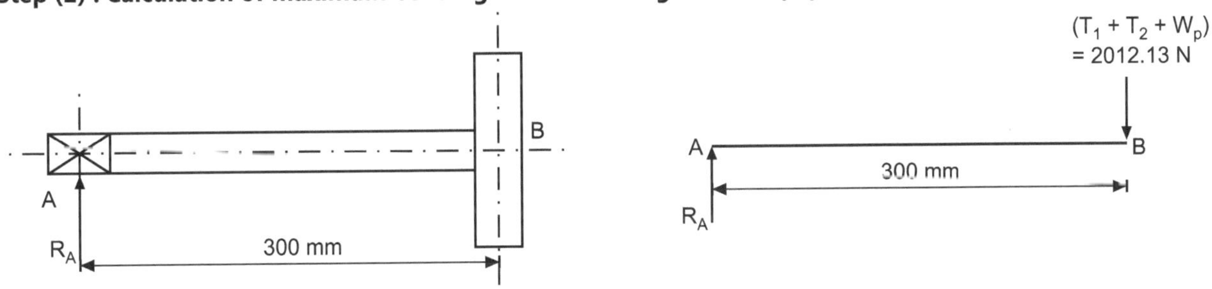


**(a) Arrangement** **(b) Vertical load diagram**

**Fig. 3.11**

Ratio of belt tensions is given by following relation,

$$\frac{T_1}{T_2} = e^{\mu\theta} = e^{(0.3 \times \pi)} = 2.566$$

$$\therefore \quad \mathbf{T_1 = 2.566\ T_2}$$

Torque transmitted by pulley,

$$T = (T_1 - T_2) \times R_p$$

$$\therefore \quad 79.58 \times 10^3 = (2.566\ T_2 - T_2) \times 100 \qquad [\therefore T_1 = 2.566\ T_2]$$

$$\therefore \quad \mathbf{T_2 = 508.17\ N}$$

$$\therefore \quad \mathbf{T_1} = 2.566\ T_2 = 2.566 \times 508.17 = \mathbf{1303.96\ N}$$

Therefore, total load acting at point 'B' = $T_1 + T_2 + W_P$ = 1303.96 + 508.17 + 200 = **2012.13 N**

Considering the shaft as a cantilever beam, we know that, bending moment at free end 'B' is zero. $\therefore$ $M_B = 0$.

Bending moment at fixed end, i.e. point 'A' is given by,

$$M_A = (T_1 + T_2 + W_P) \times 300 = 2012.13 \times 300 = \mathbf{603639\ N.mm}$$

$\therefore$ Maximum bending moment occurs at point A.

$$\therefore \quad M = M_A = \mathbf{603639\ N.mm}$$

**Step (3) : Calculation of diameter of shaft (d) :**

According to maximum shear stress theory,

$$T_e = \sqrt{M^2 + T^2} = \frac{\pi}{16} \cdot \tau_{max} \cdot d^3$$

$$\therefore \quad \sqrt{(603639)^2 + (79.58 \times 10^3)^2} = \frac{\pi}{16} \times 35 \times d^3$$

$$\therefore \quad d = \mathbf{44.58\ mm \cong 46\ mm\ (say)}$$

**Numerical Type No. 6 : "Shaft with Two Pulleys Subjected to Both Twisting and Bending Moments"**

**Problem 3.18 :** *A shaft made of mild steel is required to transmit 100 kW at 300 r.p.m. The supported length of the shaft is 3 metres. It carriers two pulleys each weighing 1500 N supported at a distance of 1 metre from both ends. Find the diameter of the shaft. Take $f_s = 60$ N/mm², $f_b = 90$ N/mm².*

**Solution : Given data :** $P = 100$ kW $= \mathbf{100 \times 10^3}$ **W**, $N = 300$ r.p.m., $W_C = W_D = 1500$ N, $f_s = \tau =$ **60 N/mm²**, $f_b = \sigma_b =$ **90 N/mm²**

**Procedure : Step (1) : To calculate torque transmitted by shaft (T) :**

We have,
$$P = \frac{2\pi NT}{60}$$

$$\therefore \quad T = \frac{P \times 60}{2\pi N} = \frac{100 \times 10^3 \times 60}{2\pi \times 300} = 3183.09 \text{ N.m} = \mathbf{3183.09 \times 10^3 \text{ N.mm}}$$

**Step (2) : To calculate maximum bending moment acting on shaft (M) :**

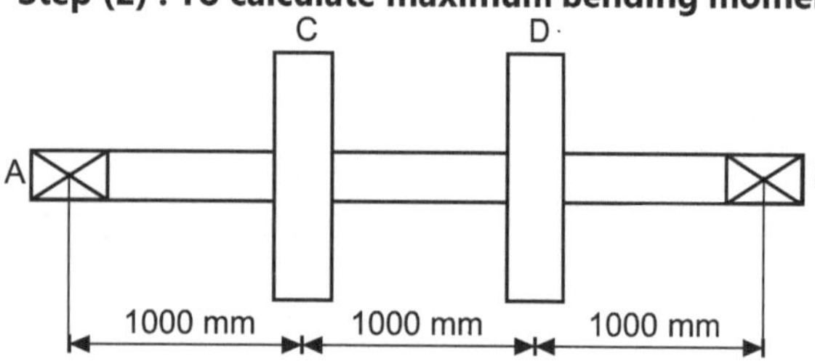


**(a) Arrangement**

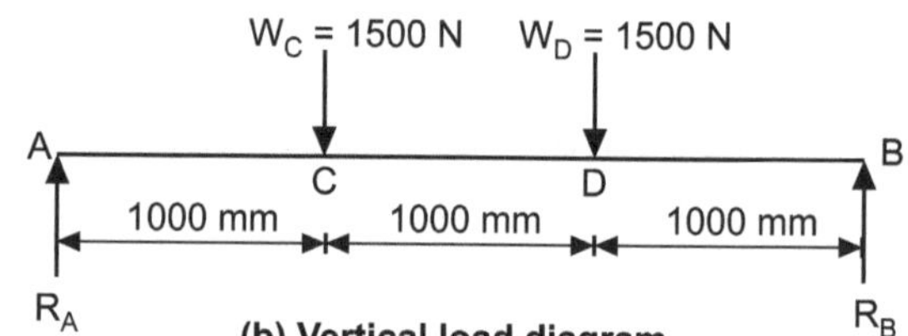


**(b) Vertical load diagram**

**Fig. 3.12**

We know, $\quad \Sigma F_y = 0$

$\therefore \quad \mathbf{R_A + R_B} = 1500 + 1500 = \mathbf{3000\ N}$ ... (1)

Also, $\quad \Sigma M_A = 0$

$\therefore \quad 1500 \times 1000 + 1500 \times 2000 = R_B \times 3000$

$\therefore \quad \mathbf{R_B = 1500\ N}$

From equation (1), $\quad R_A + 1500 = 3000$

$\therefore \quad \mathbf{R_A = 1500\ N}$

Considering the shaft as simply supported beam, we know that, bending moments at end points are zero.

$\therefore \quad M_A = 0$ and $M_B = 0$

Bending moment at point C = $\mathbf{M_C} = R_A \times 1000 = 1500 \times 1000 = \mathbf{15 \times 10^5}$ **N.mm**

Bending moment at point D = $\mathbf{M_D} = R_B \times 1000 = 1500 \times 1000 = \mathbf{15 \times 10^5}$ **N.mm**

$\therefore$ Maximum bending moment = $\mathbf{M = 15 \times 10^5}$ **N.mm**

**Step (3) : To calculate diameter of shaft (d) :**

Since the shaft is subjected to combined twisting and bending moments, it must be designed on the basis of maximum shear stress theory and maximum normal stress theory.

**(a) According to maximum shear stress theory,**

$$\sqrt{M^2 + T^2} = \frac{\pi}{16} \tau_{max} \cdot d^3$$

$$\therefore \quad \sqrt{(15 \times 10^5)^2 + (3183.09 \times 10^3)^2} = \frac{\pi}{16} \times 60 \times d^3$$

$$\therefore \quad d = \mathbf{64.84\ mm} \qquad \text{... (2)}$$

**(b) According to maximum normal stress theory,**

$$\frac{1}{2}\left[M + \sqrt{M^2 + T^2}\right] = \frac{\pi}{32} \times \sigma_{b\,max} \times d^3$$

$$\frac{1}{2} \times \left[15 \times 10^5 + \sqrt{(15 \times 10^5)^2 + (3183.09 \times 10^3)^2}\right] = \frac{\pi}{32} \times 90 \times d^3$$

$$\therefore \quad d = \mathbf{65.73\ mm} \qquad \text{... (3)}$$

- For safe design of shaft, we will take larger value of diameter by comparing equations (2) and (3).

$$\therefore \quad \mathbf{d} = 65.73 \text{ mm} \cong \mathbf{66\ mm\ (say)}$$

**Problem 3.19 :** *Fig. 3.13 shows a shaft of uniform diameter supported in two self aligning bearings at C and D. Two pulleys A and B are mounted having belt tensions 1800 N and 800 N respectively. The shaft transmits 10 kW power at 300 r.p.m. The pulleys weigh 200 N and 400 N respectively. Take safe working stress as 40 MPa.*

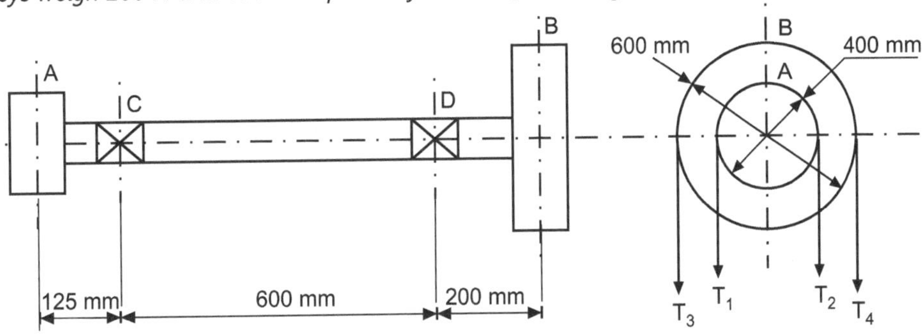


**Fig. 3.13**

**Solution : Given data : P** = 10 kW = **$10 \times 10^3$ W**, N = 300 r.p.m., $W_A$ = 200 N, $W_B$ = 400 N, $\tau$ = 40 N/mm$^2$,

$D_A$ = diameter of pulley 'A' = 400 mm, $\therefore$ $\mathbf{R_A} = \frac{D_A}{2} = \frac{400}{2}$ = **200 mm,**

$D_B$ = Diameter of pulley 'B' = 600 mm, $\therefore$ $\mathbf{R_B} = \frac{D_B}{2} = \frac{600}{2}$ = **300 mm.**

**Procedure : Step (1) : To calculate torque transmitted by shaft (T) :**

We have, Power transmitted, $P = \frac{2\pi N\ T_{mean}}{60}$

$\therefore$ Torque, $\mathbf{T} = \frac{P \times 60}{2\pi N} = \frac{10 \times 10^3 \times 60}{2 \times \pi \times 300}$ = 318.31 N-m = **$318.31 \times 10^3$ N-mm.**

**Step (2) : To calculate maximum bending moment acting on shaft (M) :**

Belt tension in pulley A = 1800 N

$\therefore$ **Load acting at point A** = 1800 + $W_A$
= 1800 + 200
= **2000 N**

Belt tension in pulley B = 800 N

$\therefore$ **Load acting at point B** = 800 + $W_B$
= 800 + 400
= **1200 N**

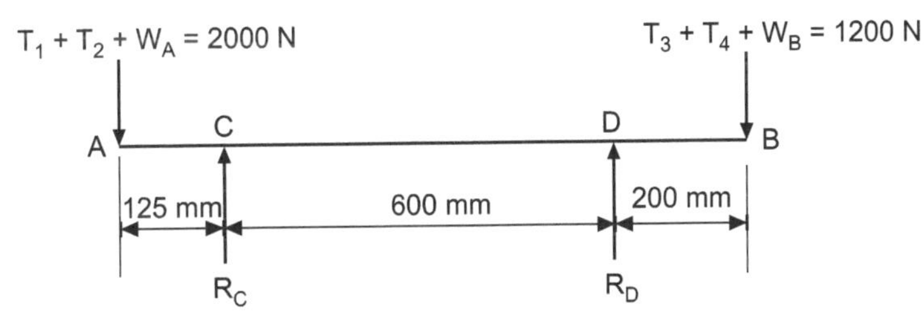


**Fig. 3.14 : Vertical load diagram**

Considering shaft as an overhanging beam, we know that, bending moments at extreme ends are zero.

$\therefore$ $M_A = 0$ and $M_B = 0$

Now, Bending moment at point 'C' = $\mathbf{M_C}$ = 2000 × 125 = **$250 \times 10^3$ N-mm**

And Bending moment at point 'D' = $\mathbf{M_D}$ = 1200 × 200 = **$240 \times 10^3$ N-mm**

Thus, maximum bending moment occurs at point C.

$\therefore$ $M_{max} = M_C = \mathbf{M}$ = **$250 \times 10^3$ N-mm**

**Step (3) : To calculate diameter of shaft (d) :**

As the shaft is subjected to both twisting and bending moment, then according to maximum shear stress theory, the equivalent twisting moment is given by,

$$T_e = \frac{\pi}{16} \cdot \tau_{max} \cdot d^3 = \sqrt{M^2 + T^2}$$

$$\therefore \quad \frac{\pi}{16} \times 40 \times d^3 = \sqrt{(250 \times 10^3)^2 + (318.31 \times 10^3)^2}$$

$$\therefore \quad \mathbf{d = 37.21\ mm \cong 38\ mm\ (say)}$$

**Numerical Type No. 7 : "Shafts Subjected to Both Twisting and Bending Moments with Shock Loading"**

**Problem 3.20 :** *Shaft is transmitting 25 kW at 1500 r.p.m, when it is subjected to bending moment of magnitude 150 N-m due to gear mounted on it. Find the diameter of shaft. Take shock and fatigue factors, $k_m$ = 1.5 and $k_t$ = 1.2. Take yield strength of material as 300 MPa.*

**Solution : Given data : P** = 25 kW = $\mathbf{25 \times 10^3}$ **W**, N = 1500 r.p.m., M = 150 N.m, $\sigma_y$ = 300 MPa, $k_m$ = 1.5, $k_t$ = 1.2.

**Procedure : Step (1) : To calculate torque transmitted (T) and Bending moment (M) :**

We have, power transmitted, $P = \frac{2\pi N\ T_{mean}}{60}$

$$\therefore \quad \text{Torque, } \mathbf{T} = \frac{P \times 60}{2\pi N} = \frac{25 \times 10^3 \times 60}{2 \times \pi \times 1500} = 159.15 \text{ N-m} = \mathbf{159.15 \times 10^3 \text{ N-mm}}$$

Bending moment due to gear = **M** = 150 N-m = $\mathbf{150 \times 10^3}$ **N-mm** (Given)

**Step (2) : To calculate diameter of shaft (d) :**

Permissible bending stress can be calculated by **assuming factor of safety as 5.**

$$\therefore \quad \sigma_b = \frac{\sigma_y}{F.O.S.} = \frac{300}{5} = \mathbf{60 \text{ N/mm}^2.}$$

As the shaft is subjected to both twisting and bending moment, then according to maximum normal stress theory, the equivalent bending moment is given by,

$$M_e = \frac{1}{2}\{(k_m M) + \sqrt{(k_m M)^2 + (k_t T)^2}\} = \frac{\pi}{32} \times \sigma_{b\,max} \times (d)^3$$

$$\therefore \quad \frac{1}{2}\{1.5 \times 150 \times 10^3 + \sqrt{(1.5 \times 150 \times 10^3)^2 + (1.2 \times 159.15 \times 10^3)^2}\} = \frac{\pi}{32} \times 60 \times (d)^3$$

$\therefore$ **d** = 35.34 mm = **36 mm (say)**

---

**Problem 3.21 :** *A mild steel shaft transmits 30 kW at 300 r.p.m. It carries a central load of 1000 N and is simply supported between the bearings 3 m apart. Determine the size of the shaft, if allowable shear stress and tensile stress are 42 MPa and 56 MPa respectively. What change would you observe in the diameter of the shaft, if same shaft is subjected to suddenly applied load ? Take $k_m = k_t$ = 2.5.*

**Solution : Given data : P** = 30 kW = $\mathbf{30 \times 10^3}$ **W**, N = 300 r.p.m., W = 1000 N, **L** = 3 m = **3000 mm**, $\tau$ = 42 N/mm², $\sigma_t$ = 56 N/mm², $k_m = k_t$ = 2.5.

**Procedure : Step (1) : To calculate torque transmitted by shaft (T) :**

We have, $P = \frac{2\pi N\ T_{mean}}{60}$

$$\therefore \quad \mathbf{T} = T_{mean} = \frac{P \times 60}{2\pi \times N} = \frac{30 \times 10^3 \times 60}{2\pi \times 300} = 954.93 \text{ N·m} = \mathbf{954.93 \times 10^3 \text{ N·mm}}$$

**Step (2) : To calculate maximum bending moment acting on shaft (M) :**

As the shaft is centrally loaded between two end bearings, the support reactions are,

$$\mathbf{R_A = R_B} = \frac{W}{2} = \frac{1000}{2} = \mathbf{500 \text{ N}}$$

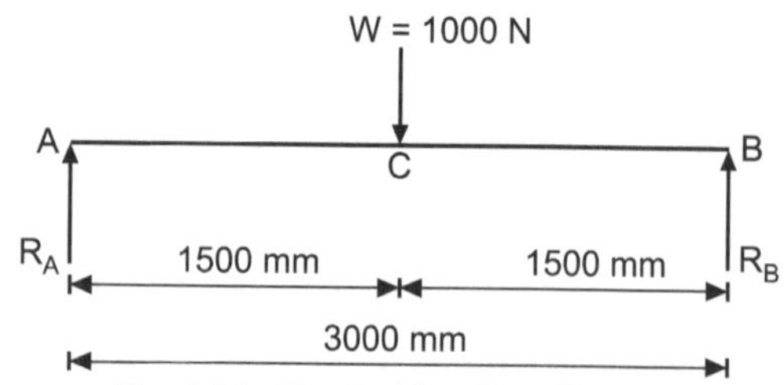

**Fig. 3.15 : Vertical loading diagram**

Considering shaft as simply supported beam, we know that, bending moment at end points is zero.

$$\therefore \quad M_A = 0, \qquad M_B = 0.$$

$$\therefore \quad \text{Bending moment,} \quad M_C = \mathbf{M} = R_A \times 1500 = 500 \times 1500 = \mathbf{750 \times 10^3 \text{ N·mm}}$$

**Step (3) : To calculate diameter of shaft (d) :**

**(a) Considering maximum shear stress theory,**

$$T_e = \sqrt{M^2 + T^2} = \frac{\pi}{16}\tau_{max} \cdot d^3$$

$$\therefore \quad \sqrt{(750 \times 10^3)^2 + (954.93 \times 10^3)^2} = \frac{\pi}{16} \times 42 \times d^3$$

$$\therefore \quad \mathbf{d} = 52.80 \text{ mm} \cong \mathbf{54 \text{ mm (say)}} \qquad \dots (1)$$

If the same shaft is subjected to suddenly applied load, then

$$T_e = \sqrt{(k_m \times M)^2 + (k_t \times T)^2} = \frac{\pi}{16} \tau_{max} \cdot d^3$$

$$\therefore \quad \sqrt{(2.5 \times 750 \times 10^3)^2 + (2.5 \times 954.93 \times 10^3)^2} = \frac{\pi}{16} \times 42 \times d^3$$

$$\therefore \quad d = 71.66 \text{ mm} \cong \mathbf{72\ mm\ (say)} \qquad \ldots (2)$$

**(b) Considering maximum normal stress theory,**

$$M_e = \frac{1}{2}\left[M + \sqrt{M^2 + T^2}\right] = \frac{\pi}{32} \sigma_{max} \cdot d^3$$

$$\therefore \quad \frac{1}{2} \times \left[750 \times 10^3 + \sqrt{(750 \times 10^3)^2 + (954.93 \times 10^3)^2}\right] = \frac{\pi}{32} \times 56 \times d^3$$

$$d = 56.31 \text{ mm} \cong \mathbf{58\ mm\ (say)} \qquad \ldots (3)$$

And for sudden loading,

$$M_e = \frac{1}{2}\left[k_m \cdot M + \sqrt{(k_m + M)^2 + (k_t \times T)^2}\right] = \frac{\pi}{32} \sigma_{max}\, d^3$$

$$\therefore \quad \frac{1}{2} \times \left[2.5 \times 750 \times 10^3 + \sqrt{(2.5 \times 750 \times 10^3)^2 + (2.5 \times 954.93 \times 10^3)^2}\right] = \frac{\pi}{32} \times 56 \times d^3$$

$$d = 76.43 \text{ mm} \cong \mathbf{78\ mm\ (say)} \qquad \ldots (4)$$

Thus, by comparing values of diameter in equations (1) and (2) as well as in (3) and (4), we observed that, the required diameter of shaft increases, if it is subjected to suddenly applied load conditions.

## Numerical Type No. 8 : "Comparison of Hollow Shaft and Solid Shaft"

**Problem 3.22 :** *Compare the weight, strength and stiffness of a hollow shaft of the same external diameter as that of solid shaft. Inside diameter of hollow shaft is 0.75 times the external diameter. Both shafts have same material and length.*

**Solution : Given data :** $d = d_o$, $\mathbf{d_i = 0.75 \times d_o}$; $\therefore$ $\mathbf{k} = \frac{d_i}{d_o} = \mathbf{0.75}$, $L_h = L_s$.

As same material is used for solid and hollow shaft, the properties like weight density, permissible stress values and modulus of rigidity will be same.

**Procedure : (i) Comparison of weight :**

We have, Weight = Weight density × Volume = Weight density × (Area × Length)

For solid shaft, $\mathbf{W_s}$ = **Weight density × (π/4) ×** $\mathbf{d^2 \times L_s}$ ... (1)

For hollow shaft, $W_h$ = Weight density × (π/4) × $(d_o^2 - d_i^2) \times L_h$

$\therefore$ $W_h$ = Weight density × (π/4) × $d_o^2 (1 - k^2) \times L_h$ $\left[\because k = \frac{d_i}{d_o}\right]$

$\therefore$ $W_h$ = Weight density × (π/4) × $d_o^2 \;[1 - (0.75)^2] \times L_h$

$\therefore$ $\mathbf{W_h}$ = **0.4375 × Weight density × (π/4) ×** $\mathbf{d_o^2 \times L_h}$ ... (2)

Dividing (2) by (1), $$\frac{W_h}{W_s} = \frac{0.4375 \times \text{Weight density} \times (\pi/4) \times d_o^2 \times L_h}{\text{Weight density} \times (\pi/4) \times d^2 \times L_s} = \mathbf{0.4375} \qquad (\because d = d_o \text{ and } L_h = L_s)$$

**(ii) Comparison of strength :**

Strength of shaft is given by,

For solid shaft, $$\mathbf{T_s = \frac{\pi}{16} \times \tau_s \times d^3} \qquad \ldots (3)$$

For hollow shaft, $$T_h = \frac{\pi}{16} \times \tau_h \times d_o^3\,(1 - k^4) = \frac{\pi}{16} \times \tau_h \times d_o^3\,[1 - (0.75)^4] = \mathbf{0.6836 \times \pi/16 \times \tau_h \times d_o^3} \qquad \ldots (4)$$

Dividing equation (4) by (3), we get,

$$\frac{T_h}{T_s} = \frac{0.6836 \times (\pi/16) \times \tau_h \times d_o^3}{(\pi/16) \times \tau_s \times d^3} = \mathbf{0.6836} \qquad (\because \tau_h = \tau_s \text{ and } d = d_o)$$

**(iii) Comparison of stiffness :** By basic torsion equation, we know that,

$$\frac{T}{J} = \frac{G\theta}{L}$$

$$\therefore \quad \frac{T}{\theta} = \frac{GJ}{L}$$

The term $\left(\frac{T}{\theta}\right)$ is called as stiffness of shaft.

$$\therefore \quad \text{Stiffness, } S = \frac{GJ}{L}$$

For solid shaft, $$\mathbf{S_s = \frac{G \times \pi/32 \times d^4}{L_s}} \quad \ldots (5)$$

For hollow shaft, $$S_h = \frac{G \times \pi/32 \times [(d_o)^4 - (d_i)^4]}{L_h} = \frac{G \times \pi/32 \times d_o^4 [1 - k^4]}{L_h} = \frac{G \times \pi/32 \times d_o^4 \times [1 - (0.75)^4]}{L_h}$$

$$\therefore \quad \mathbf{S_h = \frac{0.6836 \times G \times \pi/32 \times d_o^4}{L_h}} \quad \ldots (6)$$

Dividing equation (6) by (5), we get,

$$\frac{\mathbf{S_h}}{\mathbf{S_s}} = \frac{(0.6836 \times G \times \pi/32 \times d_o^4)/L_h}{(G \times \pi/32 \times d^4)/L_s} = \mathbf{0.6836} \qquad [\because d_o = d \text{ and } L_h = L_s]$$

**Note :** If $d_i = 0.5\ d_o$ is given in the numerical keeping other data same, then, answers will be, $\frac{W_h}{W_s} = 0.75$, $\frac{T_h}{T_s} = 0.9375$; $\frac{S_h}{S_s} = 0.9375$

---

**Problem 3.23 :** *The internal diameter of hollow shaft is 2/3rd of its external diameter. Compare strength and stiffness of hollow shaft with a solid shaft having same weight and material.*

**Solution : Given data :** $d_i = (2/3) \times d_o$; $\therefore$ $\mathbf{k} = d_i/d_o = \mathbf{0.67}$, $W_h = W_s$.

**Procedure :** We have,

$$\text{Weight} = \text{Weight density} \times \text{Volume} = W \times A \times L$$

For solid shaft, $$\mathbf{W_s = \text{Weight density} \times (\pi/4) \times d^2 \times L_s} \quad \ldots (1)$$

For hollow shaft, $$W_h = \text{Weight density} \times (\pi/4) \times (d_o^2 - d_i^2) \times L_h$$

$$\therefore \quad W_h = \text{Weight density} \times (\pi/4) \times d_o^2\ (1 - k^2) \times L_h$$

$$\therefore \quad W_h = \text{Weight density} \times (\pi/4) \times d_o^2\ [1 - (0.67)^2] \times L_h$$

$$\therefore \quad \mathbf{W_h = 0.5511 \times \text{Weight density} \times (\pi/4) \times d_o^2 \times L_h} \quad \ldots (2)$$

Given that, $W_s = W_h$; Therefore, on equating (1) and (2),

$$\therefore \quad \text{Weight density} \times (\pi/4) \times d^2 \times L_s = 0.5511 \times \text{Weight density} \times (\pi/4) \times d_o^2 \times L_h$$

As material for both shafts is same, weight density will be same.

$$\therefore \quad d^2 = 0.5511 \times d_o^2 \qquad [\text{Assume, } L_h = L_s]$$

$$\therefore \quad d = \mathbf{0.7423 \times d_o} \quad \ldots (3)$$

**(i) Comparison of strength :**

Strength of shaft is given by,

For solid shaft, $\mathbf{T_s} = \pi/16 \times \tau_s \times d^3 = \pi/16 \times \tau_s \times (0.7423 \times d_o)^3 = \mathbf{0.409 \times \pi/16 \times \tau_s \times d_o^3}$ [From equation (3)] ... (4)

For hollow shaft, $\mathbf{T_h} = \pi/16 \times \tau_h \times d_o^3 \cdot (1 - k^4) = \pi/16 \times \tau_h \times d_o^3 \times [1 - (0.67)^4] = \mathbf{0.7984 \times \pi/16 \times \tau_h \times d_o^3}$ ... (5)

Dividing equation (5) by (4), we get,

$$\frac{\mathbf{T_h}}{\mathbf{T_s}} = \frac{0.7984 \times \pi/16 \times \tau_h \times d_o^3}{0.409 \times \pi/16 \times \tau_s \times d_o^3} = \mathbf{1.952} \qquad (\because \text{For same material, } \tau_h = \tau_s)$$

**(ii) Comparison of stiffness :**

From basic torsion equation,

$$\frac{T}{J} = \frac{G\theta}{L}$$

$$\therefore \quad \frac{T}{\theta} = \frac{GJ}{L} = S \qquad \left[\because \text{Stiffness, } S = \frac{T}{\theta}\right]$$

For solid shaft,

$$\mathbf{S_s} = \frac{G \times \pi/32 \times d^4}{L_s} = \frac{G \times \pi/32 \times (0.7423 \times d_o)^4}{L_s} = \frac{\mathbf{0.3036 \times G \times \pi/32 \times d_o^4}}{\mathbf{L_s}} \qquad \dots (6)$$

For hollow shaft,

$$\mathbf{S_h} - \frac{G \times \pi/32 \times [(d_o)^4 - (d_i)^4]}{L_h} = \frac{G \times \pi/32 \times d_o^4\,[1 - k^4]}{l_h} = \frac{\mathbf{0.7984 \times G \times \pi/32 \times d_o^4}}{\mathbf{L_h}} \qquad \text{[Put } k = 0.67] \dots (7)$$

On dividing equation (7) by equation (6), we get,

$$\frac{\mathbf{S_h}}{\mathbf{S_s}} = \frac{(0.7984 \times G \times \pi/32 \times d_o^4)/L_h}{(0.3036 \times G \times \pi/32 \times d_o^4)/L_s} = \mathbf{2.6297} \qquad [\because L_h = L_s]$$

---

**Problem 3.24 :** *For an axial flow compressor, design the shaft from the following data :*

*Torque on shaft : 1000 kN-mm, Bending moment on shaft : 250 N-m, Allowable shear stress : 50 N/mm²*

*Load is applied suddenly, for which,* $k_b = 1.5$ *and* $k_t = 1$.

*Find : (a) Diameter of shaft considering it to be solid.*

*(b) Diameter of hollow shaft, if inner diameter is 0.5 times outer diameter.*

*Calculate % saving in material in case of hollow shaft.*

**Solution : Given data : T** = 1000 kN-mm = **1000 × 10³ N-mm**, **M** = 250 N-m = **250 × 10³ N-mm**, $\tau = 50\ N/mm^2$, $k_b = k_m = 1.5$ and $k_t = 1$.

**Procedure : Case I : For solid shaft :**

$$\text{Equivalent twisting moment} = T_e = \sqrt{(k_m M)^2 + (k_t T)^2} = \frac{\pi}{16} \times \tau_{max} \times (d)^3$$

$$\therefore \quad \sqrt{(1.5 \times 250 \times 10^3)^2 + (1 \times 1000 \times 10^3)^2} = \frac{\pi}{16} \times 50 \times (d)^3$$

$$\therefore \quad \mathbf{d} = 47.73 \text{ mm} \cong \mathbf{48\ mm\ (say)}$$

**Case II : For hollow shaft :** $d_i = 0.5\, d_o$; $\therefore$ $\mathbf{k} = d_i/d_o = \mathbf{0.5}$

$$\text{Equivalent twisting moment} = T_e = \sqrt{(k_m M)^2 + (k_t T)^2} = \frac{\pi}{16} \times \tau_{max} \times (d_o)^3\, \{1 - (k)^4\}$$

$$\therefore \quad \sqrt{(1.5 \times 250 \times 10^3)^2 + (1 \times 1000 \times 10^3)^2} = \frac{\pi}{16} \times 50 \times (d_o)^3\, \{1 - (0.5)^4\}$$

$$\therefore \quad \mathbf{d_o} = 48.77 \text{ mm} \cong \mathbf{50\ mm} \text{ and } \mathbf{d_i} = 0.5 \times 50 = \mathbf{25\ mm}$$

**Percentage saving in material, if hollow shaft is used :**

$$\text{Weight of solid shaft} = \text{Weight density} \times \text{Volume} = W_s = \text{Weight density} \times \frac{\pi}{4} \times d^2 \times L_s$$

$$\therefore \quad \mathbf{W_s} = \text{Weight density} \times \frac{\pi}{4} \times (48)^2 \times L_s = \mathbf{2304 \times Weight\ density \times \frac{\pi}{4} \times L_s} \qquad \dots (1)$$

$$\text{Similarly, Weight of hollow shaft} = W_h = \text{Weight density} \times \frac{\pi}{4} \times (d_o^2 - d_i^2) \times L_h$$

$$\therefore \quad W_h = \text{Weight density} \times \frac{\pi}{4} \times d_o^2\,(1 - k^2) \times L_h$$

$$\therefore \quad \mathbf{W_h} = \text{Weight density} \times \frac{\pi}{4} \times (50)^2 \times [1 - (0.5)^2] \times L_h = \mathbf{1875 \times Weight\ density \times \frac{\pi}{4} \times L_h} \qquad \dots (2)$$

Dividing (2) by (1),

$$\frac{W_h}{W_s} = \frac{1875 \times \text{Weight density} \times (\pi/4) \times L_h}{2304 \times \text{Weight density} \times (\pi/4) \times L_s} = 0.8138 \qquad (\text{Assume } L_h = L_s)$$

$\therefore$ $\mathbf{W_h = 0.8138\ W_s}$

**Percentage saving in material** $= \dfrac{W_s - W_h}{W_s} \times 100 = \dfrac{W_s - 0.8138\ W_s}{W_s} \times 100$

$\therefore$ Percentage saving in material = **18.62%**

## 3.2 INTRODUCTION TO KEY

- **A key** can be defined as, "*a machine element, which is used to connect the transmission shaft to the rotating machine element like pulley, gear, sprocket or flywheel*".

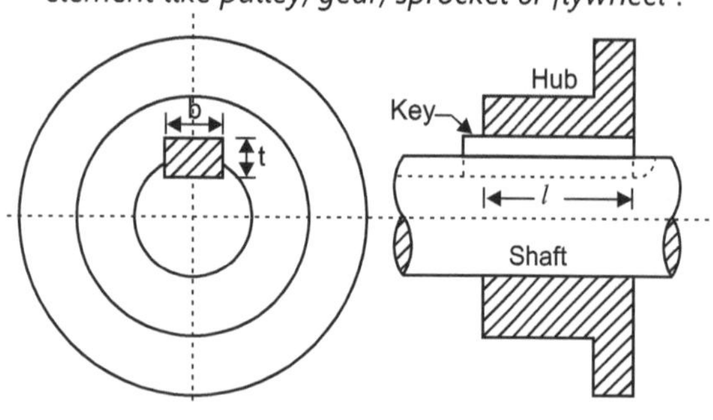


Fig. 3.16

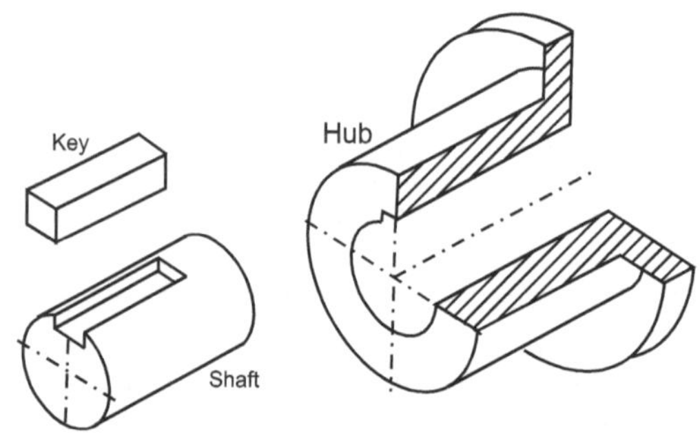


Fig. 3.17

- Key is inserted between the shaft and hub or boss of pulley to connect them together in order to prevent relative motion between them.
- It is always inserted parallel to axis of shaft.
- It is used as temporary fastening.
- **A keyway** is a slot or recess in a shaft and hub of pulley to accommodate key.

### 3.2.1 Functions of Key, Advantages and Disadvantages of Keyed Joints

**Functions of Key :**

Two basic functions of key are as follows :

(1) Primary function of key is **'to transmit the torque from shaft to the hub of mating element such as pulley, gear etc. and vice versa'**.

(2) Secondary function of key is **'to prevent relative rotational motion between the shaft and mating machine element like gear or pulley'**. In simple words, speed of shaft with key inserted should be equal to speed of mating machine part, like pulley or gear mounted on shaft. In most of the cases, key also prevents axial motion between two elements, except in case of feather key or splined connection.

**Advantages of Keyed Joints :** 1. Compact. 2. No slip, while transmitting power. 3. Low cost.

**Disadvantages of Keyed Joints :**

1. Temporary connection.
2. Weak and fails quickly.
3. No protection against overload.
4. Shaft gets weakened, due to keyway (a slot).

### 3.2.2 Types of Keys

**1. Sunk Keys :**

- A sunk key is a key, in which, half the thickness of key fits into the keyway on shaft and the remaining half thickness fits into the keyway on hub machine part such as pulley, gear etc. Therefore, keyways are required on shaft as well as on hub of the mating element. Depth of keyway is uniform inside the shaft, but is tapering in the hub.
- Sunk key is suitable for heavy duty application, since, there is no possibility of key to slip around the shaft. It means a positive drive.

**Types of Sunk Keys :**

**(a) Rectangular sunk key :**

- For rectangular sunk key, usual proportions are,

  Width (w) $= \dfrac{d}{4}$ and

  Thickness of key (t) $= \dfrac{2w}{3} = \dfrac{d}{6}$.

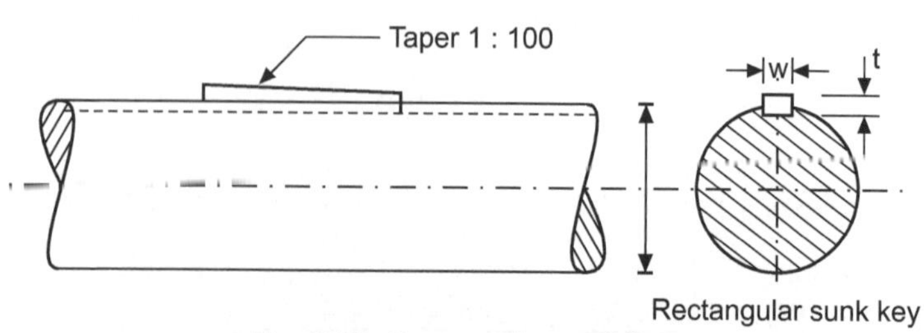


**Fig. 3.18 : Rectangular sunk key**

- It has a rectangular cross-section and has taper on one face.

**Application :** For gears and pulleys mounted on shaft.

**(b) Square sunk key :**

- In square sunk key, width (w) is equal to thickness (t).
- Except this, rectangular and square sunk keys are approximately similar.
- Usual proportions are, $w = t = \frac{d}{4}$.

**(c) Gib head key :**

- It is a **rectangular sunk key** with a head at one end called as ***gib head***.
- It is usually provided for easy removal of key.
- Usual proportions are, $w = \frac{d}{4}$ and $t = \frac{2w}{3} = \frac{d}{6}$.

**Advantages of Gib Headed Key :**

- (i) The taper surface results in wedge action and increases frictional force and the tightness of the joint.
  (ii) The taper surface is provided for easy removal of the key, particularly with gib-head.
- However, machining taper on the surface increases cost.

**Application :**

- It is used, where key is to be removed frequently.

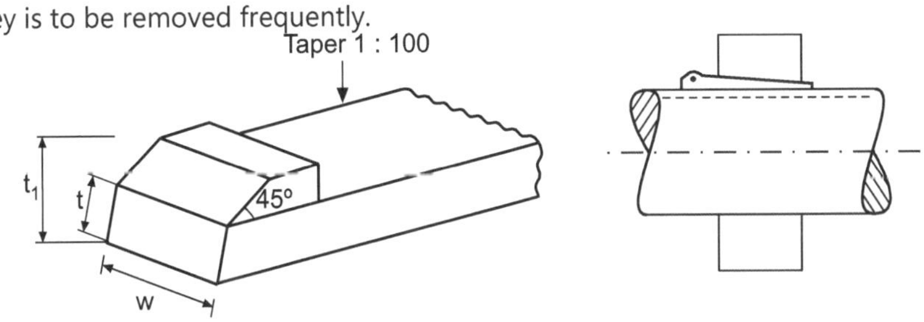


**Fig. 3.19 : Gib headed key**

**(d) Feather key :**

- A feather key is a parallel key, which is fixed either to the shaft or to the hub. It permits relative axial movement between them.
- It is a particular type of sunk key having uniform width and height.
- There are number of methods to fix the key to shaft or hub of mating machine part.
- Fig. 3.20 shows a feather key, which is fixed to the shaft by means of two cap-screws, having countersunk heads.
- There is a clearance fit between the key and the keyway in the hub. Therefore, the hub is free to slide over the key. At the same time, there is no relative rotational movement between shaft and hub. Therefore, feather key transmits torque and at the same time, it permits some axial movement of hub.

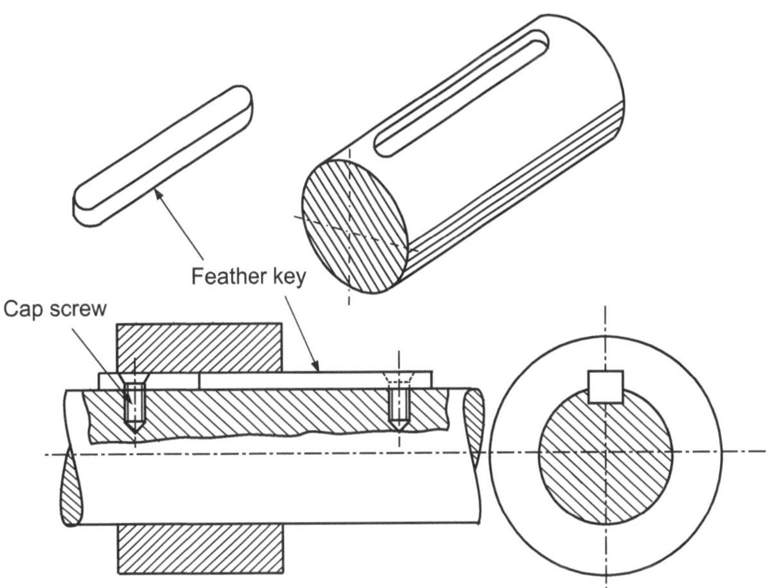


**Fig. 3.20 : Feather key**

**Applications :**

1. Feather keys are used, where the parts mounted on shaft are required to slide along the shaft, such as clutches or gear shifting devices.
2. It is an alternative arrangement to splined connection. For example : Machine tools.

**(e) Woodruff key :**

- A woodruff key is a sunk key, in the form of an almost semi-circular disk of uniform thickness as shown in Fig. 3.21.
- Keyway in the shaft is in the form of a semi-circular recess with the same curvature, as that of the key.
- Bottom portion of woodruff key fits into circular keyway of shaft.
- Keyway in the hub is made in the usual manner. The projecting part of woodruff key fits into the keyway in the hub. Once placed in position, the woodruff key can tilt itself on the shaft for self alignment.

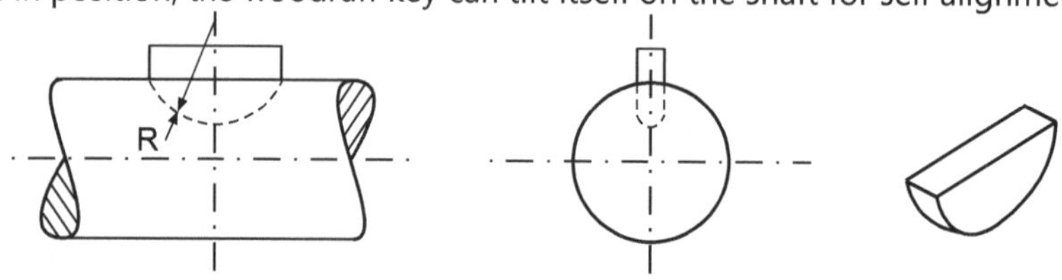


**Fig. 3.21 : Woodruff key**

**Advantages of Woodruff Key :**

(i) Woodruff key can be used on tapered shaft, because it can align itself by slight rotation in the seat.
(ii) Extra depth of key in the shaft prevents its tendency to slip over the shaft.
(iii) It permits easy removal of pulleys from shaft.

**Disadvantages of Woodruff Key :**

(i) Extra depth of keyway in the shaft increases stress concentration and reduces its strength.
(ii) It does not permit axial movement between shaft and hub.

**Applications :**

(i) Woodruff key is used on tapered shafts in machine tools and automobiles.
(ii) It is used for transmitting moderate torques, especially in automotive engineering.

**2. Saddle Keys :**

- A saddle key fits in the keyway of hub only. Here, keyway is absent on the shaft.
- There are two types of saddle keys,

**(a) Flat saddle key :**

- It is a taper key, which fits in the keyway of hub. It is flat on the shaft.
- It is likely to slip round the shaft under load.

**(b) Hollow saddle key :**

- It is a taper key, which fits in the keyway of hub.
- Bottom of key is shaped as a concave surface, so that, it can fit the curved surface of the shaft.

**Applications :**

(i) They are suitable for light duty or low power transmission, as the power is transmitted due to friction.
(ii) Saddle key is used as temporary fastening in fixing and setting eccentric parts, cams etc.

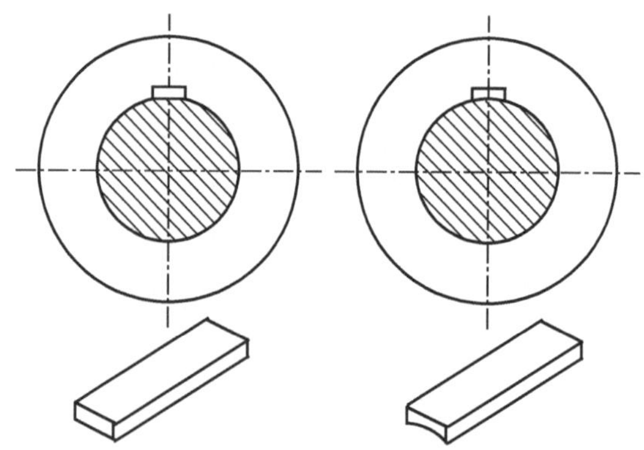

**(a) Flat saddle key** **(b) Hollow saddle say**

**Fig. 3.22 : Saddle key**

**3. Round keys :**

- They are circular in section and fit into the holes drilled partly into the hub.

**Application :**

- They are used for low power drive.

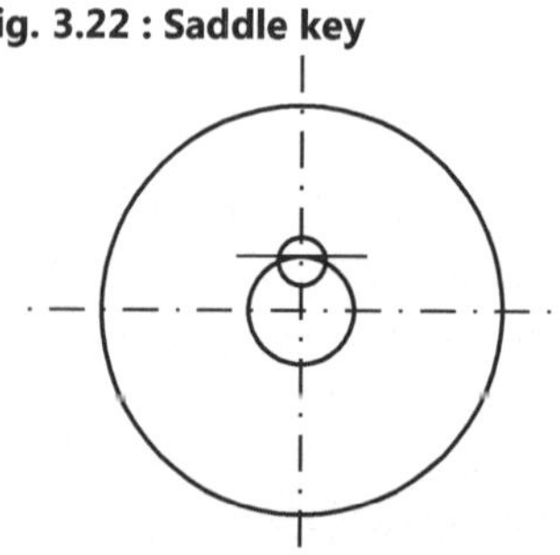

**Fig. 3.23 : Round keys**

**4. Spline :**

- A spline is a group of keyways cut on the periphery of shaft. The remaining portion of the shaft material between the keyway forms a number of equidistant keys in-built with the shaft.
- In other words, keys are made integral with the shaft, which can be fitted in the keyways broached (formed) in the hub. Such shafts are called as *splined shafts*.
- They usually have four, ten or sixteen splines.
- Projections on the splined shaft engage with corresponding recesses in a splined hub.

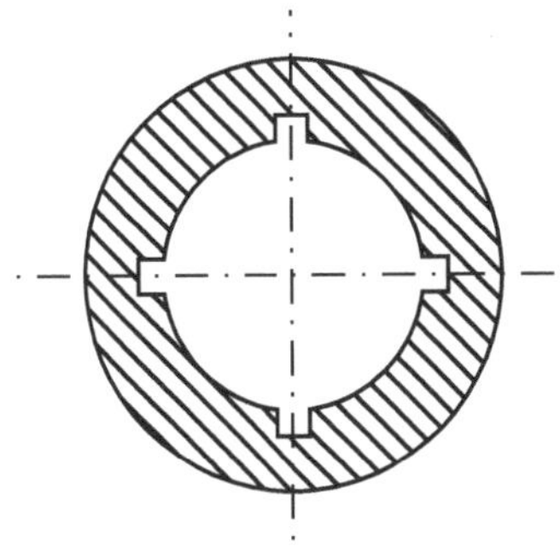

**Fig. 3.24 : Splines**

**Applications :**

(i) These splines are used for power transmission of very high order and also provide axial movement between shaft and mounted member.

(ii) Practical applications of splines may be seen in gear shifting mechanism used in automobile gear boxes.

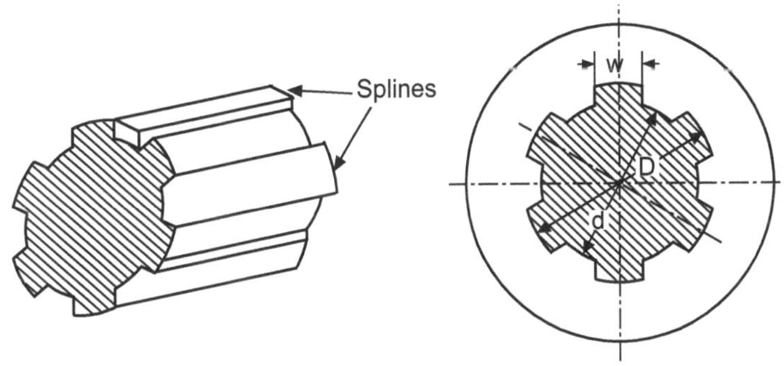


**Fig. 3.25 : Splined shaft**

### 3.2.2.1 Factors to be Considered While Selecting the Type of Key

1. Power to be transmitted.
2. Tightness of fit.
3. Stability of connection.
4. Cost.

### 3.2.2.2 Considerations in Design of Key

(1) Power to be transmitted.

(2) Tightness of fit.

(3) Stability of connection.

(4) Cost.

(5) Crushing failure of key.

(6) Shearing failure of key.

(7) Material of key and shaft should be chosen, such that, key is weaker than shaft.

## 3.2.3 Design of Sunk Key

**Strength of a Sunk Key or Different Types of Failure while Designing a Sunk Key :**

Consider a rectangular key connecting the shaft and hub.

Let,
T = Torque transmitted by shaft in 'N.mm'
F = Tangential force acting at the circumference of shaft in 'N'
d = Diameter of shaft in 'mm'
$l$ = Length of key in 'mm'
w = Width of key in 'mm'
t = Thickness of key in 'mm'
$\tau$ and $\sigma_{ck}$ = Permissible shear and crushing stress of key material used in 'N/mm$^2$'

A little consideration will show that, the **key may fail either in shearing or crushing** due to power transmitted by the shaft.

(1) Considering shearing of key, the shear stress induced can be given by,

$$\text{Shear stress} = \frac{\text{Tangential shear force acting on shaft}}{\text{Area resisting shear}}$$

$$\therefore \quad \tau = \frac{F}{l \times w}$$ [Refer Fig. 3.26 (a)]

$$\therefore \quad F = l \cdot w \cdot \tau$$

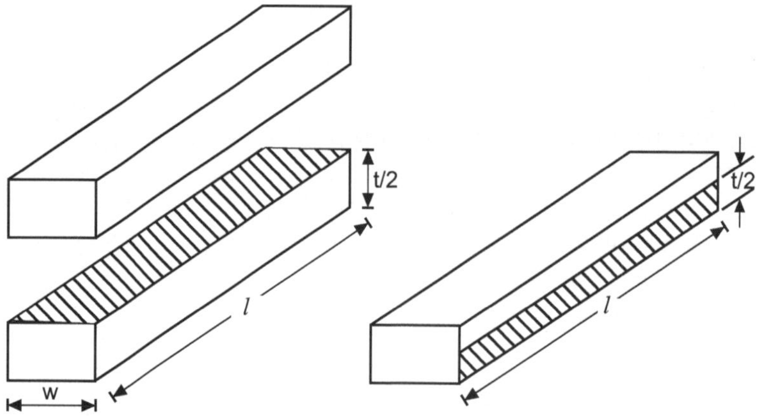


**(a) Area under shear** **(b) Area under crushing**

**Fig. 3.26 : Rectangular sunk key**

Therefore, torque transmitted,

$$\mathbf{T} = F \times \frac{d}{2} = \boldsymbol{l} \cdot \mathbf{w} \cdot \boldsymbol{\tau} \cdot \frac{\mathbf{d}}{\mathbf{2}} \quad \text{(3.15)}$$

(2) Considering crushing of key, the crushing stress induced can be given by,

$$\text{Crushing stress} = \frac{\text{Tangential crushing force acting on shaft}}{\text{Area resisting crushing}}$$

$$\therefore \quad \sigma_{ck} = \frac{F}{l \times \frac{t}{2}} \quad \text{[Refer Fig. 3.26 (b)]}$$

$$\therefore \quad F = l \cdot \frac{t}{2} \cdot \sigma_{ck}$$

Therefore, torque transmitted,

$$\mathbf{T} = F \times \frac{d}{2} = \boldsymbol{l} \cdot \frac{\mathbf{t}}{\mathbf{2}} \cdot \boldsymbol{\sigma}_{\mathbf{ck}} \cdot \frac{\mathbf{d}}{\mathbf{2}} \quad \text{... (3.16)}$$

**Note :** Design procedure for square sunk key is exactly similar to design procedure of rectangular sunk key.

### 3.2.3.1 Relation between Stresses for a Rectangular Key Equally Stronger in Crushing and Shearing

Consider a rectangular key connecting the shaft and hub.

Let,
- T = Torque transmitted by shaft
- F = Tangential force acting at the circumference of shaft
- d = Diameter of shaft
- $l$ = Length of key
- w = Width of key
- t = Thickness of key

Let, $\tau$ and $\sigma_{ck}$ be the permissible shear and crushing stresses for key material respectively.

A little consideration will show that, due to power transmitted by the shaft, the key may fail either due to shearing or crushing.

**Considering shearing of key, the tangential shear force acting on the shaft is,**

$$F = \text{Area resisting shearing} \times \text{Shear stress}$$

$$\therefore \quad F = (l \times w) \cdot \tau$$

$$\therefore \quad \text{Torque transmitted,} \quad \mathbf{T} = F \times \frac{d}{2} = \boldsymbol{l} \cdot \mathbf{w} \cdot \boldsymbol{\tau} \cdot \frac{\mathbf{d}}{\mathbf{2}} \quad \text{... (3.17)}$$

**Considering crushing of key, the tangential crushing force acting on the shaft is,**

$$F = \text{Area resisting crushing} \times \text{Crushing stress} = \left(l \cdot \frac{t}{2}\right) \cdot \sigma_{ck}$$

$$\therefore \quad \text{Torque transmitted,} \quad \mathbf{T} = F \times \frac{d}{2} = \boldsymbol{l} \cdot \frac{\mathbf{t}}{\mathbf{2}} \cdot \boldsymbol{\sigma}_{\mathbf{ck}} \cdot \frac{\mathbf{d}}{\mathbf{2}} \quad \text{... (3.18)}$$

**If key is equally stronger in shearing and crushing**, then on equating equations (3.17) and (3.18), we get,

$$l \cdot w \cdot \tau \cdot \frac{d}{2} = l \cdot \frac{t}{2} \cdot \sigma_{ck} \cdot \frac{d}{2}$$

$$\therefore \quad \frac{\tau}{\sigma_{ck}} = \frac{t}{2w}$$

But, for rectangular key, we can put $w = \frac{d}{4}$ and $t = \frac{d}{6}$ in the above equation.

$$\therefore \quad \frac{\tau}{\sigma_{ck}} = \frac{\frac{d}{6}}{2 \times \frac{d}{4}} = \frac{1}{3}$$

$$\therefore \quad \boldsymbol{\sigma_{ck} = 3\tau}$$

Thus, for a **rectangular key** equally stronger in crushing and shearing, **'crushing stress is thrice the shear stress'**.

### 3.2.3.2 Relation between Stresses for a Square Key Equally Stronger in Crushing and Shearing

Considering a square key connecting the shaft and hub.

Let,
- $T$ = Torque transmitted by shaft
- $F$ = Tangential force acting at the circumference of shaft
- $d$ = Diameter of shaft
- $l$ = Length of key
- $w$ = Width of key
- $t$ = Thickness of key

For square key, we have, $w = t$

Let $\tau$ and $\sigma_{ck}$ be the permissible shear and crushing stresses for key material respectively.

A little consideration will show that due to power transmitted by shaft, the key may fail either due to shearing or crushing.

**Considering shearing of key, the tangential shearing force acting on the shaft is,**

$$F = \text{Area resisting shearing} \times \text{Shear stress} = l \cdot w \cdot \tau$$

$$\therefore \quad \text{Torque transmitted} = \mathbf{T} = F \times \frac{d}{2} = \boldsymbol{l \cdot w \cdot \tau \cdot \frac{d}{2}} \qquad \text{... (3.19)}$$

**Considering crushing of key, the tangential crushing force acting on the shaft is,**

$$F = \text{Area resisting crushing} \times \text{Crushing stress} = l \cdot \frac{t}{2} \cdot \sigma_{ck}$$

$$\therefore \quad \text{Torque transmitted} = \mathbf{T} = F \times \frac{d}{2} = \boldsymbol{l \cdot \frac{t}{2} \cdot \sigma_{ck} \cdot \frac{d}{2}} \qquad \text{... (3.20)}$$

**If key is equally stronger in shearing and crushing**, then on equating equations (3.19) and (3.20), we get,

$$l \cdot w \cdot \tau \cdot \frac{d}{2} = l \cdot \frac{t}{2} \cdot \sigma_{ck} \cdot \frac{d}{2}$$

$$\therefore \quad \frac{\tau}{\sigma_{ck}} = \frac{t}{2w}$$

$$\therefore \quad \frac{\tau}{\sigma_{ck}} = \frac{1}{2} \text{ or } \boldsymbol{\sigma_{ck} = 2\tau} \qquad \text{[For square key, } w = t\text{]}$$

Thus, for a **square key**, equally stronger in crushing and shearing, **'crushing stress is twice the shearing stress**.

## 3.2.4 Differentiate between Key and Cotter

| Key | Cotter |
|---|---|
| (1) Key is driven parallel to axis of shaft. | (1) Cotter is normally driven at right angle to axis of connected parts (circular rods). |
| (2) Key is subjected to crushing stress and torsional shear stress. | (2) Cotter is subjected to crushing stress and transverse shear stress. |
| (3) Key resists shear over longitudinal section. | (3) Cotter resists shear over transverse section. |

### 3.2.5 Effect of Keyways on Strength of Shaft

- Keyway is a slot machined either on the shaft or in the hub to accommodate the key.
- It is cut by vertical or horizontal milling cutter.
- A little consideration will show that, **'rectangular slot cut as keyway'** will **reduce load carrying capacity of shaft**.
- This is due to two reasons :
  (1) Stress concentration near the corners of keyway and (2) Reduction in cross-sectional area of shaft.
- In other words, torsional strength of shaft is reduced.
- Following relation of reduction factor is used to analyze the weakening effect of keyway given by H. F. Moore.

$$e = 1 - 0.2\,(w/d) - 1.1(h/d) \quad \ldots (3.21)$$

Where,

$e$ = Shaft strength factor = $\dfrac{\text{Strength of shaft with keyway}}{\text{Strength of shaft without keyway}}$

$d$ = Diameter of shaft,

$w$ = Width of keyway,

$h$ = Depth of keyway = $\dfrac{1}{2} \times$ Thickness of key = $\dfrac{t}{2}$

- It is usually assumed that, strength of keyed shaft is 75% of solid shaft (i.e. without keyway).
- Thus, after finding out dimensions of key, the reduction factor 'e' is calculated using the above equation (3.21). **If value of calculated 'e' is less than 0.75, design is safe.**
- **Note :** In case of **'Hollow saddle key'** and **'Tangent key'**, there is no effect of keyway on strength of shaft.

**Numerical Type No. 9 : "Design of Key"**

**Problem 3.25 :** *Design a rectangular key for shaft of 50 mm diameter. The permissible stresses for key material are 40 N/mm² in shear and 70 N/mm² in crushing.*

**Solution : Given data :** $d = 50$ mm, $\tau = 40$ N/mm², $\sigma_{ck} = 70$ N/mm²

**Procedure :** We know that, torque transmitted by shaft is given by,

$$T = \frac{\pi}{16} \times \tau_s \times d^3 = \frac{\pi}{16} \times 40 \times (50)^3 = \mathbf{1.03 \times 10^6\ N\text{-}mm}$$

We can find two dimensions of key, i.e. **width** and **thickness** by using empirical relations.

By empirical relations, $\mathbf{w}$ = **Width of key** $= \dfrac{d}{4} = \dfrac{50}{4} = \mathbf{12.5\ mm}$

$\mathbf{t}$ = **Thickness of key** $= \dfrac{d}{6} = \dfrac{50}{6} = \mathbf{8.333\ mm}$

To find **length of key**, we consider shear failure and crushing failure in the following way.

**(1) Considering the key under shear failure :**

We have, $T = l \times w \times \tau_k \times \dfrac{d}{2}$

$\therefore \quad 1.03 \times 10^6 = l \times 12.5 \times 40 \times \dfrac{50}{2}$

$\therefore \quad l$ = Length of key = **82.4 mm** $\quad \ldots (1)$

**(2) Considering the key under crushing failure :**

We have, $T = l \times \dfrac{t}{2} \times \sigma_{ck} \times \dfrac{d}{2}$

$\therefore \quad 1.03 \times 10^6 = l \times \dfrac{8.333}{2} \times 70 \times \dfrac{50}{2}$

$\therefore \quad l$ = length of key = **141.26 mm** $\quad \ldots (2)$

- For safe design, we always take larger value of any dimension obtained by considering all types of failures. On comparing (1) and (2), **Length of key = $l$ = 141.26 mm ≅ 142 mm (say).**

---

**Problem 3.26 :** *A belt pulley is fastened to a 90 mm diameter shaft running at 300 r.p.m. by means of a key 20 mm wide and 140 mm long. Allowable stresses for shaft and key material are 40 N/mm² in shear and 100 N/mm² in crushing. Find power transmitted and depth of key required.*

**Solution : Given data :** $d = 90$ mm, $l = 140$ mm, $\tau = 40$ N/mm², $w = 20$ mm, $\sigma_{ck} = 100$ N/mm², $N = 300$ r.p.m.

**Procedure :** Consider shearing failure of key,

$$\text{Torque transmitted, } \mathbf{T} = l \cdot w \cdot \tau \cdot \frac{d}{2} = 140 \times 20 \times 40 \times \frac{90}{2} = \mathbf{5040 \times 10^3\ N\text{-}mm = 5040\ N\text{-}m}$$

**(1) Power transmission capacity :**

$$\mathbf{P} = \frac{2\pi NT}{60} = \frac{2\pi \times 300 \times 5040}{60} = 158.336 \times 10^3 \text{ Watt} = \mathbf{158.336\ kW}$$

**(2) Thickness of key :**

Consider crushing failure of key,

$$T = l \times \frac{t}{2} \times \sigma_{ck} \times \frac{d}{2}$$

$$\therefore \quad 5040 \times 10^3 = 140 \times \frac{t}{2} \times 100 \times \frac{90}{2}$$

$$\therefore \quad t = \mathbf{16\ mm}$$

---

**Problem 3.27 :** *A shaft of 30 mm diameter is transmitting power at a maximum shear stress of 80 MPa. If a pulley is connected to the shaft by means of a key, find the dimensions of key, so that, stress in key is not to exceed 50 MPa and length of key is 4 times the width.*

**Solution : Given data :** d = 30 mm,

$$\tau_{shaft} = 80 \text{ MPa (i.e. N/mm}^2)$$

$$\tau_{key} = 50 \text{ MPa (i.e. N/mm}^2)$$

$$l = 4 \times w$$

**Procedure :** Torque transmitted by shaft can be obtained by,

$$T = \frac{\pi}{16} \tau d^3$$

$$\therefore \quad T = \frac{\pi}{16} \times 80 \times (30)^3 \qquad [\therefore \tau_{shaft} = 80 \text{ N/mm}^2]$$

$$\therefore \quad \mathbf{T = 424115\ N.mm}$$

If there is no loss of power transmission, we know that, torque transmitted by key will be equal to torque transmitted by shaft.

$$\therefore \quad T_{key} = T_{shaft} = \mathbf{424115\ N.mm}$$

**Design of key (i.e. to determine dimensions of key) :**

**(i) Width of key can be calculated by considering shearing failure of key.**

We have, $$T_{key} = l \times w \times \tau_{key} \times \frac{d}{2}$$

$$\therefore \quad 424115 = (4 \times w) \times w \times 50 \times \frac{30}{2} \qquad [\text{Given : } l = 4 \times w \text{ and } \tau_{key} = 50 \text{ N/mm}^2]$$

$$\therefore \quad \mathbf{w} = 11.889 \text{ mm} \cong \mathbf{12\ mm\ (say)}$$

**(ii) Length of key** can be determined by, $\boldsymbol{l} = 4 \times w = 4 \times 12 = \mathbf{48\ mm}$

**(iii) Thickness of key** can be determined by using empirical relations given as under.

(1) If key is rectangular in cross-section, then

$$t = \frac{2w}{3} = \frac{2 \times 12}{3} = \mathbf{8\ mm}$$

(2) If key is square in cross-section, then

$$t = w = \mathbf{12\ mm}$$

---

## 3.3 INTRODUCTION TO COUPLINGS

- Shafts are usually available upto 7 metres length for easy handling and transportation. Therefore, when power is to be transmitted at a greater distance more than 7 metres, two lengths of shaft or more are required to be connected by means of coupling.
- Shafts connected by coupling can be disengaged only after dismantling the coupling.

### 3.3.1 Functions of Coupling

Shaft couplings are used in machinery for different purposes, such as :

1. For provision of connecting shaft units, which are manufactured separately. For example : A motor shaft and A generator shaft.
2. For provision facility of disconnections for repairs or maintenance.
3. To compensate for misalignment of connected shafts.
4. To introduce mechanical flexibility (such as in case of flexible coupling.)
5. To reduce transmission of shock loads from one shaft to another shaft.
6. To provide protection against overloads.
7. To change vibration characteristics of rotating members, whenever required.

### 3.3.2 Requirements or Pre-requisites of a Good Coupling

1. Coupling should be easy to connect or disconnect.
2. Coupling should transmit 100% power from one shaft to the other shaft without any power losses.
3. Coupling should hold the shafts in perfect alignment.
4. Coupling should reduce the transmission of shock loads from one shaft to another shaft.
5. Coupling should not have projected parts.

#### 3.3.2.1 Factors to be Considered, While Selecting a Coupling

1. Misalignment of shafts.
2. Required torque and desired speed.
3. Operating conditions.
4. Protection against overload.
5. Durability and reliability.
6. Cyclic operations.
7. Direction of rotation.

### 3.3.3 Types of Couplings

Shaft couplings are classified as,

1. **Rigid Coupling :**
   - It is used to **connect two shafts**, which are in **perfect alignment**.
   - It consists of two hubbed flanges, one keyed to the driving shaft and other to driven shaft. The two flanges are bolted together.
   - **Applications of rigid coupling :**
     (i) When the required length of shaft is too long and it is difficult to manufacture such long shaft, then two or more shafts of suitable length are connected with the help of sleeve or muff coupling.
     (ii) For motor-pump sets, motor-gear box sets etc.
   - **Types of rigid coupling :**
     (a) Sleeve or muff coupling, (b) Split muff or clamp coupling, (c) Flange coupling.
2. **Flexible Coupling :**
   - It is used to **connect two shafts** having both **lateral and angular misalignments**.
   - Its construction is similar to a rigid coupling, except that, a resilient element, such as, rubber bush is inserted between the coupled parts.
   - Rubber bush is provided with brass lining to avoid excessive wear. These bushes absorb the misalignment between the shafts.
   - **Types of flexible coupling :**
     (a) Bushed pin type coupling, (b) Universal coupling, (c) Oldham coupling.

## 3.3.4 Applications of Different Types of Shaft Couplings

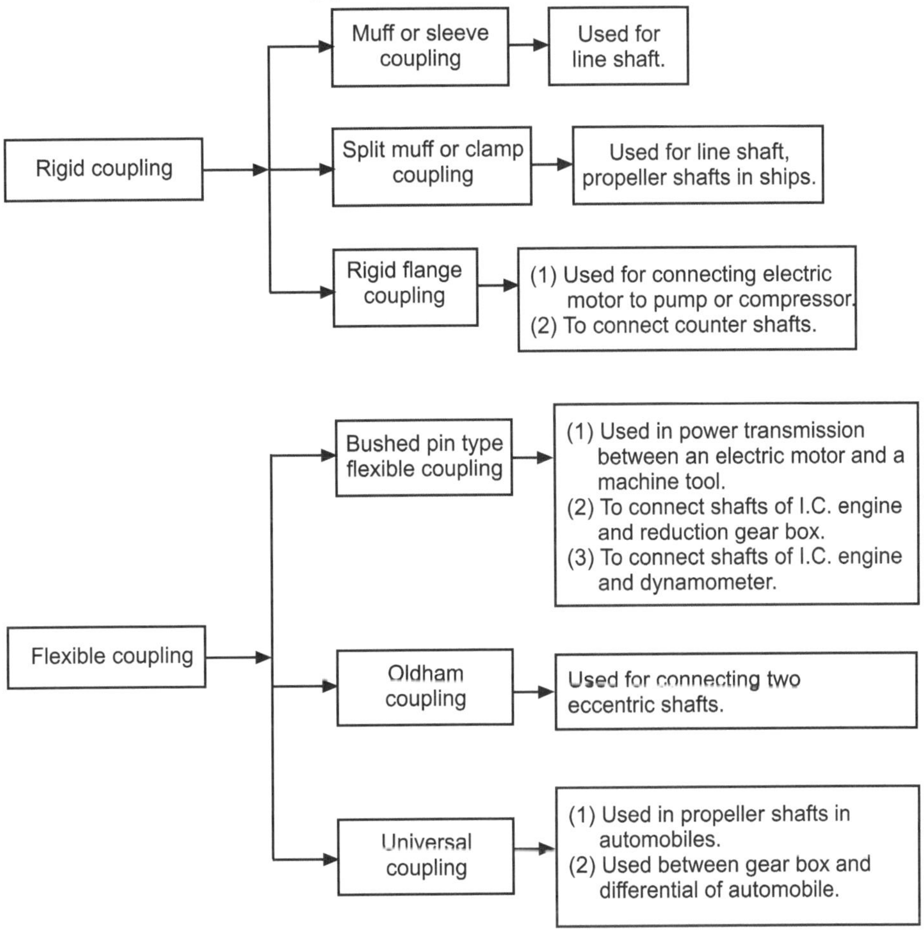


**Chart showing detailed classification of couplings with applications of each type**

## 3.3.5 Design Procedure of Sleeve or Muff Coupling

- Sleeve or muff coupling consists of a hollow cylinder called as *sleeve or muff* having internal diameter equal to diameter of shafts.
- Sleeve or muff is fitted at the ends of shafts to be connected with the help of gib headed key. Thus, the power is transmitted through the sleeve and key.
- It is generally used with small diameter shafts.

**Step I : Design of shaft :** Considering the shaft under twisting moment, $T = \frac{\pi}{16} \cdot \tau \cdot d^3$

From above expression, diameter of the shaft can be determined.

**Step II : Design of sleeve :**

Dimensions of sleeve are calculated by using standard proportions given below.

| | | | |
|---|---|---|---|
| Outer diameter of sleeve, $D$ | = | $(2d + 13)$ | in 'mm' |
| Inner diameter of sleeve | = | Diameter of shaft = $d$ | in 'mm' |
| Length of sleeve, $L$ | = | $3.5 \times d$ | in 'mm' |
| where, $d$ | = | diameter of shaft | in 'mm' |
| Let, $T$ | = | Torque transmitted by coupling | in 'N·mm' |
| $\tau$ | = | Permissible shear stress of sleeve material | in 'N/mm$^2$' |

We can write, $$T = \frac{\pi}{16} \times \tau \times D^3 (1 - k^4) \qquad \left[\text{Where, } k = \frac{d}{D}\right]$$

From the above expression, outer diameter of the sleeve (D) calculated by using standard proportions can be checked to find out, whether torsional shear stress induced is less than the permissible shear stress. If yes, design of sleeve is said to be safe.

**Step III : Design of key :** Key may be designed using the following empirical relations.

$$\text{Width of key} = w = \frac{d}{4}$$

$$\text{Thickness of key} = t = \frac{2w}{3} = \frac{d}{6}$$

Length of key must be at least equal to length of sleeve.

The coupling key is usually made in two parts, so that, length of key in each shaft,

$$l = \frac{L}{2}$$

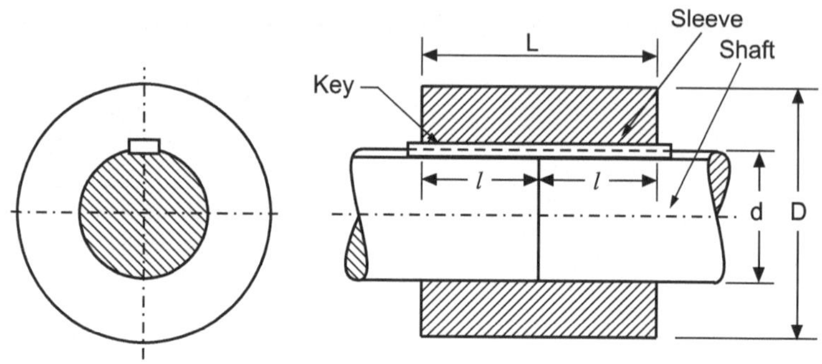


**Fig. 3.27 : Muff or Sleeve coupling**

After determining the dimensions of key, the crushing stress and shear stress induced in the key may be checked by using following design equations of key :

$$\text{Torque transmitted,} \quad T = F \times \frac{d}{2} = l \times w \times \tau \times \frac{d}{2} \quad \text{... (In shear)}$$

$$\text{Torque transmitted,} \quad T = F \times \frac{d}{2} = l \times \frac{t}{2} \times \sigma_{ck} \times \frac{d}{2} \quad \text{... (In crushing)}$$

**Numerical Type No. 10 : "Muff Coupling"**

**Problem 3.28:** *Design a Muff coupling for connecting two shafts transmitting 50 kW at 450 r.p.m. The material for shaft and key is same, for which, allowable shearing and crushing stresses are 40 MPa and 80 MPa respectively.*

**Solution : Given data : P** = 50 kW = $\mathbf{50 \times 10^3}$ **W**, N = 450 r.p.m., $\tau$ = 40 N/mm$^2$, $\sigma_{ck}$ = 80 N/mm$^2$

**Procedure :**

**Step I :** Power transmitted by shaft,

$$P = \frac{2\pi NT}{60}$$

$$\therefore \quad \text{Torque, } \mathbf{T} = \frac{P \times 60}{2\pi N} = \frac{50 \times 10^3 \times 60}{2 \times \pi \times 450} = 1061.03 \text{ N-m} = \mathbf{1061.03 \times 10^3 \text{ N-mm}}$$

We know that, torque transmitted by shaft is given by,

$$T = \frac{\pi}{16} \times \tau \times d^3$$

$$\therefore \quad 1061.03 \times 10^3 = \frac{\pi}{16} \times 40 \times d^3$$

$\therefore$ Diameter of shaft, $\mathbf{d}$ = 51.31 mm ≅ **52 mm (say)**

Diameter of shaft is equal to inner diameter of sleeve.

$\therefore$ Inner diameter of sleeve = d = 52 mm.

**Step II : Design of sleeve :** Dimensions of sleeve are calculated by using standard proportions given below.

Outer diameter of sleeve, D = 2d + 13 = 2 × 52 + 13 = **117 mm**

Length of sleeve, L = 3.5 d = 3.5 × 52 = **182 mm**

We have,
$$k = \frac{d}{D} = \frac{52}{117} = \mathbf{0.4444}$$

If T = Torque transmitted by the coupling,

and $\tau$ = Permissible shear stress for sleeve material,

then,
$$T = \frac{\pi}{16} \times \tau \times D^3 (1 - k^4)$$

$$\therefore \quad \tau_{ind} = \frac{T \times 16}{\pi \times D^3 (1 - k^4)} = \frac{1061.03 \times 10^3 \times 16}{\pi \times (117)^3 \times [1 - (0.4444)^4]}$$

$$\therefore \quad \boldsymbol{\tau_{ind}} = \mathbf{3.51 \text{ N/mm}^2 < 40 \text{ N/mm}^2} \text{ (Permissible shear stress)}$$

As induced shear stress is less than given permissible shear stress, design is safe.

**Step III : Design of key :** Key may be designed using the following empirical relations.

$$\text{Width of key (w)} = \frac{d}{4} = \frac{52}{4} = \mathbf{13\ mm}$$

$$\text{Thickness of key (t)} = \frac{2w}{3} = \frac{d}{6} = \frac{52}{6} = \mathbf{8.67\ mm}$$

Length of key must be at least equal to length of sleeve. The coupling key is usually designed into two parts.

$$\therefore \quad \text{Length of key } (l) = \frac{L}{2} = \frac{182}{2} = \mathbf{91\ mm}$$

Checking the key under shear failure, we have,

$$\text{Torque transmitted, T} = l \times w \times \tau_{ind} \times \frac{d}{2}$$

$$\therefore \quad 1061.03 \times 10^3 = 91 \times 13 \times \tau_{ind} \times \frac{52}{2}$$

$$\therefore \quad \tau_{ind} = \mathbf{34.496\ N/mm^2 < 40\ N/mm^2} \text{ (Permissible shear stress)}$$

As induced shear stress in key is less than permissible value, design of key is safe.

Checking the key under crushing failure, we have,

$$T = l \times \frac{t}{2} \times \sigma_{ck} \times \frac{d}{2}$$

$$\therefore \quad 1061.03 \times 10^3 = 91 \times \left(\frac{8.67}{2}\right) \times \sigma_{ck_{ind}} \times \frac{52}{2}$$

$$\therefore \quad \sigma_{ck_{ind}} = \mathbf{103.45\ N/mm^2 > 80\ N/mm^2} \text{ (Permissible crushing stress)}$$

As induced crushing stress (103.45 $N/mm^2$) is more than permissible value (80 $N/mm^2$), we will modify the value of '$l$' by keeping the value of permissible crushing stress as follows :

$$T = l \times \frac{t}{2} \times \sigma_{ck} \times \frac{d}{2}$$

$$\therefore \quad 1061.03 \times 10^3 = l \times \left(\frac{8.67}{2}\right) \times 80 \times \left(\frac{52}{2}\right)$$

$$\therefore \quad l = 117.67 \text{ mm} \cong \mathbf{118\ mm\ (say)}$$

### 3.3.5.1 Advantages of Muff Coupling

(1) It is the simplest form of coupling and easy to construct.

(2) It has no projection parts except the key head.

(3) External surface of sleeve is smooth.

### 3.3.5.2 Disadvantages of Muff Coupling

(1) It is difficult to assemble or dismantle, since the sleeve has to be either shifted over the shaft or ends of shafts to connect them together.

(2) It cannot tolerate mis-alignment between the axes of shafts to be connected together.

(3) It cannot absorb shocks and vibrations.

(4) It requires more axial space as compared to flange coupling.

### 3.3.5.3 Clearance in Muff coupling

- Small clearance is provided at joining surface of muff coupling to create a pressure between the surface of shaft and coupling.
- Pressure created by tightening the bolts causes friction on the surface of shaft, which transmits the required torque.

## 3.3.6 Flange Coupling

**Introduction :**

- A flange coupling is a type of rigid coupling. It consists of two cast iron flanges, one mounted on the driving shaft and other on the driven shaft.
- Faces of couplings are turned (or finished) in such a way that, they become perpendicular to axes of shafts.

- Two flanges are connected together by means of four or six bolts arranged on a circle concentric with axes of shafts. Power is transmitted from the driving shaft to left side flange through the key. It is then transmitted from left side flange to right side flange through the bolts. Finally, power is transmitted from right side flange to the driven shaft through the key.
- There are two types of rigid flange couplings.
  1. Unprotective or unprotected type flange coupling.
  2. Protective or protected type flange coupling.
- Fig. 3.28 shows an unprotective or unprotected type of flange coupling. Revolving bolt heads and nuts are dangerous to the operator and may lead to accidents.

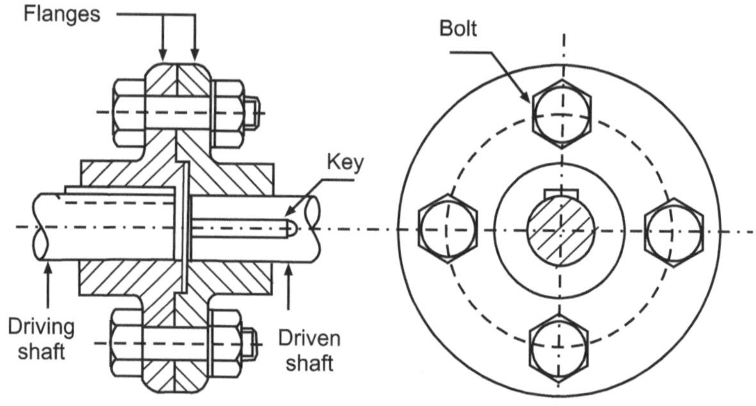


**Fig. 3.28 : Unprotective type flange coupling**

- Fig. 3.29 shows protective or protected type of flange coupling. It consists of two hubs keyed on the ends of two shafts.
- The hubs extend into the flanges, whose faces are brought together and connected by means of bolts. Bolts are guarded and protected by giving projections at the flanges.

**Design Procedure of Protected Type Flange Coupling :**

Let,
- $d$ = Diameter of shaft or inner diameter of hub
- $D$ = Outer diameter of hub
- $d_c$ = Core diameter of bolts
- $D_1$ = Pitch circle diameter (P.C.D.) of bolts
- $n$ = Number of bolts
- $t_f$ = Thickness of flange

Let $\tau_s$, $\tau_b$, $\tau_k$ and $\tau_f$ be the permissible shear stresses for shaft, bolt, key and flange material, and $\sigma_{ckb}$ and $\sigma_{ckk}$ be the permissible crushing stresses for bolt and key material respectively.

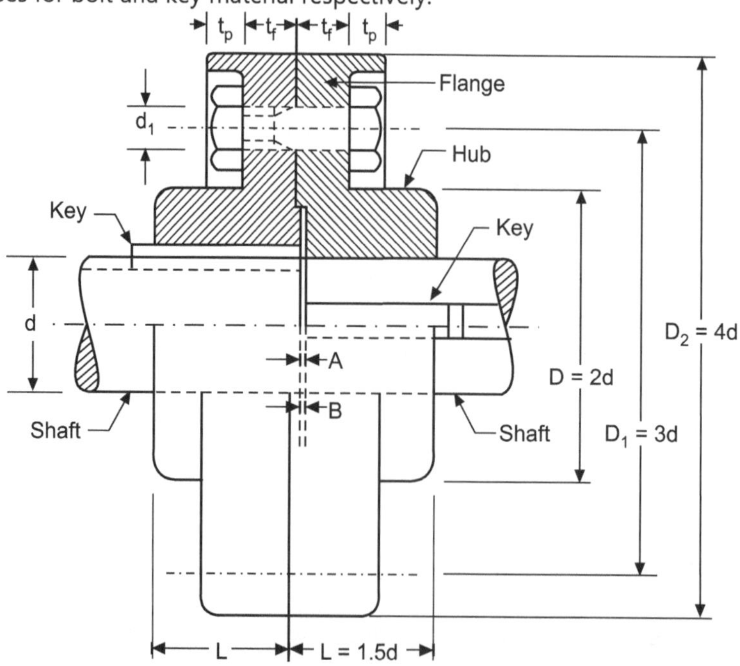


**Fig. 3.29 : Protective type flange coupling**

**Step I : Design of shaft :** Considering the shaft under twisting moment, we have, $T = \frac{\pi}{16} \tau d^3$

From the above design equation, diameter of shaft (d) is determined.

**Step II : Design of hub :** Usual proportions are,

$$D = \text{Outer diameter of hub} = 2d, \; L = \text{Length of hub} = 1.5\, d, \quad k = \frac{d}{D}$$

Considering hub as a hollow shaft, transmitting the same torque as that of shaft,

$$T = \frac{\pi}{16} \times \tau \times D^3 (1 - k^4) \qquad \text{(where, } k = d/D\text{)}$$

By putting the value of D in above expression, check whether shear stress induced in hub is less than its permissible value. If yes, the design of hub is said to be safe.

**Step III : Design of key :** Key may be designed using the following empirical relations.

$$\text{Width of key} = w = \frac{d}{4}$$

$$\text{Thickness of key} = \frac{2w}{3} = \frac{d}{6}$$

Length of key must be at least equal to length of hub. $\therefore l = L$

After determining the dimensions of key, the crushing stress and shear stress induced in the key may be checked by using design equations of key, given below.

$$\text{Torque transmitted,} \quad T = F \times \frac{d}{2} = l \times w \times \tau_k \times \frac{d}{2} \qquad \text{... (In shearing)}$$

$$\text{And} \quad T = F \times \frac{d}{2} = l \times \frac{t}{2} \times \sigma_{ckk} \times \frac{d}{2} \qquad \text{... (In crushing)}$$

**Step IV : Design of flange :** While transmitting the torque, flange may undergo shear failure.

Torque transmitted, $T$ = Circumference of hub × Thickness of flange × Shear stress × Radius of hub

$$\therefore \quad T = (\pi \times D) \times t_f \times \tau_f \times \frac{D}{2}$$

Taking $t_f = \frac{d}{2}$, shear stress induced in the flange ($\tau_f$) may be checked.

(**Note :** For marine type of flange coupling, $t_f = d/3$).

**Step V : Design of bolts :** Bolts are subjected to shear stress due to torque transmitted.

$$\text{Load on each bolt} = \frac{\pi}{4} \times (d_c)^2 \times \tau_b$$

$$\therefore \quad \text{Total load on all bolts} = n \times \frac{\pi}{4} \times (d_c)^2 \times \tau_b$$

$$\therefore \quad \text{Torque transmitted} = \text{Force (or Load)} \times \text{Radius} = n \times \frac{\pi}{4} \times (d_c)^2 \times \tau_b \times \frac{D_1}{2}$$

where, $D_1$ = Pitch circle diameter of bolts = 3d.

And, Number of bolts,

$$n = 3, \text{ for 'd' upto 40 mm}$$
$$= 4, \text{ for 'd' upto 100 mm}$$
$$= 6, \text{ for 'd' upto 180 mm}$$

Keeping the values of $D_1$ and n in above equation, we can find core diameter of bolt ($d_c$).

Other proportions are, $D_2 = 4 \times d$, and

Thickness of protective circumferential flange = $t_p$ = 0.25 d.

### Numerical Type No. 11 : "Protective Type Flange Coupling"

**Problem 3.29 :** *In a rigid flange coupling to transmit 20 kW at 700 r.p.m., the flanges are of cast iron and other parts of C-40. Assume stresses for C.I., Tensile = 28 N/mm$^2$, Compressive = 60 N/mm$^2$, Shear = 10 N/mm$^2$*

*Stresses for C-40, Tensile = Compressive = 110 N/mm$^2$, Shear = 40 N/mm$^2$, Bearing = 25 N/mm$^2$.*

*Determine: Diameter of shaft, diameter of bolts, if 6 bolts are used and various flange dimensions.*

**Solution : Given data :** $P = 20 \text{ kW} = \mathbf{20 \times 10^3 \ W}$, $N = 700$ r.p.m.

$\tau_s = 40 \text{ MPa} = \mathbf{40 \ N/mm^2} = \tau_k = \tau_b$, $\sigma_{ckk} = 110 \text{ N/mm}^2$

$\tau_f = \tau_{ci} = 10 \text{ N/mm}^2$, $n = 6$

**Procedure :** Power transmitted by steel shafts,

$$P = \frac{2\pi NT}{60}$$

$$\therefore \quad \text{Torque, } \mathbf{T} = \frac{P \times 60}{2\pi N} = \frac{20 \times 10^3 \times 60}{2\pi \times 700} = 272.84 \text{ N-m} = \mathbf{272.84 \times 10^3 \ N\text{-}mm}$$

We know that, torque transmitted by shaft is given by,

$$T = \frac{\pi}{16} \times \tau_s \times d^3$$

$$\therefore \quad 272.84 \times 10^3 = \frac{\pi}{16} \times 40 \times d^3$$

$$\therefore \quad \text{Diameter of shaft, } d = 32.63 \text{ mm} \cong \mathbf{34 \ mm \ (say)}$$

**Step I : Design of hub :** Usual proportions are,

$D$ = Outer diameter of hub = 2d = 2 × 34 = **68 mm**

$L$ = Length of hub = 1.5 × d = 1.5 × 34 = **51 mm**

$$k = \frac{d}{D} = \frac{34}{68} = \mathbf{0.5}$$

Considering hub as a hollow shaft transmitting the same torque as that of shaft, we have,

$$T = \frac{\pi}{16} \times \tau_{ci} \times D^3 (1 - k^4)$$

$$\therefore \quad 272.84 \times 10^3 = \frac{\pi}{16} \times \tau_{ci} \times (68)^3 \times [1 - (0.5)^4)]$$

$$\therefore \quad \tau_{ci} = \mathbf{4.71 \ N/mm^2 < 10 \ N/mm^2 \ (Permissible \ stress)}$$

Thus, the induced shear stress in the cast iron hub is less than the given permissible shear stress. Hence, the design is safe.

**Step II : Design of flange :** Take, $\mathbf{t_f} = \frac{d}{2} = \frac{34}{2} = \mathbf{17 \ mm}$

While transmitting the torque, the flange is under shear failure.

Torque transmitted, T = Circumference of hub × Thickness of flange × Shear stress × Radius of hub

$$\therefore \quad T = (\pi \times D) \times t_f \times \tau_f \times \frac{D}{2}$$

$$\therefore \quad 272.84 \times 10^3 = \pi \times 68 \times 17 \times \tau_f \times \frac{68}{2}$$

$$\therefore \quad \tau_f = \mathbf{2.2 \ N/mm^2 < 10 \ N/mm^2 \ (Permissible \ stress)}$$

Thus, induced shear stress is less than given permissible shear stress for flange material. Hence, the design is safe.

Other proportions are, $\mathbf{D_2}$ = 4 × d = 4 × 34 = **136 mm**

Thickness of protective circumferential flange is given by

$\mathbf{t_p}$ = 0.25 × d = 0.25 × 34 = **8.5 mm**

**Step III : Design of bolts :** Bolts are subjected to shear stress due to torque transmitted.

$$\therefore \quad \text{Load on each bolt} = \frac{\pi}{4} \times (d_c)^2 \times \tau_b$$

$$\therefore \quad \text{Total load on all bolts} = n \times \frac{\pi}{4} \times (d_c)^2 \times \tau_b$$

$$\therefore \quad \text{Torque transmitted} = \text{Load} \times \text{Radius} = n \times \frac{\pi}{4} (d_c)^2 \times \tau_b \times \frac{D_1}{2}$$

Taking $\mathbf{D_1}$ = 3 d = 3 × 34 = **102 mm** and **n** = **6** (given), the above equation gives,

$$272.84 \times 10^3 = 6 \times \frac{\pi}{4} \times (d_c)^2 \times 40 \times \frac{102}{2}$$

$$\therefore \quad d_c = \mathbf{5.327 \ mm}$$

We have, $d_c = 0.84 \times d_o$

$\therefore \quad d_o = \frac{d_c}{0.84} = \frac{5.327}{0.84} = \mathbf{6.34\ mm \cong 8\ mm\ (say)}$

$\therefore$ We can use M8 bolts. But for safety reasons, we must increase size of bolts upto M16.

---

**Problem 3.30 :** *Design and draw a protective type of cast iron flange coupling for a steel shaft transmitting 20 kW at 259 r.p.m. and having an allowable shear stress of 40 MPa. Working stress in the bolts should not exceed 90 N/mm$^2$. Assume that, same material is used for shaft and key and that the crushing stress is twice the value of its shear stress. Maximum torque is 25% greater than the full load torque. Shear stress for cast iron is 14 N/mm$^2$.*

**Solution : Given data : P** = 20 kW = $\mathbf{20 \times 10^3\ W}$, N = 259 r.p.m., $\tau_s$ = 40 N/mm$^2$, $\tau_b$ = 90 N/mm$^2$, $\tau_{ci}$ = 14 N/mm$^2$. **For key,** $\sigma_{ck} = 2\tau_s = 2 \times 40 = \mathbf{80\ N/mm^2}$

**Procedure :** Power transmitted by steel shafts = $P = \frac{2\pi N T_{mean}}{60}$

$\therefore$ Torque, $T_{mean} = \frac{P \times 60}{2\pi \times N} = \frac{20 \times 10^3 \times 60}{2\pi \times 259} = 737.39\ \text{N-m} = \mathbf{737.39 \times 10^3\ N\text{-}mm}$

But, maximum torque is 25% greater than mean torque.

$\therefore \quad T_{max} = 1.25 \times T_{mean} = 1.25 \times 737.39 \times 10^3 = \mathbf{921.73 \times 10^3\ N.mm}$

We know that, torque transmitted by shaft is given by,

$$T = \frac{\pi}{16} \times \tau_s \times d^3$$

$$\therefore \quad 921.73 \times 10^3 = \frac{\pi}{16} \times 40 \times d^3$$

$\therefore$ Diameter of shafts = d = 48.95 mm $\cong$ **50 mm (say)**

We know that, inner diameter of hub is equal to diameter of shaft.

$\therefore$ Inner diameter of hub = d = 50 mm.

**Step I : Design of hub :**

Usual proportions are,

D = Outer diameter of hub = 2 × d = 2 × 50 = **100 mm**

L = Length of hub = 1.5 × d = 1.5 × 50 = **75 mm**

$k = \frac{d}{D} = \frac{50}{100} = \mathbf{0.5}$

Considering hub as a hollow shaft, transmitting the same torque as that of shaft,

We have, $$T = \frac{\pi}{16} \times \tau_{ci} \times D^3 (1 - k^4)$$

$$\therefore \quad 921.73 \times 10^3 = \frac{\pi}{16} \times \tau_{ci} \times (100)^3 \times \{1 - (0.5)^4\}$$

$\therefore \quad \tau_{ci} = \mathbf{5\ N/mm^2 < 14\ N/mm^2}$ (Permissible shear stress)

Thus, the induced shear stress in the cast iron hub is less than the given permissible shear stress. Hence, the design is safe.

**Step II : Design of key :**

As material for shaft and key is same, we can take, $\tau$ = 40 N/mm$^2$

It is given that, crushing stress for key material is twice the shear stress.

$\therefore \quad \sigma_{ck} = 2\tau = 2 \times 40 = 80$ N/mm$^2$

Also, $\sigma_{ck} = 2\tau$ indicates square key, with w = t.

Key may be designed using the following empirical relations.

Width of key = $w = \frac{d}{4} = \frac{50}{4} = \mathbf{12.5\ mm}$

$\therefore$ Thickness of key = t = w = **12.5 mm**

Length of key must be at least equal to length of hub. $\therefore$ $l$ = **75 mm**

Checking the key under shear failure,

$$\text{Torque transmitted, } T = l \times w \times \tau \times \frac{d}{2}$$

$$\therefore \quad 921.73 \times 10^3 = 75 \times 12.5 \times \tau \times \frac{50}{2}$$

$\therefore \quad \tau = \mathbf{39.32\ N/mm^2 < 40\ N/mm^2}$ **(Permissible shear stress)**

Checking the key under crushing failure,

$$\text{Torque transmitted, } T = l \times \frac{t}{2} \times \sigma_{ck} \times \frac{d}{2}$$

$$\therefore \quad 921.73 \times 10^3 = 75 \times \left(\frac{12.5}{2}\right) \times \sigma_{ck} \times \left(\frac{50}{2}\right)$$

$$\therefore \quad \sigma_{ck} = \mathbf{78.65\ N/mm^2 < 80\ N/mm^2}\ \textbf{(Permissible crushing stress)}$$

Hence, design of key is safe.

**Step III : Design of flange :**

Take, $\mathbf{t_f} = \frac{d}{2} = \frac{50}{2} = \mathbf{25\ mm}$

While transmitting the torque, the flange is under shear. The torque transmitted is,

T = Circumference of hub × Thickness of flange × Shear stress × Radius of hub

$$\therefore \quad T = (\pi \times D) \times t_f \times \tau_f \times \frac{D}{2}$$

$$\therefore \quad 921.73 \times 10^3 = \pi \times 100 \times 25 \times \tau_f \times \frac{100}{2}$$

∴ $\tau_f$ = **2.347 N/mm²**, which is less than given permissible shear stress (14 $N/mm^2$) for flange material. So, design is safe.

**Step IV : Design of bolts :**

Let us assume, n = number of bolts = 4. [For 'd' between 40 to 100 mm, n is taken as 4]

$\mathbf{D_1}$ = diameter of bolt circle or pitch circle diameter of bolts = 3 × d = 3 × 50 = **150 mm**

Bolts are subjected to shear stress due to torque transmitted.

Load on each bolt $= \frac{\pi}{4} \cdot (d_c)^2 \cdot \tau_b$

∴ Total load on all bolts $= n \times \frac{\pi}{4} \cdot (d_c)^2 \cdot \tau_b$

We have, Torque = Load × Radius

∴ Torque transmitted $= n \times \frac{\pi}{4} \cdot (d_c)^2 \cdot \tau_b \cdot \frac{D_1}{2}$

$$\therefore \quad 921.73 \times 10^3 = 4 \times \frac{\pi}{4} \times (d_c)^2 \times 90 \times \left(\frac{150}{2}\right)$$

∴ $d_c$ = 6.59 mm.

∴ $d_o = \frac{d_c}{0.84} = \frac{6.59}{0.84} = 7.84\ \text{mm} \cong$ **8 mm (say)**

Thus, we may use M8 bolts. But, for safer side, we must increase size of bolts upto M16.

Other proportions are, $D_2$ = 4 × d = 4 × 50 = **200 mm** and

Thickness of protective circumferential flange = $\mathbf{t_p}$ = 0.25 × d = 0.25 × 50 = **12.5 mm**

### 3.3.6.1 Advantages of Flange Coupling

(1) It has high torque transmitting capacity.

(2) It is easy to assemble and dismantle.

(3) It is simple in construction and easy to manufacture.

### 3.3.6.2 Disadvantages of Flange Coupling

(1) It cannot tolerate misalignment between the axes of two shafts.

(2) It can be used only, where the motion is free from shocks and vibrations.

(3) It requires more radial space as compared to muff coupling.

## 3.3.7 Flexible Coupling and its Advantages on Flange Coupling

- Flange coupling is a rigid type of coupling. It can be used, only when, there is perfect alignment between the axes of two shafts and the motion is free from vibrations and shocks.
- **In practice**, it is **impossible** to obtain perfect **alignment** of shafts.

- **Misalignment exists due to the following reasons :**
  1. Deflection of shafts due to lateral forces.
  2. Error in shaft mounting due to manufacturing tolerances.
  3. Use of two separately manufactured units, such as, an electric motor and a worm gear box.
  4. Thermal expansion of parts.
- If rigid coupling is used in such circumstances, then misalignment causes excessive bearing reactions resulting in vibrations and wear.
- To overcome this problem, flexible couplings are used.
- A flexible coupling employs a flexible element like rubber bush between the driving and driven flanges. This flexible rubber bush not only accommodates for the misalignment, but also absorbs shocks and vibrations.

**Construction of Flexible Coupling :**

- Construction of flexible coupling is shown in Fig. 3.30. It is similar to rigid type of flange coupling except the provision of rubber bush and pins in place of bolts.
- Flexible coupling consists of two flanges, one keyed to the input shaft and the other to the output shaft. Two flanges are connected together by means of 4 or 6 pins.
- At one end, the pin is fixed to the output flange by means of a nut. Diameter of pin is enlarged in input flange, where a rubber bush is to be mounted over the pin.
- Rubber bush is provided with brass lining at the inner surface to avoid excessive wear of rubber bush.
- Power is transmitted from the input shaft (driver) to the input flange through the key. It is then transmitted from the input flange to the pin through the rubber bush. The pin then transmits the power to the output flange by shear resistance. Finally, power is transmitted from the output flange to the output shaft (driven) through the key.

## 3.3.8 Bushed Pin Type Flexible Coupling

**Introduction :**

- A bushed pin type flexible coupling as shown in Fig. 3.30. It is a modification of the rigid type of flange coupling. The coupling bolts are known as *pins*. The rubber or leather bushes are used over the pins.
- Two halves (flanges) of coupling are dissimilar in construction. A clearance of 5 mm is kept between the faces of two halves of coupling. There is no rigid connection between them and the drive takes place through the medium of compressible rubber or leather bushes.
- Flexible coupling is used to connect a machine to a prime mover for direct drive.

**Purpose of Flexible Couplings :**

(1) To prevent transmission of shock from one shaft to other.

(2) To eliminate stress reversals, when either shaft is subjected to deflection at or near the coupling.

**Design Procedure of Bushed Pin Type Flexible Coupling :**

- In designing a bushed pin type flexible coupling, the proportions of rigid type flange coupling are modified. It involves reduction of bearing pressure on the rubber or leather bushes and it should not exceed 0.5 N/mm$^2$. For this, pitch circle diameter and number of pins are increased.

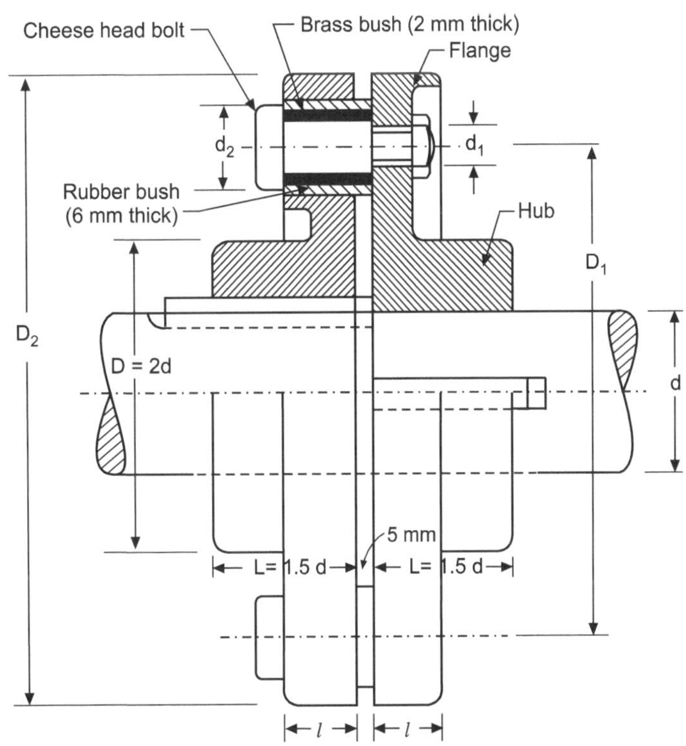


**Fig. 3.30 : Bushed-pin type flexible coupling**

Let,
$l$ = Length of bush in flange
$d_2$ = Diameter of bush
$P_b$ = Bearing pressure on the bush or pin
$d_1$ = Diameter of pins
n = Number of pins
$D_1$ = Pitch circle diameter of pins

**Step I :** Calculate diameter of shaft 'd', by using following equations.

Torque transmitted by shaft,

**(a)** $T = \frac{P \times 60}{2\pi N}$ and **(b)** $T = \frac{\pi}{16} \tau d^3$

From these expressions, find 'd'.

**Step II (a) :** Calculate the diameter of pin,

$$d_1 = \frac{0.5\ d}{\sqrt{n}}$$

where, n can be assumed depending upon number of bolts.

The calculated diameter $d_1$ is modified and a brass bush of 2 mm thickness is fitted over the enlarged portion of pin. Also, the brass bush carries a rubber bush of 6 mm thickness.

∴ Diameter of rubber bush = $d_2 = d_1 + 2 \times 2 + 2 \times 6$

Pitch circle diameter of pins = $D_1 = 2 \times d + d_2 + 2 \times 6$

**Step II (b) :** We have, bearing load acting on each pin = $P_b \times d_2 \times l$

∴ Total bearing load on all pins = $n \times P_b \times d_2 \times l$

∴ Torque transmitted by coupling = $W \times \left(\frac{D_1}{2}\right) = n \times P_b \times d_2 \times l \times \left(\frac{D_1}{2}\right)$

Using the above equation, we can determine the value of $l$.

**Step II (c) :** Calculate the direct shear stress induced in the coupling halves,

$$\tau = \frac{W}{\frac{\pi}{4}(d_1)^2}$$

Calculate the maximum bending stress induced in the pin,

$$\sigma_b = \frac{M}{Z}$$

where, $M = W \times \left(\frac{l}{2} + 5\right)$ and $Z = \frac{\pi}{32}(d_1)^3$

- According to maximum shear stress theory,

$$\tau_{max} = \text{Maximum shear stress} = \frac{1}{2}\sqrt{\sigma_b^2 + 4(\tau)^2}$$

- According to normal stress theory,

$$\sigma_{b\,max} = \text{Maximum normal stress} = \frac{1}{2}\sigma_b + \frac{1}{2}\sqrt{\sigma_b^2 + 4(\tau)^2}$$

Check whether, the above calculated maximum stresses are within the safe limits or not.

**Step III : Design of hub :**

Let,
Diameter of hub = D = 2d
Length of hub = L = 1.5 d

Considering the hub as a hollow shaft, check the induced shear stress for hub material (cast iron).

We have, $T = \frac{\pi}{16}\tau_c D^3 (1 - k^4)$ $\left(\text{where, } k = \frac{d}{D}\right)$

Find $\tau_{c\ ind.}$

$\tau_{c\ ind.}$ should be less than $\tau_{c\ permissible}$ for safe design.

**Step IV : Design for key :**

Let, $w$ = Width of key

$t$ = Thickness of key

For rectangular key, $w = \frac{d}{4};\ t = \frac{d}{6}$ and For square key, $w = \frac{d}{4};\ t = \frac{d}{4}$

Length of key $l_1$ is slightly greater than length of hub.

$$l_1 > L$$

Now, check the induced stresses $\tau_k$ and $\sigma_{ck}$ in key by considering shearing and crushing failure of key.

$$T = w \times l_1 \times \tau_k \times \frac{d}{2} \qquad \text{(Shearing of key)}$$

Find $\tau_{ind}$. It should be less than $\tau_{k\ permissible}$.

$$T = \frac{t}{2} \times l_1 \times \sigma_{ck} \times \frac{d}{2} \qquad \text{(Crushing of key)}$$

Find $\sigma_{ck\ ind.}$. It should be less than $\sigma_{ck\ permissible}$.

**Step V : Design for flange :**

Thickness of flange, $t_f = 0.5\ d$

Check the induced shear stress in the flange by considering the flange at the junction of hub in shear.

$$T = (\pi D t_f) \times \tau_c \times \frac{D}{2}$$

Find $\tau_{c\ induced}$. It should be less than $\tau_{c\ permissible}$ for safe design of flange.

**Numerical Type No. 12 : "Bushed Pin Type Flange Coupling"**

**Problem 3.31 :** *Design a bushed pin type flexible coupling for connecting a motor shaft to pump shaft for the following service conditions :*

*Power to be transmitted = 40 kW; Speed of motor shaft = 1000 r.p.m.; Diameter of motor shaft = 50 mm; Diameter of pump shaft = 45 mm; Allowable shear stress for C.I. = 15 MPa. The bearing pressure in the rubber bush and allowable shear stress in the pins are to be limited to 0.45 N/mm² and 25 N/mm² respectively.*

*The allowable shear stress and crushing stress for shaft and key material is 42 MPa and 84 MPa respectively.*

**Solution : Given data :** $P = 40\ kW = \mathbf{40 \times 10^3\ W}$, $N = 1000$ r.p.m., $d = 50$ mm, $d_p = 45$ mm, $\tau_{ci} = 15\ N/mm^2$, $P_b = 0.45\ N/mm^2$, $\tau = 25\ N/mm^2$.

**Procedure :**

**Step I :**

We have, power transmitted, $P = \frac{2\pi NT}{60}$

$\therefore$ Torque, $\mathbf{T} = \frac{P \times 60}{2\pi N} = \frac{40 \times 10^3 \times 60}{2 \times \pi \times 1000} = 381.97\ \text{N-m} = \mathbf{381.97 \times 10^3\ N\text{-}mm}$

**Step II (a) :** Let n = number of pins = 6.

Calculate the diameter of pin as,

$$\mathbf{d_1} = \frac{0.5\ d}{\sqrt{n}} = \frac{0.5 \times 50}{\sqrt{6}} = \mathbf{10.206\ mm}$$

In order to permit the bending stress induced in the pin due to compressibility of brass bush, let us modify, diameter of pin from 10.206 mm to **20 mm**. This diameter is threaded and secured in right half coupling.

Let us take, diameter of the enlarged portion in the left half of coupling, $d_1$ = **24 mm.**

A brass bush of 2 mm thickness is fitted over the enlarged portion of pin. Also, the brass bush carries a rubber bush of 6 mm thickness.

$\therefore$ Diameter of rubber bush = $d_2 = d_1 + 2 \times 2 + 2 \times 6 = 24 + 4 + 12 =$ **40 mm**, and

Pitch circle diameter of pins = $D_1 = 2 \times d + d_2 + 2 \times 6 = 2 \times 50 + 40 + 12 =$ **152 mm**

**Step II (b) :**

We have, bearing load acting on each pin, $\mathbf{W} = P_b \times d_2 \times l = 0.45 \times 40 \times l = \mathbf{18 \times \mathit{l}}$

$\therefore$ Total bearing load on all pins = $n \times W = 6 \times (18 \times l)$

$\therefore$ Torque transmitted by coupling, $T = n \times W \times \frac{D_1}{2}$

$\therefore \quad 381.97 \times 10^3 = 6 \times (18 \times l) \times \frac{152}{2}$

$\therefore \quad$ **$l$ = 46.54 mm**

$\therefore \quad$ W = $18 \times l = 18 \times 46.54$ = **837.72 N**

**Step II (c) :**

Calculate the direct shear stress induced in the coupling halves (taking minimum pin diameter).

$$\tau = \frac{W}{(\pi/4) \times (d_1)^2} = \frac{837.72}{\pi/4 \times (20)^2} = \mathbf{2.67\ N/mm^2}$$

Calculate the maximum bending stress induced in the pin,

$$\sigma_b = \frac{M}{Z} = \frac{W \times (l/2 + 5)}{\pi/32 \times (d_1)^3} = \frac{837.72 \times (\frac{46.54}{2} + 5)}{\pi/32 \times (20)^3} = \mathbf{30.15\ N/mm^2}$$

According to maximum shear stress theory,

$$\tau_{max} = \text{Maximum shear stress} = \frac{1}{2}\sqrt{\sigma_b^2 + 4(\tau)^2} = \frac{1}{2}\sqrt{(30.15)^2 + 4 \times (2.67)^2} = \mathbf{15.31\ N/mm^2}$$

According to maximum normal stress theory,

$$\sigma_{b\,max} = \text{Maximum normal stress} = \frac{1}{2}\sigma_b + \frac{1}{2}\sqrt{\sigma_b^2 + 4(\tau)^2}$$

$$\boldsymbol{\sigma_{b\,max}} = \frac{1}{2} \times 30.15 + \frac{1}{2}\sqrt{(30.15)^2 + 4 \times (2.67)^2} = \mathbf{30.38\ N/mm^2}$$

As the above maximum stresses are within the safe limits, design is safe.

**Step III : Design for hub :**

Diameter of hub, $\quad D = 2d = 2 \times 50$ = **100 mm**

Length of hub, $\quad L = 1.5\,d = 1.5 \times 50$ = **75 mm**

Considering hub as a hollow shaft, check the induced shear stress in the hub.

$$T = \frac{\pi}{16} \cdot \tau_{ci} \cdot D^3 \cdot (1 - k^4)$$

$$\therefore \quad 381.97 \times 10^3 = \frac{\pi}{16} \times \tau_{ci} \times 100^3\,[1 - (0.5)^4] \qquad (\because k = d/D = 0.5)$$

$\therefore \quad \tau_{c\,ind}$ = **2.075 N/mm² < 15 N/mm²** (Permissible shear stress)

As induced shear stress is less than permissible stress, design is safe.

## 3.3.8.1 Advantages of Bushed Pin Type Flexible Coupling

(1) It is simple in construction.

(2) It can be used for transmitting high torques.

(3) It prevents transmission of shock from one shaft to the other shaft.

(4) It absorbs vibrations.

(5) It can tolerate 0.5 mm of lateral misalignment or 1.5° angular misalignment.

## 3.3.8.2 Disadvantages of Bushed Pin Type Flexible Coupling

(1) High cost.

(2) It requires more radial space as compared to other type of couplings.

## 3.3.8.3 Reasons of Placing Coupling Near to Bearing

Coupling is provided near the bearing due to following reasons :

(1) It gives minimum vibrations.

(2) Bending load on shaft can be minimized.

(3) It increases power transmission stability.

(4) It avoids deflection of shaft.

### 3.3.8.4 Couplings Suggested for Specific Cases

| Sr. No. | Case | Coupling suggested |
|---|---|---|
| 1. | Shafts having parallel axes with small distance apart. | Oldham's coupling. |
| 2. | Shafts with intersecting axes. | Universal coupling. |
| 3. | Shafts having perfect alignment. | Rigid couplings like sleeve or muff and flange coupling. |
| 4. | Shafts having both lateral and angular misalignment. | Flexible coupling. |
| 5. | Overhead crane shafts. | Protected type flange coupling. |
| 6. | High speed pump and motor assembly. | Flexible coupling. |

## 3.3.9 Comparison between Rigid and Flexible Coupling

| Sr. No. | Comparative Point | Rigid Coupling | Flexible Coupling |
|---|---|---|---|
| 1. | **Purpose** | It is used to connect two shafts, which are in perfect alignment. | It is used to connect two shafts, having both lateral and angular misalignments. |
| 2. | **Shock and vibrations** | Due to absence of torsional flexibility, they cannot damp out vibrations. | Due to torsional flexibility, they are able to damp out vibrations to certain extent. |
| 3. | **Rotational speed** | Relatively low. | Relatively high. |
| 4. | **Alignment** | Rigid coupling cannot tolerate any misalignment between two shafts. | Flexible coupling can tolerate small amount of misalignment between two shafts. |
| 5. | **Deflection** | Less. | More. |
| 6. | **Cost** | Less expensive. | More expensive. |
| 7. | **Suitability** | They are suitable for the applications like electric motor with pump or compressor. | They are suitable for high power transmission applications like electric motor and machine tool, Shaft of I.C. engine with reduction gear box or dynamometer. |

## 3.3.10 Difference between Coupling and Clutch

| Coupling | Clutch |
|---|---|
| 1. It is used to make a permanent or semi-permanent connection between two shafts. | 1. It permits rapid connection or disconnection as per the need of driver or operator. |
| 2. Example: To connect an electric motor with a machine like pump. | 2. Example: To connect or disconnect the engine shaft (driver shaft) to a driven shaft. |

**Theory Questions for Practice**

1. Discuss the various types of shafts and standard sizes of transmission shafts.
2. What types of stresses are induced in shafts?
3. What do you understand by torsional rigidity and lateral rigidity?
4. "A hollow shaft has greater strength and stiffness than solid shaft of equal weight". Justify.
5. What is a key? State its function.
6. How keys are classified? Draw neat sketches of different types and state their applications.
7. What is the effect of keyway cut into the shaft?
8. Explain the strength equations of rectangular sunk key.
9. What are functions of coupling?
10. Sketch a protective type flange coupling and explain its design.
11. Distinguish between rigid and flexible couplings.
12. Describe design procedure of bush-pin type flexible coupling with neat sketch.

## Numerical Problems for Practice

1. A mild steel shaft is transmitting 12.5 kW at 300 r.p.m. and is supported in two bearings 750 mm apart, to a pulley of 450 mm diameter fixed at a distance of 200 mm from one end. The belt passing over the pulley is vertical and tension ratio is 2. The safe working stresses for the shaft is $F_t$ = 70 N/mm$^2$ and $f_s$ = 56 N/mm$^2$. Design the diameter of shaft. The pulley weights 600 N. (**Ans.** d = 52 mm)
2. A mild steel shaft is supported on two bearings 1 metre apart. It transmits 15 kW at 300 r.p.m. to a pulley of diameter 200 mm mounted at a distance of 300 mm from one end. The belt passing over the pulley is vertical and the ratio of belt tension is 2 : 1. Pulley weighs 500 N. Design suitable diameter of shaft with following permissible stresses as $\sigma_t$ = 70 N/mm$^2$, $\sigma_s$ = 56 N/mm$^2$. (**Ans.** d = 49.1467 ≅ 50 mm (say))
3. A shaft is required to transmit 100 kW at 1440 r.p.m. The outside diameter of shaft should not exceed 50 mm and maximum shear stress is 70 N/mm$^2$. Find the dimensions of hollow and solid shaft, which would just meet these requirements and compare the weights of the two. (**Ans.** $W_H$ = 0.75 $W_s$, $d_i$ = 44.36 mm, d = 36.40 mm)
4. A line shaft is to transmit 25 kW at 150 r.p.m. It is driven by a motor placed directly under it by means of belt running on the 1000 mm diameter pulley keyed to the end of the shaft. The ratio of tight side to slack side tensions is 2.5 and the centre of pulley overhangs at 150 mm beyond the center line of end bearing. Determine diameter of shaft, if shear stress is 56 N/mm$^2$ and weight of pulley is 1.6 kN. (**Ans.** d = 59 mm)
5. Design of the diameter of a rotating shaft, which is subjected to a maximum torque of 200 N-m and bending moment of 350 N-m. Both the torque and bending moment are applied suddenly for following conditions:
   (a) Yield point value in shear = 300 N/mm$^2$.
   (b) Factor of safety = 3.
   (c) $K_b$ = 2 and $K_t$ = 1.5 are combined shock and fatigue load factor and twisting respectively. (**Ans.** d = 36 mm)
6. A hollow shaft is required to transmit 600 kW at 110 r.p.m. The maximum torque being 20% greater than the mean. The shear stress is not to exceed 63 MPa and twist in a length of 3 m not to exceed 1.4°. Find external diameter of the shaft, if ratio of internal diameter to external diameter is 3/8. Take modulus of rigidity 84 GPa. (**Ans.** $d_o$ = 176.2 mm)
7. A pulley is keyed to a shaft of diameter 100 mm by means of key of 20 mm width and 150 mm length. The allowable shear stress is 40 N/mm$^2$ and crushing stress is 100 N/mm$^2$. The shaft is running at 250 r.p.m. Find the power transmission capacity and (depth) thickness of the key. (**Ans.** P = 157.079 kW; t = 16 mm)
8. Design a cast iron protective flange coupling to connect the shafts in order to transmit 7.3 kW at 500 r.p.m.
   (a) Allowable shear stress for shaft, bolt and key = 42 N/mm$^2$.
   (b) Allowable crushing stress for bolt and key = 82 N/mm$^2$.
   (c) Allowable shear stress for C.I. Flange = 8 N/mm$^2$. (**Ans.** d = 26 mm, w = 6.5 mm, $l$ = 41 mm, M8 bolts)
9. A rigid C.I. flange coupling transmits 15 kW at 400 r.p.m. The maximum torque transmitted is 25% greater than the full load torque. Find the diameter and number of bolts required if :
   Shear stress for shaft = 40 N/mm$^2$.
   Shear stress for bolt = 30 N/mm$^2$. (**Ans.** M16 bolts, n = 4)
10. A sleeve coupling is to connect two shafts. The power transmitted is 38 kW at 200 r.p.m. If shear stress for the shaft and key material is 60 N/mm$^2$ and for muff 10 N/mm$^2$. Find : (i) diameter of shaft, (ii) diameter of sleeve and (iii) key dimensions. (**Ans.** d = 54 mm, D = 108 m, L = 81 mm, w = 13.5 mm, t = 9 mm)
11. Design muff coupling, which is used to connect two shafts transmitting 30 kW at 300 r.p.m. The material for the shaft and key is plain carbon steel having allowable shear and crushing stresses as 35 MPa and 70 MPa respectively. The material for the muff is C.I., for which, allowable shear stress is assumed as 15 MPa. (**Ans.** D = 117 mm, L = 182 mm, w = t = 13 mm, $l$ = 91 mm)

❑❑❑

*Chapter* **4**

# DESIGN OF POWER SCREWS AND SPUR GEAR

## 4.1 INRODUCTION TO POWER SCREWS

- **Power screws (translation screws)** are used to transmit power by converting rotary motion into linear motion (translatory motion).
- **For example :**
  1. In *lead screw of lathe*, the rotary motion is available, but the tool has to be advanced (travelled) axially in the direction of cut against cutting resistance of material.
  2. In *screw jack*, a small force (effort) applied in horizontal plane, is used to raise or lower a large load.
  3. In *machine vices*, *testing machines*, mechanical *presses* etc.
- **Main components of power screw** are : (i) Screw and (ii) Nut. Torque is applied to any one (generally screw), causing it to rotate and move either itself or the other element in axial direction.
- One holding arrangement is necessary to hold either screw or nut in its place.
- **Depending upon the holding arrangement, power screws operate in two different ways :**
  (a) In case of lead screw of lathe, the screw rotates in its bearing, while the nut has axial motion.
  (b) In case of screw jack or machine vice, the nut is kept stationary and screw moves in axial direction.

### 4.1.1 Material Used for Screw and Nut

- In case of square threads, the nut is usually made of soft material, so that, it can be replaced, if worn out.

| Particular | Suitable Materials |
|---|---|
| 1. Screw | Steel |
| 2. Nut | Phosphor bronze or Brass. |

### 4.1.2 Advantages of Power Screws

1. Large load carrying capacity.
2. Compact construction due to small overall dimensions.
3. Simple to design and manufacture.
4. Large mechanical advantage (i.e. M.A.), which means that, a load of 15 kN can be raised by applying an effort as small as 400 N.
5. Provides precisely controlled and highly accurate linear motion required in machine tool applications.
6. Low cost and more reliability due to less number of parts.
7. Smooth and noiseless service without any maintenance.

### 4.1.3 Disadvantages of Power Screws

1. Very poor efficiency as low as 40%. So, power screw is not preferred in continuous power transmission elements like machine tools except lead screw of lathe.
2. Rapid wear of screw and nut due to high friction in threads.

### 4.1.4 Applications of Power Screws

- Power screws are mainly used for intermittent motion i.e. occasionally required for lifting the load such as car, bus or actuating any mechanism.
- Various types of power screws with practical application for each type are as under.

| Type of Power Screw | Application |
|---|---|
| 1. Screw jack | To raise the load. |
| 2. Lead screw of lathe | To have accurate motion in machining operations. |
| 3. Vice | To clamp a workpiece. |
| 4. Universal testing machine (U.T.M.) | To load specimen. |

### 4.1.5 Terminology of Power Screw

1. **Nominal diameter or Major diameter :**
   - It is defined as, "*the largest diameter of the screw*". It is also called as *major diameter.*
   - It is denoted by letter $\mathbf{d_o}$.
2. **Core diameter or Root diameter or Core diameter :**
   - It is defined as, "*the smallest diameter of screw thread*". It is also called as *minor diameter.*
   - It is denoted by letter $\mathbf{d_c}$.
3. **Pitch :**
   - It is defined as, "*the distance from a point on one thread to the corresponding point on adjacent thread*".
   - *Pitch (i.e. distance) is measured parallel to thread axis.*
   - It is denoted by **p**.
   - For square threads, $\mathbf{d_c = d_o - p}$ ... (4.1)

     i.e. $\mathbf{d_o = d_c + p}$ ... (4.2)
4. **Mean diameter or Pitch diameter :**
   - It is defined as, "*the average of nominal diameter $d_o$ and core diameter $d_c$*".
   - It is also known as *pitch diameter*. It is denoted by letter **d**.
   - Mathematically, Mean diameter, $\mathbf{d = \dfrac{d_o + d_c}{2}}$ ... (4.3)

   Mean diameter can also be expressed as,

   (a) $d = \dfrac{d_o + d_c}{2}$

   $\therefore \quad d = \dfrac{d_o + (d_o - p)}{2}$ [From equation 4.1]

   $\therefore \quad d = \dfrac{2d_o - p}{2}$

   $\therefore \quad \mathbf{d = d_o - \dfrac{p}{2}}$ ... (4.4)

   (b) $d = \dfrac{d_o + d_c}{2}$

   $\therefore \quad d = \dfrac{(d_c + p) + d_c}{2}$ [From equation 4.2]

   $\therefore \quad d = \dfrac{2d_c + p}{2}$

   $\therefore \quad \mathbf{d = d_c + \dfrac{p}{2}}$ ... (4.5)

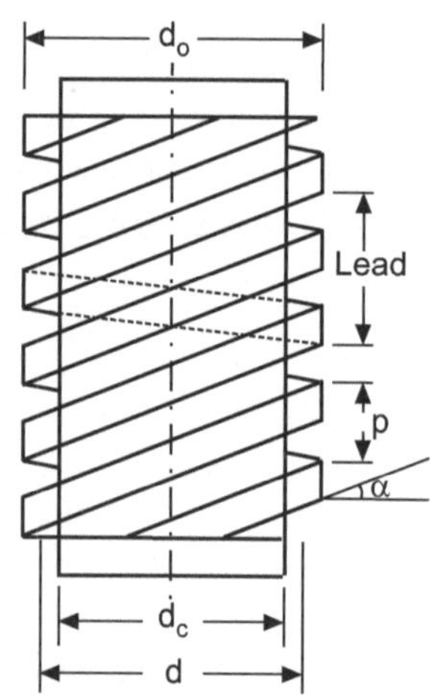

**Fig. 4.1 : Terminology of power screw**

5. **Lead :**
   - Lead is defined as, "*the distance, through which, a screw advances (travels/moves) axially in one revolution of nut*".
   - It is denoted by letter '***l***'.

**Important Note :**
- For single start threaded screw, lead is equal to pitch, whereas for double start threaded screw, it is equal to twice the pitch and so on.
- For multiple threaded screws, Lead = Number of starts × Pitch

6. **Helix angle or Lead angle :**
   - Helix angle is defined as, "*the angle made by helix of thread with a plane perpendicular to the axis of screw*".
   - It is also called as *lead angle*. It is denoted by **α**.

   Mathematically, $\tan\alpha = \dfrac{\text{Lead}}{\pi d}; \quad \therefore \alpha = \tan^{-1}\left(\dfrac{\text{Lead}}{\pi d}\right)$ ... (4.6)
7. **Right hand and Left hand threads :**
   - When the axis of screw is vertical and if the thread slopes upward from left to right, it is known as **right hand thread**.

- When the axis of screw is vertical and if the thread slopes upward from right to left, it is known as **left hand thread**.
- For right hand threads, if the right hand fingers are kept in the direction of rotation of nut, the thumb will indicate the direction of advancement of nut. Whereas for left hand threads, if the left hand fingers are kept in the direction of rotation of nut, the thumb will indicate the direction of advancement of nut.
- Right hand threads are most commonly used in *power screws, lifting and clamping devices*.

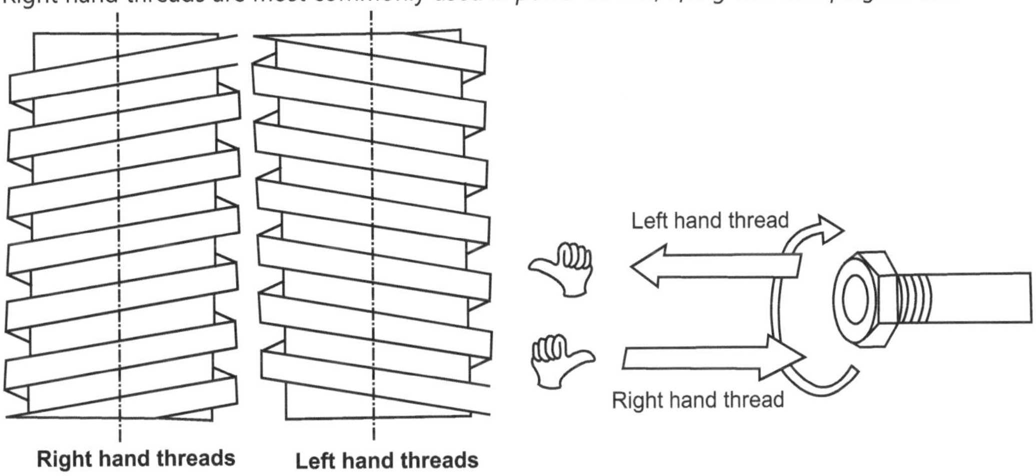


Fig. 4.2 : Right hand and left hand threads

## 4.2 VARIOUS THREAD PROFILES USED FOR POWER SCREWS

1. **Square Thread (Square thread profile) :**

- Square thread has a zero profile thread angle. It has flanks perpendicular to each other.
- Square threads are used, when force is to be applied in *both directions*.
- As they transmit power without any side thrust, they are used in *screw jacks and clamping devices*. Refer Fig. 4.3 (a).

2. **Trapezoidal Threads :**

- In trapezoidal threads, the thread angle is 30°. Refer Fig. 4.3 (b).
- They are manufactured on thread milling machine with the help of multipoint cutting tool, because machining with multipoint cutting tool is economical as compared to single point cutting tool used to machine square threads.
- They are used in *lead screw* and other power transmission devices used in *machine tools*.

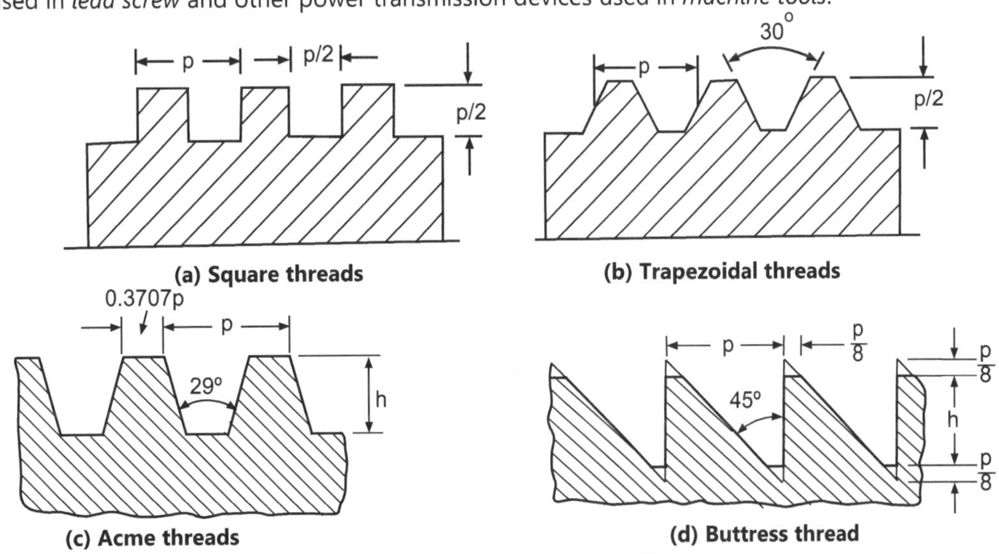


Fig. 4.3 : Types of thread profiles

3. **Acme Threads :**

- Acme thread is a special type of trapezoidal thread.
- Both acme and trapezoidal threads are identical in all respects except the **thread angle**.
- Thread angle in Acme threads is 29°, which is 30° in trapezoidal threads. Refer Fig. 4.3 (c).

- Acme threads are used for *lead screw* and other power transmission devices used in *machine tools*.
- Acme threads can be cut by dies, therefore, they are easy to manufacture.
- Efficiency of acme threads is less than square thread, because of slight slope on the flanks.

**4. Buttress Threads :**

- In case of buttress threads, the flanks are inclined at angle of 45°. Refer Fig. 4.3 (d).
- As they take unidirectional thrust, they are used in *vices and clamping devices*, where force is to be applied in *one direction only*.
- Buttress threads are stronger than any other type of threads.

### 4.2.1 Advantages, Disadvantages and Applications of Different Thread Profiles

**1. Square Threads :**

**Advantages :**

(a) Maximum efficiency.

(b) Minimum radial or bursting pressure on nut. As there is no side thrust, the motion of nut is uniform, which increases the life of nut.

**Disadvantages :**

(a) Low strength as compared to other thread profiles.

(b) Engagement and disengagement is difficult.

(c) Do not permit the use of split nut to compensate for wear.

(d) Difficult to manufacture.

(e) Machining with single point cutting tool is an expensive operation.

**Applications :**

(a) Screw jacks. (b) Mechanical presses. (c) Clamping devices. (d) Feed screws.

**2. Trapezoidal Threads :**

**Advantages :**

(a) Trapezoidal threads permit the use of split nut, which can compensate the wear.

(b) Due to larger cross-section at the root, trapezoidal threads are stronger than square threads against shear failure.

**Disadvantages :**

(a) Lower efficiency than square threads, due to slope given to flanks (sides).

(b) Slope on the flanks (sides) introduces some bursting pressure on the nut.

**Applications :**

(a) Lead screw of machine tools. (b) Bench vices.

**3. Acme Threads :**

- Since Acme thread profile is a special type of trapezoidal thread profile, its advantages, disadvantages and applications are same as those of trapezoidal threads.
- Only difference is in thread angle. In case of trapezoidal thread profile, thread angle is 30°, which is 29° in Acme thread profile.

**4. Buttress Threads :**

- Buttress threads are used, where heavy axial force acts along the screw axis in one direction only.

**Advantages :**

(a) Due to largest cross-section at the roots, buttress threads are stronger than any other type of threads.

(b) They combine the advantages of square threads (high efficiency) and V-threads (high strength).

(c) They can be economically manufactured on thread milling machine.

**Disadvantage :**

(a) They can be used to transmit power in only one direction.

**Applications :**

(a) Machine vice. (b) Clamping devices.

### 4.2.1.1 Difference between Square Threads and Acme Threads

| Square threads | Acme threads |
|---|---|
| 1. Flanks are perpendicular to each other. | 1. Roots and crests are straight and angle made by flanks of thread is 29°. |
| 2. They are used for power transmission in either direction. | 2. They are used for power transmission in only one direction. |
| 3. More efficiency. | 3. Less efficiency. |
| 4. They are manufactured on lathe machine by single point cutting tool. So difficult to manufacture on mass scale. | 4. They are cut by dies and therefore, easy to manufacture. |

### 4.2.1.2 Difference between V-threads and Square Threads

| Sr. No. | Comparative Point | V-threads | Square threads |
|---|---|---|---|
| 1. | **Thread angle** | 60° | 0° |
| 2. | **Strength** | Thickness of V-threads at the base is equal to 'p', therefore, they are stronger than square threads. So less chances of thread failure. | Thickness of square threads at the base is equal to 'p/2', therefore, they are weaker than 'V'-threads. So more chances of thread failure. |
| 3. | **Load carrying capacity** | Due to more strength, load carrying capacity is more. | Due to comparatively low strength, load carrying capacity is less. |
| 4. | **Efficiency** | Less due to high friction. | High due to less friction. |
| 5. | **Radial or bursting pressure** | Due to thread angle (60°), the nut is subjected to radial or bursting pressure. | Due to zero thread angle (0°), radial or bursting pressure is minimum or absent. |
| 6. | **Manufacturing ease** | Easy to manufacture on mass scale, as they can be easily cut on thread milling machine. | Difficult to manufacture on mass scale, as they are manufactured on lathe machine by single point cutting tool. |
| 7. | **Cost** | Cheaper due to low production cost. | Costlier due to high production cost. |
| 8. | **Applications** | *In case of fasteners*, due to high strength, low cost and easy availability. | *In case of power screws*, due to high power transmission efficiency. |

### 4.2.1.3 Advantages and Disadvantages of Square Threads over 'V' Threads

**Advantages of Square Threads over V-threads :**

1. **Radial or Bursting pressure :** Due to zero thread angle in square threads, radial or bursting pressure is absent. But, due to thread angle of 60° in V-threads, the nut is subjected to radial or bursting pressure.
2. **Efficiency :** Transmission efficiency of square threads is greatest due to zero thread angle. Whereas, V threads are not used for power transmission due to low efficiency because of 60° thread angle.

**Disadvantages of Square Threads as Compared to V-threads :**

1. **Strength :** Thickness of V-threads at the base is equal to 'p', whereas thickness of square threads at the base is equal to 'p/2'. Therefore, square threads are weaker than 'V' threads.
2. **Manufacturing :** Square threads are difficult to manufacture on mass scale, as they are manufactured on lathe machine by single point cutting tool. Whereas, 'V'-threads are easy to manufacture, since they can be easily cut on thread milling machine.
3. **Cost :** Square threads are costlier than 'V'-threads due to high production cost.
4. **Load carrying capacity :** Load carrying capacity of square threads is less as compared to 'V'-threads, due to low strength.

### 4.2.1.4 Advantages and Disadvantages of 'V' Threads

**Advantages of 'V' Threads over Square Threads:**

1. **Strength:** As the thickness of V-threads at the base is equal to 'p' and the thickness of square threads at the base is equal to 'p/2', therefore 'V' threads are stronger than square threads. So there are very less chances of thread failure.

2. **Application :** Due to high friction, V-threads offer greater resistance to unscrewing and screwing. Therefore, they are used as fasteners. But, square threads are not used for fastening purpose, because of less friction involved.
3. **Load carrying capacity:** In case of 'V' threads, load carrying capacity is more due to more strength as compared to square threads having low strength.
4. **Mass production :** As compared to square threads, 'V' threads are easy to manufacture on mass scale, as they can be easily cut on thread milling machine.
5. **Cost :** 'V' threads are cheaper due to low production cost. In case of square threads, production cost is more due to necessity of machining with a single point cutting tool, which is an expensive operation.

**Disadvantages of 'V' Threads as compared to Square Threads:**

1. **Efficiency :** Efficiency of 'V' threads is less due to high friction as compared to square threads having less friction.
2. **Radial or bursting pressure :** In case of 'V' threads, the nut is subjected to radial or bursting pressure due to 60° thread angle. This bursting pressure is minimum or absent in square threads due to zero (0°) thread angle.
3. **Power transmission :** 'V' threads are not suitable for power transmission due to high profile angle, which reduces the transmission efficiency. But, square threads are commonly used for power transmission, because their zero profile angle gives greatest power transmission efficiency.

#### 4.2.1.5 Reasons for using Square Threads in Power Transmission

- Square threads are preferred for power screws due to following reasons,
  1. Efficiency of power screws depends upon the profile angle. As profile angle is zero, the square threads have the *greatest efficiency*.
  2. They produce *minimum bursting pressure* on the nut.
  3. Square threads *transmit power* without any side thrust in *both directions*.
  4. More power transmission efficiency, due to *less friction*.
  5. They are *smooth* and *noiseless* in operation.

#### 4.2.1.6 Reasons for using 'V-Threads' as Fasteners

- Following factors permit the use of V-threads as fasteners.
  1. 'V' threads are most suitable and commonly used for fastening purposes.
  2. 'V' threads offer greater resistance at the time of unscrewing and screwing due to high friction involved. Therefore, they are used in fasteners.

     For example: Screw and nut assembly.
  3. High strength.
  4. Low cost.
  5. Easy manufacturing.
  6. Large load carrying capacity.
- But, V-threads are not suitable for power transmission, because of high profile angle (60°), which reduces the power transmission efficiency.

### 4.2.2 Multiple Threaded Screws

- Multiple threaded screws are also known as **multistart threads.**
- **Multiple threaded screws** are used in certain applications, where **higher travelling speed** is required or comparatively **large axial movement** is required without much reduction in root area of threads.
- In other words, multistart threads are used, when large lead with fine threads or high efficiency is required. They are also called as *multiple start screws*.

  **For example :** Double start, Triple start or Quadruple start screws.
- These screws have two or more threads cut side by side, around the rod.
- Refer the Fig. 4.4, where pitch is denoted by p and lead is denoted by *l*.

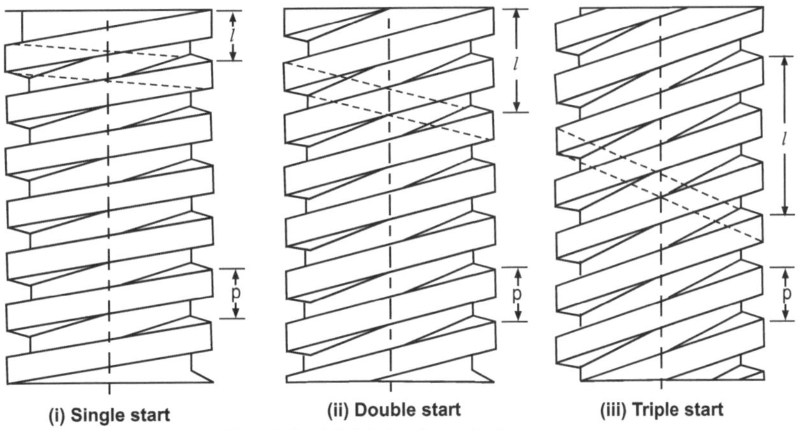


**Fig. 4.4 : Multiple threaded screws**

1. **Single Start Thread :**

- It is a single helical groove thread cut on the screws.
- For single start threads,

Lead = Axial distance travelled by screw after completing one revolution = Pitch

$\therefore \quad l = p$ (For single start thread)

2. **Double Start Thread :**

- When two threads are cut with equal spacing on the screw, threads are known as *double start threads*.
- For Double start threads,

Lead = Axial distance travelled by screw after completing one revolution

$\therefore \quad l = 2 \times \text{Pitch} = 2 \times p$ (For two start or double start thread)

Similarly, for Triple start threads, Lead = $3 \times$ pitch = $3 \times p$

for Quadruple start threads, Lead = $4 \times$ pitch = $4 \times p$

for Multiple start threads, Lead = $n \times$ pitch = $n \times p$

where, n = Number of starts.

#### 4.2.2.1 Advantages of Multiple Threaded Screw

1. Increased travelling speed due to large axial movement per revolution of screw.
2. Efficiency of multi-threaded screw is more than single threaded screw due to increase in helix angle.

#### 4.2.2.2 Disadvantages of Multiple Threaded Screw

1. Lower mechanical advantage as compared to single threaded screw.
2. Chances of loosing the property of self locking.

## 4.3 EXPRESSION FOR TORQUE REQUIRED TO RAISE THE LOAD

- Load (W) to be raised or lowered is placed on the head or top of the square threaded rod. This square threaded rod is rotated by applying effort (P) at the end of the lever for raising (lifting) the load (W).
- If one complete turn of screw is unwound and developed, it will form an inclined plane, as shown in Fig. 4.5 (a) and (b).

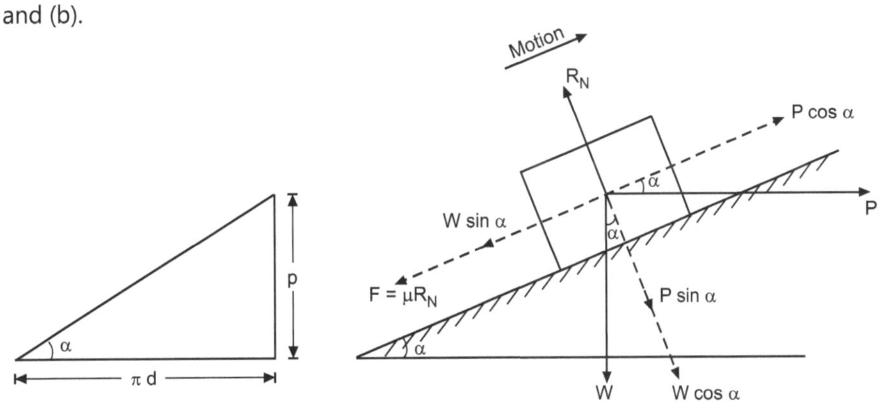


**(a) Development of screw** **(b) Forces acting on screw, while raising the load**

**Fig. 4.5**

Let, $p$ = Pitch of screw in 'mm'

$d$ = Mean diameter of screw in 'mm'

$\alpha$ = Helix angle in 'degrees' or 'radians'

$P$ = Effort applied at the circumference of screw in 'N'

$W$ = Load to be lifted in 'N'

$\mu$ = Coefficient of friction = $\tan \phi$

- From the development of screw as shown in Fig. 4.5 (a), we get,

$$\tan \alpha = \frac{p}{\pi d}$$

- Since the load is being lifted, therefore, the force of friction ($F = \mu \cdot R_N$) will act downward.
- Resolving the forces parallel to the plane (along the plane), we have,

$$\Sigma F_x = 0$$

$$\therefore \quad P \cos \alpha - W \sin \alpha - \mu R_N = 0$$

$$\therefore \quad P \cos \alpha = W \sin \alpha + \mu R_N \qquad \dots (4.7)$$

- Resolving the forces perpendicular to the plane, we have,

$$\Sigma F_y = 0$$

$$\therefore \quad R_N - P \sin \alpha - W \cos \alpha = 0$$

$$\therefore \quad R_N = P \sin \alpha + W \cos \alpha \qquad \dots (4.8)$$

Substituting the value of $R_N$ from equation (4.8) in the equation (4.7), we have,

$$P \cos \alpha = W \sin \alpha + \mu (P \sin \alpha + W \cos \alpha)$$

$$\therefore \quad P \cos \alpha = W \sin \alpha + \mu \cdot P \sin \alpha + \mu \cdot W \cos \alpha$$

$$\therefore \quad P \cos \alpha - \mu \cdot P \sin \alpha = W \sin \alpha + \mu \cdot W \cos \alpha$$

$$\therefore \quad P (\cos \alpha - \mu \cdot \sin \alpha) = W (\sin \alpha + \mu \cdot \cos \alpha)$$

$$\therefore \quad P = W \times \frac{(\sin \alpha + \mu \cdot \cos \alpha)}{(\cos \alpha - \mu \cdot \sin \alpha)}$$

By putting the value of $\mu$ = Coefficient of friction = $\tan \phi$ in above equation, we have,

$$P = W \times \frac{(\sin \alpha + \tan \phi \cdot \cos \alpha)}{(\cos \alpha - \tan \phi \cdot \sin \alpha)} = W \times \frac{(\sin \alpha \cos \phi + \sin \phi \cos \alpha)}{(\cos \alpha \cos \phi - \sin \phi \sin \alpha)}$$

$$\therefore \quad P = W \times \frac{\sin (\alpha + \phi)}{\cos (\alpha + \phi)} = W \tan (\alpha + \phi)$$

- Torque required to overcome friction between screw and nut,

$$T_1 = P \times \frac{d}{2} = W \cdot \tan (\alpha + \phi) \times \frac{d}{2} \text{ in 'N.mm'} \qquad \dots (4.9)$$

- It is desired that, load should not rotate with screw. Therefore, a thrust collar/bearing is provided at the top of screw.
- Torque required to overcome the friction at the collar will be,

$$T_2 = \frac{2}{3} \times \frac{\mu_1 \times W \times (R_1^3 - R_2^3)}{(R_1^2 - R_2^2)} \quad \text{(Uniform pressure condition)} \qquad \dots (4.10)$$

or

$$T_2 = \mu_1 \times W \times R \quad \text{(Uniform wear condition)} \qquad \dots (4.11)$$

where, $R_1$ = Radius of head, $R_2$ = Radius of pin, and $R$ = Mean radius = $\frac{R_1 + R_2}{2}$,

and $\mu_1$ = Coefficient of friction between screw and collar (bearing)

$\therefore$ **Total torque required to raise the load will be, $T = T_1 + T_2$** in 'N.mm' or 'N.m'.

## 4.4 EXPRESSION FOR TORQUE REQUIRED TO LOWER THE LOAD

- Load (W) to be lowered is placed on the head or top of the square threaded rod. This square threaded rod is rotated by applying effort (P) at the end of lever for lowering the load.
- If one complete turn of screw is unwound and developed, it will form an inclined plane, as shown in Fig. 4.6.

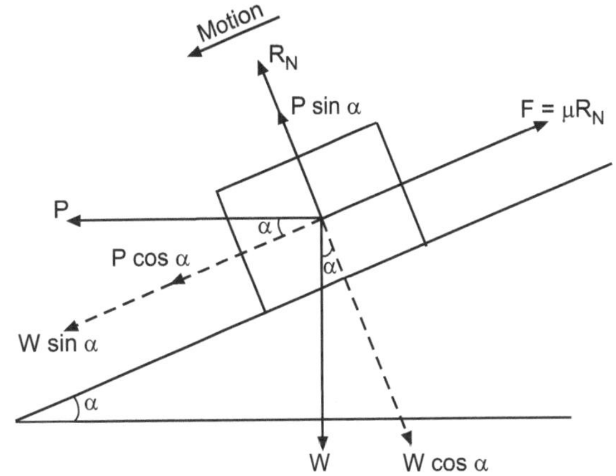


**Fig. 4.6: Forces acting on the screw, while lowering the load**

- Resolving the forces along the plane (i.e. parallel to plane),

$$\sum F_x = 0$$

$$\therefore \quad -P\cos\alpha - W\sin\alpha + \mu R_N = 0$$

$$\therefore \quad P\cos\alpha + W\sin\alpha = \mu R_N \qquad \ldots (4.12)$$

- Resolving the forces perpendicular to the plane,

$$\sum F_y = 0$$

$$\therefore R_N + P\sin\alpha - W\cos\alpha = 0$$

$$\therefore \quad R_N = W\cos\alpha - P\sin\alpha \qquad \ldots (4.13)$$

Substituting the value of $R_N$ from equation (4.13) in the equation (4.12),

$$P\cos\alpha + W\sin\alpha = \mu(W\cos\alpha - P\sin\alpha)$$

$$\therefore \quad P = W \times \frac{(\mu\cos\alpha - \sin\alpha)}{(\cos\alpha + \mu\sin\alpha)}$$

By putting the value of $\mu$ = Coefficient of friction = $\tan\phi$ in above equation, we get,

$$P = W \times \frac{(\tan\phi\cos\alpha - \sin\alpha)}{(\cos\alpha + \tan\phi\sin\alpha)} = W \times \frac{(\sin\phi\cos\alpha - \sin\alpha\cos\phi)}{(\cos\alpha\cos\phi + \sin\phi\sin\alpha)}$$

$$\therefore \quad P = W \times \frac{\sin(\phi - \alpha)}{\cos(\phi - \alpha)} = W \times \tan(\phi - \alpha)$$

- Torque required to overcome friction between screw and nut,

$$T_1 = P \times \frac{d}{2} = W \cdot \tan(\phi - \alpha) \times \frac{d}{2} \qquad \text{in 'N.mm'} \qquad \ldots (4.14)$$

- It is desired that, load should not rotate with screw. Therefore, a thrust collar is provided at the top of screw.
- Torque required to overcome the friction at the collar will be,

$$T_2 = \frac{2}{3} \times \frac{\mu_1 \times W \times (R_1^3 - R_2^3)}{(R_1^2 - R_2^2)} \qquad \text{(Uniform pressure condition)}$$

or $$T_2 = \mu_1 \times W \times R \qquad \text{(Uniform wear condition)}$$

where $R_1$ = Radius of head, $R_2$ = Radius of pin, and R = Mean radius = $\frac{(R_1 + R_2)}{2}$,

and $\mu_1$ = Coefficient of friction between screw and collar (bearing).

$\therefore$ **Total torque required to lower the load will be, $T = T_1 + T_2$**

- In both cases (Raising and lowering the load), the torque applied at the end of lever/handle of length 'L' is given by,

$$T = P_1 \times L$$

where, $P_1$ = Effort applied at the end of handle or lever to rotate the screw.

## 4.5 DIFFERENCE BETWEEN SCREW FRICTION AND COLLAR FRICTION

| Screw friction | Collar friction |
|---|---|
| 1. It is the friction between screw and nut. | 1. It is the friction between screw and collar. |
| 2. Torque required to overcome screw friction is,<br>(a) $T_1 = W \tan(\alpha + \phi) \times \frac{d}{2}$<br>... Raising the load<br>(b) $T_1 = W \tan(\phi - \alpha) \times \frac{d}{2}$<br>... Lowering the load. | 2. Torque required to overcome the collar friction is,<br>(a) $T_2 = \frac{2}{3} \times \frac{\mu_1 \times W \times (R_1^3 - R_2^3)}{(R_1^2 - R_2^2)}$ (Uniform pressure condition)<br>Where, $R_1$ and $R_2$ are outer and inner radii of collar respectively.<br>(b) $T_2 = \mu_1 \times W \times R$ (Uniform wear condition)<br>Where, R is mean radius of collar $= \frac{R_1 + R_2}{2}$, and<br>$\mu_1$ = coefficient of friction between screw & collar surface. |

## 4.6 VARIOUS TYPES OF STRESSES INDUCED IN POWER SCREW

1. **Direct compressive stress due to axial load W :**

- $$\sigma_c = \frac{W}{\frac{\pi}{4} \cdot d_c^2}$$

2. **Torsional shear stress :**

- We know that, the torque transmitted by screw is given by,

$$T = \frac{\pi}{16} \times \tau \times d_c^3$$

$\therefore$ Torsional shear stress, $$\tau = \frac{16\,T}{\pi \times d_c^3}$$

- When screw is subjected to both stresses simultaneously, i.e. direct stress (compressive type) and torsional shear stress, then design must be based on,

(a) Maximum shear stress theory, according to which, maximum shear stress on minimum cross-section is,

$$\tau_{max} = \frac{1}{2}\sqrt{\sigma_c^2 + 4\,\tau^2}$$

(b) Maximum normal stress theory, according to which, maximum normal stress on minimum cross-section is,

$$\sigma_{c\,max} = \frac{1}{2}\sigma_c + \frac{1}{2}\sqrt{\sigma_c^2 + 4\,\tau^2}$$

3. **Transverse shear stress due to axial load :**

- Transverse shear stress for screw,

$$\tau_{screw} = \frac{\text{Load}}{\text{Area}} = \frac{W}{\pi\, d_c \cdot n \cdot t}$$

- Transverse shear stress for nut,

$$\tau_{nut} = \frac{\text{Load}}{\text{Area}} = \frac{W}{\pi\, d_o \cdot n \cdot t}$$

4. **Bearing pressure :**

- In order to reduce the wear of screw and nut, bearing pressure on thread surfaces must lie within permissible limit. It is given by,

$$P_b = \frac{\text{Load}}{\text{Bearing area}} = \frac{W}{\frac{\pi}{4}(d_o^2 - d_c^2)\,n} \quad \text{OR} \quad \frac{W}{\pi\, d \cdot n \cdot t}$$

where $t$ = Thickness or Width of screw $= \frac{p}{2}$ ... (4.15)

$n$ = Number of threads of screw in contact with the nut

$$\therefore \quad n = \frac{\text{Height of nut}}{\text{Pitch of threads}} = \frac{H}{p} \quad \text{... (4.16)}$$

## 4.7 EFFICIENCY OF SCREW

- We know that, actual effort required at the circumference of screw to lift the load is given by,

$$P = W \tan(\alpha + \phi) \quad \text{in 'N'} \qquad \ldots (4.17)$$

where, $W$ = Load to be lifted in 'N'

$\phi$ = Angle of friction in 'degrees'

$\alpha$ = Helix angle in 'degrees'

- If there is no friction between the screw and nut, then $\phi = 0°$.

So, the above equation (4.17) reduces to give ideal effort as,

$$P_o = W \tan \alpha \quad \text{in 'N'} \qquad \ldots (4.18)$$

Ideal effort means the effort applied, when friction is absent.

- **Efficiency of screw** is defined as, '*the ratio of ideal effort to actual effort*'.

∴ Efficiency of screw is given by,

$$\eta = \frac{P_o}{P} = \frac{W \tan \alpha}{W (\alpha + \phi)} = \frac{\tan \alpha}{\tan (\alpha + \phi)}$$

- Therefore, we conclude that, efficiency of screw is independent of load to be lifted (W), but it depends upon helix angle ($\alpha$).

### 4.7.1 Overall Efficiency of Screw Jack

- Total torque required to raise the load is the algebraic sum of

1. Torque required to overcome friction between screw and nut ($T_1$), and
2. Torque required to overcome collar or bearing friction ($T_2$).

∴ Total torque, $T = T_1 + T_2$

where, $T_1 = W \tan(\alpha + \phi) \cdot \frac{d}{2}$

and $T_2 = \mu_1 \cdot W \cdot R$ (Assuming uniform wear condition)

$= \frac{2}{3} \mu_1 \cdot W \cdot \frac{(R_1^3 - R_2^3)}{(R_1^2 - R_2^2)}$ (Assuming uniform pressure condition)

where, $R_1$ = Outer radius of collar = Radius of head ($R_3$) of screw jack

$R_2$ = Inner radius of collar = Radius of pin ($R_4$) of screw jack

$R$ = Mean radius = $\frac{R_1 + R_2}{2}$

- **Overall efficiency** is defined as, "**ratio of ideal torque to actual torque**".

Mathematically,

$$\eta_o = \frac{\text{Ideal torque}}{\text{Actual torque}} = \frac{T_o}{T} = \frac{T_o}{W \tan(\alpha + \phi) \cdot \frac{d}{2} + \mu_1 \cdot W \cdot R} \qquad \ldots (4.19)$$

- In case of ideal torque, friction will be absent. Therefore, $\phi = 0$ and $\mu_1 = 0$. Thus, ideal torque is given by,

$$T_o = W \tan(\alpha + \phi) \cdot \frac{d}{2} = W \tan(\alpha + 0) \cdot \frac{d}{2} + 0 = W \tan \alpha \cdot \frac{d}{2}$$

- Equation (4.19) becomes,

$$\eta_o = \frac{W \tan \alpha \cdot \frac{d}{2}}{W \tan(\alpha + \phi) \cdot \frac{d}{2} + \mu_1 WR} \qquad \ldots (4.20)$$

**OR**

$$\eta_o = \frac{T_o}{T} = \frac{T_o}{T_1 + T_2}$$

### 4.7.2 How to Improve Efficiency of Power Screws ?

- We have, Efficiency of screw $= \eta = \frac{\tan \alpha}{\tan(\alpha + \phi)}$

where, $\alpha$ is the helix angle.

- Thus, efficiency of screw can be improved by increasing the helix angle.
- The efficiency is maximum for a helix angle ranging between 40° to 45°.
- Figure 4.7 shows graph of helix angle versus efficiency of screw.

**Conclusion of Graph :**

1. Efficiency is very less, when the value of $\alpha$ is close to 0° or 90°.
2. Efficiency increases rapidly upto helix angle of 20° and remains more or less constant for the value of helix angle between 30° to 60°.
3. The efficiency is maximum for a helix angle in the range of 40° to 45°.
4. But, when the helix angle is increased beyond 60°, the efficiency drops, because the normal thread force increases, thereby increasing the force of friction.

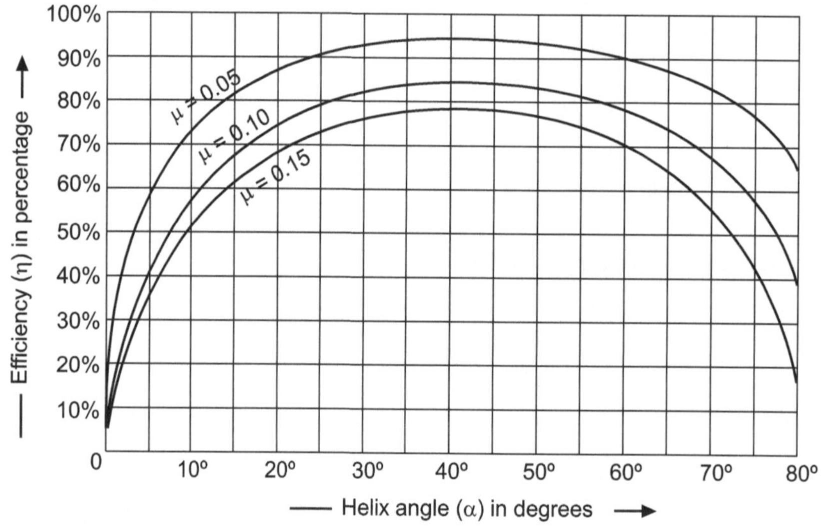


**Fig. 4.7 : Graph of 'Efficiency v/s Helix Angle'**

## 4.7.3 Self-Locking and Overhauling of Screw

- We know that, the effort required at circumference of screw to *lower the load* is,

$$P = W \tan(\phi - \alpha)$$

- Torque required to lower the load is,

$$T = P \times \frac{d}{2} = W \tan(\phi - \alpha) \times \frac{d}{2}$$

- In the above expression, if $\phi < \alpha$, then torque required to lower the load will be negative.
- In other words, the load will start moving downward without application of any torque. Such a condition is called as *overhauling*.
- However, if $\alpha < \phi$, then torque required will be positive, indicating that some effort is required to be applied to lower the load. Such a condition is called as *self-locking*.
- Efficiency of screw $= \eta = \dfrac{\tan \alpha}{\tan(\alpha + \phi)}$ ... (4.21)
- For self locking of screw, the condition is $\alpha \leq \phi$.

  $\therefore$ By substituting $\alpha = \phi$ in equation (4.21), efficiency for self-locking screw becomes,

$$\eta = \frac{\tan \phi}{\tan 2\phi} = \frac{\tan \phi}{\frac{2 \tan \phi}{1 - \tan^2 \phi}} = \frac{1 - \tan^2 \phi}{2} = \frac{1}{2} - \frac{\tan^2 \phi}{2} = 0.5 - \frac{\tan^2 \phi}{2}$$

- In terms of percentage,

$$\eta = \left(0.5 - \frac{\tan^2 \phi}{2}\right) \times 100 = \left[50 - \left(\frac{100 \tan^2 \phi}{2}\right)\right] \%$$

$$\eta = [50 - 50 \tan^2 \phi] \%$$

- This means that, efficiency will always come less than 50% for self locking screw.

- *Therefore, condition for self-locking is $\phi > \alpha$, and efficiency to be less than 50%.*
- Whereas, $\phi < \alpha$ is the condition for overhauling.
- *Also, if efficiency is more than 50%, the screw is said to be overhauling.*

## 4.7.4 Application of Overhauling Screw

- **Hand press** is an application of overhauling screw.
- In case of hand presses, the torque required to lower the load should be negative. It is possible only, when $\phi < \alpha$.
- Thus, the load will start moving downward without any application of torque. Such a condition is known as overhauling of screws.

**Important Formulae**

(1) Nominal diameter = Core diameter + Pitch

$$d_o = d_c + p$$

(2) Mean diameter, $d = \dfrac{d_o + d_c}{2}$

$$d = d_o - \frac{p}{2}$$

$$d = d_c + \frac{p}{2}$$

(3) Coefficient of friction between screw and nut = $\mu = \tan\phi$.

(4) Coefficient of friction between screw and collar = $\mu_1$.

(5)
$$\tan\alpha = \frac{\text{Lead}}{\pi \cdot d}$$

where, Lead = Pitch 'p' [For single start square thread]

$= 2 \times p$ [For 2-start or double start square thread]

$= 3 \times p$ [For 3-start or triple start square thread]

$= 4 \times p$ [For quadruple start square thread]

(6) (i) Torque required to overcome friction between screw and nut at the time of lifting the load :

$$T_1 = W\tan(\alpha + \phi) \times \frac{d}{2}$$

(ii) Torque required to overcome friction between screw and nut at the time of lowering the load

$$T_1 = W\tan(\phi - \alpha) \cdot \frac{d}{2}$$

(7) Torque required to overcome collar friction :

(i) $T_2 = \mu_1 WR$ ... [Assuming uniform wear condition]

(ii) $T_2 = \dfrac{2}{3}\mu_1 \cdot W \cdot \left[\dfrac{R_1^3 - R_2^3}{R_1^2 - R_2^2}\right]$ ... [Assuming uniform pressure condition]

where,

$W$ = Load in 'N'

$\alpha$ = Helix angle in 'degrees'

$\phi$ = Angle of friction in 'degrees'

$R_1$ = External or outer radius of collar in 'mm'

$R_2$ = Internal or inner radius of collar in 'mm'

$R$ = Mean radius of collar = $\dfrac{R_1 + R_2}{2}$ in 'mm'

$\mu_1$ = Coefficient of collar friction/bearing friction

(8) Power required to rotate screw,

$$P = \frac{2\pi NT}{60} \quad \text{in 'Watts'}$$

where,

$N$ = Number of revolutions made by screw per minute (r.p.m.)

$T$ = Torque in 'N.m'

(9) $\begin{bmatrix}\text{Number of revolutions} \\ \text{made by screw} \\ \text{per minute}\end{bmatrix} = N = \dfrac{\text{Linear speed or Cutting speed}}{\text{Pitch of screw}}$ in 'r.p.m.'

(10) Efficiency of screw,

$$\eta = \frac{\text{Ideal torque }(T_o)}{\text{Actual torque }(T)} = \frac{T_o}{T_1 + T_2}$$

(a) When collar/bearing friction is neglected (i.e. $T_2 = 0$), then

$$\eta = \frac{\tan\alpha}{\tan(\alpha + \phi)}$$

(b) When collar/bearing friction is considered, then

$$\eta = \frac{T_o}{T_1 + T_2} = \frac{W \tan\alpha \times \frac{d}{2}}{W \tan(\alpha + \phi) \times \frac{d}{2} + \mu_1 \cdot W \cdot R}$$

The above equation is also called as formula to calculate **overall efficiency of screw jack ($\eta_o$).**

**Note :** To calculate overall efficiency, the same expression is used i.e. 10 (b), where $\eta$ is nothing but $\eta_o$.

(11) Stresses induced in various parts of screw jack are :

**(i) For screw :**

(a) Compressive stress, $\sigma_c = \dfrac{W}{\frac{\pi}{4} d_c^2}$

(b) Torsional shear stress, $\tau = \dfrac{16T_1}{\pi d_c^3}$

(c) Principal stresses :

(i) Maximum shear stress = $\tau_{max} = \dfrac{1}{2} \times \sqrt{\sigma_c^2 + 4\tau^2}$

(ii) Maximum normal stress = $\sigma_{c\,max} = \dfrac{1}{2}\sigma_c + \dfrac{1}{2} \times \sqrt{\sigma_c^2 + 4\tau^2}$

(d) Transverse shear stress for screw, $\tau_{screw} = \dfrac{W}{\pi d_c \cdot n \cdot t}$

**(ii) For nut :**

(a) Bearing pressure, $P_b = \dfrac{W}{\frac{\pi}{4} \cdot (d_o^2 - d_c^2) \times n}$

(b) Transverse shear stress for nut, $\tau_{nut} = \dfrac{W}{\pi \cdot d_o \cdot n \cdot t}$

(12) Height of nut, $H = n \times p$

(13) Thickness of thread, $t = \dfrac{p}{2}$.

**Numerical Type No. 1 : "Neglecting Bearing or Collar Friction"**

**Problem 4.1 :** *A lead screw of lathe has square threads of 24 mm outside diameter and 5 mm pitch. In order to drive tool carriage, the screw exerts an axial pressure of 2.5 kN. Find the efficiency of single threaded screw and power required to drive the screw, if it rotates at 300 r.p.m. Neglect the bearing friction. Assume coefficient of friction for screw thread as 0.12.*

**Solution : Given data :** $d_o$= 24 mm, p = 5 mm, **W** = 2.5 kN = $2.5 \times 10^3$ N = **2500 N**, $\mu$ = 0.12, N = 300 r.p.m.

**Procedure :** We have,

$$\mathbf{d_c} = d_o - p = 24 - 5 = \mathbf{19\ mm}$$

$$\therefore \quad \mathbf{d} = \frac{d_o + d_c}{2} = \frac{24 + 19}{2} = \mathbf{21.5\ mm}$$

$\therefore \quad \tan\alpha = \frac{p}{\pi d} = \frac{5}{\pi \times 21.5} = \mathbf{0.07402}$

$\therefore \quad \alpha = \tan^{-1}(0.07402) = \mathbf{4.233°}$

Also, $\mu = \tan\phi = 0.12$

$\therefore \quad \phi = \tan^{-1}(0.12) = \mathbf{6.843°}$

**Note for students :** Always try to take at least 3 digits after dot, while calculating angles $\alpha$ and $\phi$ for accurate answers.

Torque required to overcome the friction between screw and nut is given by,

$$T_1 = P \times \frac{d}{2} = W \tan(\alpha + \phi) \times \frac{d}{2} = 2500 \times \tan(4.233 + 6.843) \times \frac{21.5}{2}$$

$\therefore \quad \mathbf{T_1} = 5260.97 \text{ N-mm} \cong \mathbf{5.261 \text{ N-m}}$

As we have to neglect the bearing or collar friction, the torque required ($T_2$) to overcome friction between screw and collar bearing will be zero. It means that, $T_2 = 0$.

$\therefore$ Total torque required to raise the load is,

$$\mathbf{T} = T_1 + T_2 = 5.261 + 0 = \mathbf{5.261 \text{ N-m}}$$

**1. Efficiency of the screw :**

$$\eta = \frac{\tan\alpha}{\tan(\alpha + \phi)} = \frac{\tan(4.233)}{\tan(4.233 + 6.843)} \times 100 = \mathbf{37.81\%}$$

**2. Power required to rotate the screw :**

$$P = \frac{2\pi NT}{60} = \frac{2 \times \pi \times 300 \times 5.261}{60} = 165.28 \text{ W} = \mathbf{0.165 \text{ kW}}$$

---

**Problem 4.2 :** *An electric motor driven power screw moves a nut in a horizontal plane against a force of 100 kN at a speed of 360 mm/min. Screw has pitch 8 mm on a major diameter of 50 mm. Coefficient of friction is 0.15.*

1. *Estimate power of the motor.*
2. *State whether the screw is self-locking or over-hauling ?*

**Solution : Given Data :** $d_o = 50$ mm, $p = 8$ mm,

$\mathbf{W} = 100 \text{ kN} = \mathbf{100 \times 10^3 \text{ N}}$, $\mu = 0.15$, Linear speed = 360 mm/min.

**Procedure :**

We have, $\mathbf{d_c} = d_o - p = 50 - 8 = \mathbf{42 \text{ mm}}$

And, $\mathbf{d} = \frac{d_o + d_c}{2} = \frac{50 + 42}{2} = \mathbf{46 \text{ mm}}$

We know that, screw travels upwards or downwards by a distance equal to pitch.

Therefore, Speed = Pitch of screw × Number of revolutions made by screw per minute

$$\therefore \quad \begin{bmatrix} \text{Number of revolutions} \\ \text{made by screw} \\ \text{per minute} \end{bmatrix} = N = \frac{\text{Linear speed or Cutting speed}}{\text{Pitch of screw}} \quad \text{in 'r.p.m.'}$$

$$\therefore \quad \mathbf{N} = \frac{\text{Speed}}{\text{Pitch of screw}} = \frac{360 \text{ mm/min}}{8 \text{ mm}} = \mathbf{45 \text{ r.p.m.}}$$

We know, $\tan\alpha = \frac{\text{Lead}}{\pi d} = \frac{p}{\pi d}$

$$\therefore \quad \boldsymbol{\alpha} = \tan^{-1}\left(\frac{p}{\pi d}\right) = \tan^{-1}\left(\frac{8}{\pi \times 46}\right) = \mathbf{3.1685°}$$

Also, $\mu = \tan\phi$

$\therefore \quad \boldsymbol{\phi} = \tan^{-1}(\mu) = \tan^{-1}(0.15) = \mathbf{8.5307°}$

Torque required to overcome the friction between screw and nut is given by,

$$T_1 = P \times \frac{d}{2} = W \times \tan(\alpha + \phi) \cdot \frac{d}{2}$$

$$\therefore \quad \mathbf{T_1} = 100 \times 10^3 \times \tan(3.1685 + 8.5307) \times \frac{46}{2} = \mathbf{476.27 \times 10^3 \text{ N.mm} = 476.27 \text{ N.m}}$$

As we have to neglect the bearing or collar friction, the torque ($T_2$) required to overcome friction between screw and collar will be, $\mathbf{T_2 = 0}$.

∴ Total torque required to raise the load, $T = T_1 + T_2$

∴ $\mathbf{T} = 476.27 + 0 = \mathbf{476.27\ N.m}$

**(1) Power of the motor :**

Power required to rotate the screw is given by,

$$P = \frac{2\pi NT}{60} = \frac{2 \times \pi \times 45 \times 476.27}{60} = 2244.36\ W = \mathbf{2.244\ kW}$$

**(2) Self locking or Overhauling :**

Efficiency of the screw is,

$$\eta = \frac{\tan \alpha}{\tan (\alpha + \phi)} = \frac{\tan (3.1685)}{\tan (3.1685 + 8.5307)} = \mathbf{0.2673 = 26.73\%}$$

As the efficiency is less than 50%, therefore screw is **self-locking**.

---

**Problem 4.3 :** *A double start square threaded power screw of nominal diameter 100 mm and pitch 12 mm is to be used to raise the load of 300 kN. The coefficient of friction at screw thread is 0.15. Neglect collar friction. Calculate :*

1. *Torque required to raise the load.*
2. *Efficiency of the screw.*

**Solution : Given data :** $d_o = 100$ mm, p = 12 mm, **W** = 300 kN = $\mathbf{300 \times 10^3\ N}$, $\mu = 0.15$

**Procedure :** For double start square thread,

Lead of screw, $\boldsymbol{l} = 2 \times p = 2 \times 12 = \mathbf{24\ mm}$

Also, $d_c = d_o - p = 100 - 12 = \mathbf{88\ mm}$

∴ Mean diameter, $\mathbf{d} = \frac{d_o + d_c}{2} = \frac{100 + 88}{2} = \mathbf{94\ mm}$

We have, $\tan \alpha = \frac{\text{lead}}{\pi d}$

∴ $\alpha = \tan^{-1}\left(\frac{24}{\pi \times 94}\right) = \mathbf{4.6463°}$

Also, $\tan \phi = \mu$

∴ $\phi = \tan^{-1}(\mu) = \tan^{-1}(0.15) = \mathbf{8.5307°}$

**(1) Torque required to raise the load :**

Torque required to overcome the friction between screw and nut,

$$T_1 = P \times \frac{d}{2} = W \tan (\alpha + \phi) \cdot \frac{d}{2}$$

∴ $\mathbf{T_1} = 300 \times 10^3 \times \tan (4.6463 + 8.5307) \times \frac{94}{2} = 3301154.225\ N.mm = \mathbf{3301.15 \times 10^3\ N.mm}$

Neglect the torque required to overcome collar friction, therefore, $T_2 = 0$

Total torque (T) required to raise the load = $T_1 + T_2$

∴ $\mathbf{T} = T_1 + T_2 = 3301.15 \times 10^3 + 0 = \mathbf{3301.15 \times 10^3\ N.mm}$

**(2) Efficiency of screw :**

$$\eta = \frac{\tan \alpha}{\tan (\alpha + \phi)} = \frac{\tan (4.6463)}{\tan (4.6463 + 8.5307)} = \mathbf{0.3471} \text{ or } \mathbf{34.71\%}$$

---

## Numerical Type No. 2 : " Considering Bearing or Collar Friction"

**Problem 4.4 :** *The mean diameter of a square threaded screw having pitch of 10 mm is 50 mm. A load of 20 kN is lifted to a distance of 170 mm. Find the work done in lifting load and efficiency of screw, when (a) load rotates with screw, (b) load rests on loose head, which does not rotate with the screw.*

*The internal and external diameters of bearing surface are 10 mm and 60 mm respectively and μ for all may be taken as 0.08.*

**Solution : Given data :** d = 50 mm, p = 10 mm, **W** = 20 kN = $\mathbf{20 \times 10^3\ N}$, Lift = 170 mm

$\mu = 0.08$, $\mu_1 = 0.08$, $D_1 = 60$ mm, $D_2 = 10$ mm

**Procedure :** We know, $\mathbf{d_o} = d + \frac{p}{2} = 50 + \frac{10}{2} = \mathbf{55\ mm}$

And, $\mathbf{d_c} = d_o - p = 55 - 10 = \mathbf{45\ mm}$

Also, $R_1 = \frac{D_1}{2} = \frac{60}{2}$ and $R_2 = \frac{D_2}{2} = \frac{10}{2}$

$\therefore$ $\mathbf{R_1 = 30\ mm}$ and $\mathbf{R_2 = 5\ mm}$

$\therefore$ Mean radius of collar, $\mathbf{R} = \frac{R_1 + R_2}{2} = \frac{30 + 5}{2} = \frac{35}{2} = \mathbf{17.5\ mm}$

We have, $N = \frac{\text{Lift}}{\text{Pitch}} = \frac{170}{10} = \mathbf{17\ r.p.m.}$

And, $\tan \alpha = \frac{p}{\pi d} = \frac{10}{\pi \times 50}$

$\therefore$ $\alpha = \tan^{-1}\left(\frac{10}{\pi \times 50}\right) = \tan^{-1}(0.0636) = \mathbf{3.6426°}$

Also, $\mu = \tan \phi$

$\therefore$ $\phi = \tan^{-1}(\mu) = \tan^{-1}(0.08) = \mathbf{4.5739°}$

Torque required to overcome the friction between screw and nut is given by,

$$T_1 = P \times \frac{d}{2} = W \times \tan(\alpha + \phi) \times \frac{d}{2}$$

$$\therefore \quad \mathbf{T_1} = 20 \times 10^3 \times \tan(3.6426 + 4.5739) \times \frac{50}{2} = \mathbf{72.198 \times 10^3\ N.mm}$$

**Case I :** As the load rotates with the screw, it means, no bearing or collar is provided. Therefore, neglecting the bearing or collar friction, the torque ($T_2$) required to overcome friction between screw and collar will be zero.

i.e. $T_2 = 0$

$\therefore$ Total torque required to raise the load is,

$$\mathbf{T} = T_1 + T_2 = (72.198 \times 10^3) + 0 = \mathbf{72.198 \times 10^3\ N.mm} \text{ or } \mathbf{72.198\ N.m}$$

**(1) Work done in lifting load :**

Work done in lifting load is the power required to rotate the screw, which is given by,

$$P = \frac{2\pi NT}{60} = \frac{2 \times \pi \times 17 \times 72.198}{60} = \mathbf{128.53\ W = 0.128\ kW}$$

**(2) Efficiency of the screw :**

$$\eta = \frac{\tan \alpha}{\tan(\alpha + \phi)} = \frac{\tan(3.6426)}{\tan(3.6426 + 4.5739)} = \mathbf{0.4409} \text{ or } \mathbf{44.09\%}$$

**Case II :** As the load does not rotate with the screw, it means that collar or bearing is provided on the top of screw. Therefore, we must consider the torque required to overcome the bearing or collar friction. It is given by,

$$T_2 = \mu_1 \times W \times R \quad \text{... (Assuming uniform wear condition)}$$

$$\therefore \quad \mathbf{T_2} = 0.08 \times 20 \times 10^3 \times 17.5 = \mathbf{28 \times 10^3\ N.mm}$$

$\therefore$ Total torque required to raise load is,

$$\mathbf{T} = T_1 + T_2 = (72.198 \times 10^3) + (28 \times 10^3) = \mathbf{100.198 \times 10^3\ N.mm = 100.198\ N.m}$$

**(1) Work done in lifting load :**

Work done in lifting load is the power required to rotate the screw, which is given by,

$$P = \frac{2\pi NT}{60} = \frac{2 \times \pi \times 17 \times 100.198}{60} = \mathbf{178.38\ W = 0.178\ kW}$$

**(2) Efficiency of the screw :**

Since bearing or collar is provided at the top of screw, torque ($T_2$) required to overcome collar friction must be considered. Due to this, efficiency of screw along with collar friction must be calculated, which is called as **overall efficiency of screw jack**.

$$\eta_o = \frac{\text{Ideal torque}}{\text{Actual torque}} = \frac{T_o}{T} = \frac{W \times \tan \alpha \cdot \frac{d}{2}}{T} = \frac{20 \times 10^3 \times \tan(3.6426) \times \frac{50}{2}}{100.198 \times 10^3} = \mathbf{0.3176} \text{ or } \mathbf{31.76\%}$$

**Problem 4.5 :** *A vertical screw with single start square thread of 50 mm mean diameter and 12.5 mm pitch is raised against a load of 10 kN by means of hand wheel. The boss of wheel is threaded to act as a nut. The axial load is taken up by a thrust collar, which supports the wheel boss and has mean diameter 60 mm. The coefficient of friction is 0.15 for screw and 0.18 for collar. If tangential force applied by each hand of the wheel is 100 N, find suitable diameter of hand wheel.*

**Solution : Given data :** $d = 50$ mm, $p = 12.5$ mm, $W = 10$ kN $= \mathbf{10 \times 10^3}$ **N**,

$\mu = 0.15$, $\mu_1 = 0.18$, $D = 60$ mm, $P_1 = 100$ N [Effort applied by each hand]

**Procedure :** We know,

$$\mathbf{d_o} = d + \frac{p}{2} = 50 + \frac{12.5}{2} = \mathbf{56.25\ mm},$$

$$\mathbf{d_c} = d_o - p = 56.25 - 12.5 = \mathbf{43.75\ mm},$$

$$\mathbf{R} = \frac{D}{2} = \frac{60}{2} = \mathbf{30\ mm}$$

And,

$$\tan\alpha = \frac{\text{Lead}}{\pi d} = \frac{p}{\pi d} = \frac{12.5}{\pi \times 50} = 0.07957 \quad \begin{bmatrix}\text{For single start} \\ \text{square threads}\end{bmatrix}$$

$\therefore$

$$\alpha = \tan^{-1}(0.07957) = \mathbf{4.5498°}$$

Also,

$$\mu = \tan\phi = 0.15$$

$\therefore$

$$\phi = \tan^{-1}(0.15) = \mathbf{8.5307°}$$

Torque required to overcome the friction between screw and nut is given by,

$$T_1 = P \times \frac{d}{2} = W \tan(\alpha + \phi) \times \frac{d}{2}$$

$\therefore$

$$\mathbf{T_1} = 10 \times 10^3 \times \tan(4.5498 + 8.5307) \times \frac{50}{2} = \mathbf{58.087 \times 10^3\ N.mm}$$

As the load does not rotate with the screw, the torque required to overcome the bearing or collar friction is given by,

$$T_2 = \mu_1 \cdot W \cdot R \quad \text{(Assuming uniform wear condition)}$$

$\therefore$

$$\mathbf{T_2} = 0.18 \times 10 \times 10^3 \times 30 = \mathbf{54 \times 10^3\ N.mm}$$

$\therefore$ Total torque required to raise the load is,

$$\mathbf{T} = T_1 + T_2 = 58.087 \times 10^3 + 54 \times 10^3 = \mathbf{112.087 \times 10^3\ N.mm}$$

Diameter of handle ($D_H$) may be obtained by,

Torque required = Effort applied by hand wheel × Radius of hand wheel

$\therefore$

$$T = P_1 \times \frac{D_H}{2}$$

$\therefore$

$$112.087 \times 10^3 = (2 \times 100) \times \frac{D_H}{2}$$

(Effort is taken as twice, as it is applied by each hand of wheel)

$\therefore$ Diameter of hand wheel $= D_H = \mathbf{1120.87\ mm \cong 1122\ mm\ (say)}$

---

**Problem 4.6 :** *The spindle of a screw jack has single start square thread with outside diameter 45 mm and a pitch of 10 mm. The spindle moves in a nut. The load is carried on a swivel head, but it is not free to rotate. The bearing surface of swivel has a mean diameter of 60 mm. Coefficient of friction between screw and nut is 0.2 and between the swivel head and spindle is 0.9. Calculate the load, which can be raised by effort of 100 N each applied at the end of two levers, each of effective length 350 mm. Also, determine efficiency of lifting arrangement.*

**Solution : Given data :** $d_o = 45$ mm, $p = 10$ mm, $\mu = 0.2$, $\mu_1 = 0.9$,

$D = 60$ mm, $P_1 = 100$ N (applied at end of two levers), $L = 350$ mm

**Procedure :** We know,

$$\mathbf{d} = d_o - \frac{p}{2} = 45 - \frac{10}{2} = \mathbf{40\ mm}$$

$$\mathbf{d_c} = d_o - p = 45 - 10 = \mathbf{35\ mm}$$

$$\mathbf{R} = \frac{D}{2} = \frac{60}{2} = \mathbf{30\ mm}$$

We have,

$$\tan\alpha = \frac{\text{Lead}}{\pi d} = \frac{p}{\pi d} \quad \text{[For single start square thread]}$$

$\therefore \quad \alpha = \tan^{-1}\left(\frac{p}{\pi d}\right) = \tan^{-1}\left(\frac{10}{\pi \times 40}\right) = \mathbf{4.5498°}$

Also, $\quad \mu = \tan \phi$

$\therefore \quad \phi = \tan^{-1}(\mu) = \tan^{-1}(0.2) = \mathbf{11.3099°}$

**(1) Load to be raised :**

Torque required to overcome the friction between screw and nut is given by,

$$T_1 = P \times \frac{d}{2} = W \times \tan(\alpha + \phi) \times \frac{d}{2} = W \times \tan(4.5498 + 11.3099) \times \frac{40}{2} = \mathbf{5.682\ 'W'}$$

As the load is carried on a swivel head and not free to rotate, the torque required to overcome the bearing or collar friction is given by,

$T_2 = \mu_1 \cdot W \cdot R$ (Assuming uniform wear condition)

$\therefore \quad T_2 = 0.9 \times W \times 30 = \mathbf{27\ 'W'}$ in 'N.mm'

$\therefore$ Total torque required to raise the load is,

$T = T_1 + T_2 = 5.682\ 'W' + 27\ 'W' = \mathbf{32.682\ 'W'}$ in 'N.mm'

We have, Torque required = Effort applied by each lever × Effective length of lever

$\therefore \quad T = P_1 \times L$

$\therefore \quad 32.682\ W = (2 \times 100) \times 350$ ... ($\because$ There are two levers)

$\therefore \quad W = \text{Load raised} = \mathbf{2141.85\ N}$

**(2) Efficiency of the screw :**

We have, $\quad \eta = \frac{\text{Ideal torque}}{\text{Actual torque}}$

$\therefore \quad \eta = \frac{T_o}{T} = \frac{W \times \tan\alpha \times d/2}{32.862\ W} = \frac{\tan(4.5498°) \times 40/2}{32.862} = \mathbf{0.0484 \text{ or } 4.84\%}$

---

**Problem 4.7 :** *A vertical 2-start square threaded screw of 120 mm mean diameter and 24 mm pitch, supports a vertical load of 20 kN. The axial thrust in screw is taken by collar bearing of 300 mm outside and 150 mm inside diameter. Find the force required at the end of the lever, which is 400 mm long in order to lift and lower the load. Coefficient of friction for screw and nut is 0.18 and for collar bearing is 0.25.*

**Solution : Given data :** d = 120 mm, p = 24 mm, $\mu$ = 0.18, $\mu_1$ = 0.25, **W** = 20 kN = $\mathbf{20 \times 10^3}$ **N**, L = 400 mm, $D_1$ = 300 mm, $D_2$ = 150 mm.

**Procedure :** We have, $\quad \mathbf{R_1} = \frac{D_1}{2} = \frac{300}{2} = \mathbf{150\ mm}$, and $\mathbf{R_2} = \frac{D_2}{2} = \frac{150}{2} = \mathbf{75\ mm}$,

$\therefore$ Mean radius of collar,

$$\mathbf{R} = \frac{R_1 + R_2}{2} = \frac{150 + 75}{2} = \mathbf{112.5\ mm}$$

We know that, for double start or two start square thread, **Lead** = 2 × p

We have, $\quad \tan\alpha = \frac{\text{Lead}}{\pi d} = \frac{2 \times p}{\pi d} = \frac{2 \times 24}{\pi \times 120} = 0.1273$

$\therefore \quad \alpha = \tan^{-1}(0.1273) = \mathbf{7.2561°}$

Also, $\quad \mu = \tan\phi = 0.18$

$\therefore \quad \phi = \tan^{-1}(0.18) = \mathbf{10.2039°}$

**Case I : For raising the load :**

Torque required to overcome the friction between screw and nut is given by,

$$T_1 = P \times \frac{d}{2} = W \tan(\alpha + \phi) \times \frac{d}{2}$$

$$\therefore \quad \mathbf{T_1} = 20 \times 10^3 \times \tan(7.2561 + 10.2039) \times \frac{120}{2} = \mathbf{377.437 \times 10^3\ N.mm}$$

As the load is not free to rotate, therefore, torque required to overcome the bearing or collar friction is given by,

$T_2 = \mu_1 \cdot W \cdot R$ (Assuming uniform wear condition)

$\therefore \quad \mathbf{T_2} = 0.25 \times 20 \times 10^3 \times 112.5 = \mathbf{562.5 \times 10^3\ N.mm}$

$\therefore$ Total torque required to raise the load is,

$$\mathbf{T} = T_1 + T_2 = 377.437 \times 10^3 + 562.5 \times 10^3 = \mathbf{939.937 \times 10^3\ N.mm}$$

We know that,

$$\text{Torque required} = \text{Effort applied by lever} \times \text{Effective length of lever}$$

$$\therefore \quad T = P_1 \times L$$

$$\therefore \quad 939.937 \times 10^3 = P_1 \times 400$$

$$\therefore \quad \mathbf{P_1 = 2349.84\ N}$$

**Case II : For lowering the load :**

Torque required for lowering the load to overcome the friction between screw and nut is given by,

$$T_1 = P \times \frac{d}{2} = W \tan(\phi - \alpha) \times \frac{d}{2}$$

$$\therefore \quad \mathbf{T_1} = 20 \times 10^3 \times \tan(10.2039 - 7.2561) \times \frac{120}{2} = \mathbf{61.793 \times 10^3\ N}$$

In case (I), we have already calculated the torque required to overcome the bearing or collar friction as,

$$\mathbf{T_2 = 562.5 \times 10^3\ N.mm}$$

∴ Total torque required to lower the load is given as,

$$\mathbf{T} = T_1 + T_2 = 61.793 \times 10^3 + 562.5 \times 10^3 = \mathbf{624.293 \times 10^3\ N.mm}$$

We know that,

$$\text{Torque required} = \text{Effort applied by lever} \times \text{Effective length of lever}$$

$$\therefore \quad T = P_1 \times L$$

$$\therefore \quad 624.293 \times 10^3 = P_1 \times 400$$

$$\therefore \quad \mathbf{P_1 = 1560.73\ N}$$

---

**Problem 4.8 :** *A square threaded bolt of mean diameter 24 mm and pitch 5 mm is tightened by screwing a nut, whose mean diameter of bearing surface is 50 mm. The load on the bolt is 10 kN. The coefficient of friction between nut and bolt is 0.1 and between nut and bearing surface is 0.16. Find the force required at the end of 0.5 m long spanner to tighten the nut.*

**Solution : Given data :** $d = 24$ mm, $p = 5$ mm, $D = 50$ mm; $\therefore$ $\mathbf{R} = \frac{50}{2} = \mathbf{25\ mm}$

$\mathbf{W} = 10$ kN $= \mathbf{10 \times 10^3\ N}$, $\mu = 0.1$, $\mu_1 = 0.16$, $\mathbf{L} = 0.5$ m $= \mathbf{500\ mm}$

**Procedure :**

We have, $\tan\alpha = \frac{p}{\pi d}$

$$\therefore \quad \alpha = \tan^{-1}\left(\frac{p}{\pi d}\right) = \tan^{-1}\left(\frac{5}{\pi \times 24}\right) = \mathbf{3.7939°}$$

Also, $\tan\phi = \mu$

$$\therefore \quad \phi = \tan^{-1}(\mu) = \tan^{-1}(0.1) = \mathbf{5.7106°}$$

Torque ($T_1$) required to overcome the friction,

$$T_1 = P \times \frac{d}{2} = W \tan(\alpha + \phi) \times \frac{d}{2}$$

$$\therefore \quad \mathbf{T_1} = 10 \times 10^3 \times \tan(3.7939 + 5.7106) \times \frac{24}{2} = \mathbf{20.091 \times 10^3\ N.mm}$$

Assuming uniform wear condition, torque ($T_2$) required to overcome the collar friction is given by,

$$\mathbf{T_2} = \mu_1 \cdot W \cdot R = 0.16 \times 10 \times 10^3 \times 25 = \mathbf{40 \times 10^3\ N.mm}$$

∴ Total torque required to raise the load,

$$\mathbf{T} = T_1 + T_2 = (20.091 \times 10^3) + (40 \times 10^3) = \mathbf{60.091 \times 10^3\ N.mm}$$

Now, torque required at the end of spanner,

$$T = P_1 \times L$$

where, $P_1$ is the effort or force applied at the end of spanner.

$$\therefore \quad \mathbf{P_1} = \frac{T}{L} = \frac{60.091 \times 10^3}{500} = \mathbf{120.18\ N}$$

---

**Problem 4.9 :** *A machine vice has single start square threads with 22 mm nominal diameter and 5 mm pitch. The outer and inner diameter of the friction collar are 55 mm and 45 mm respectively. The coefficients of friction for thread and collar are 0.15 and 0.17 respectively. The machinist can comfortably exert a force of 125 N on the handle of mean radius of 150 mm. Assuming uniform wear for the collar, calculate :*

1. *Clamping force developed between jaws.* 2. *Overall efficiency of clamp.*

**Solution : Given data :** For screw, $d_o$ = 22 m, p = 5 mm, $D_1$ = 55 mm, $D_2$ = 45 mm, $\mu$ = 0.15, $\mu_1$ = 0.17, $P_1$ = 125 N, $R_H$ = 150 mm.

**Procedure :** We have,

$$\mathbf{d_c} = d_o - p = 22 - 5 = \mathbf{17\ mm},$$

$$\therefore \quad \mathbf{d} = \frac{d_o + d_c}{2} = \frac{22 + 17}{2} = \mathbf{19.5\ mm},$$

Also,

$$\mathbf{R_1} = \frac{D_1}{2} = \frac{55}{2} = \mathbf{27.5\ mm} \text{ and } \mathbf{R_2} = \frac{D_2}{2} = \frac{45}{2} = \mathbf{22.5\ mm}$$

$$\therefore \quad \mathbf{R} = \text{Mean radius} = \frac{R_1 + R_2}{2} = \frac{27.5 + 22.5}{2} = \mathbf{25\ mm}$$

We have,

$$\tan \alpha = \left(\frac{\text{Lead}}{\pi d}\right) = \frac{p}{\pi d} \qquad \text{(For single start square threads)}$$

$$\therefore \quad \alpha = \tan^{-1}\left(\frac{p}{\pi d}\right) = \tan^{-1}\left(\frac{5}{\pi \times 19.5}\right) = \mathbf{4.666^\circ}$$

And

$$\tan \phi = \mu$$

$$\therefore \quad \phi = \tan^{-1}(\mu) = \tan^{-1}(0.15) = \mathbf{8.5307^\circ}$$

**(1) Clamping force (W) :**

Torque required to overcome the friction between screw and nut is given by,

$$T_1 = P \times \frac{d}{2} = W \times \tan(\alpha + \phi) \times \frac{d}{2}$$

$$\therefore \quad \mathbf{T_1} = W \times \tan(4.666 + 8.5307) \times \frac{19.5}{2} = \mathbf{2.2862\ 'W'} \qquad \textbf{in 'N.mm'}$$

Also, torque required in overcoming collar friction,

$$T_2 = \mu_1 \cdot W \cdot R = 0.17 \times W \times 25 \qquad \text{(Assuming uniform wear condition)}$$

$$\therefore \quad \mathbf{T_2} = \mathbf{4.25\ 'W'} \qquad \textbf{in 'N.mm'}$$

Total torque required to raise the load, $T = T_1 + T_2$

$$\therefore \quad \mathbf{T} = 2.2862\ 'W' + 4.25\ 'W' = \mathbf{6.5362\ 'W'} \qquad \textbf{'N.mm'} \qquad \ldots(1)$$

An effort $P_1$ is applied externally by handle having mean radius $R_H$, we can write,

Torque applied,

$$\mathbf{T} = P_1 \times R_H = 125 \times 150 = \mathbf{18750\ N.mm} \qquad \ldots(2)$$

Equating equations (1) and (2), we get,

$$6.5362\ W = 18750$$

$$\therefore \quad W = \mathbf{2868.64\ N}$$

**(2) Overall efficiency :**

$$\eta_o = \frac{T_o}{T} = \frac{W \tan \alpha \cdot \frac{d}{2}}{T} = \frac{2868.4 \times \tan(4.666^\circ) \times \frac{19.5}{2}}{18750} = \mathbf{0.1217 = 12.17\%}$$

## 4.8 DESIGN OF SCREW JACK

**Brief Description of Screw Jack :**

- Screw jack is a portable device consisting of screw mechanism used to raise or lower the load.
- There are two types of jacks – Hydraulic and Mechanical.
- Mechanical jacks can be either hand operated or motor power driven.
- Hydraulic jack consists of cylinder and piston mechanism, where the movement of piston rod is used to lower or raise the load.

**Construction of a Screw Jack :**

- Mechanical screw jack consists of a screw and nut. Nut is fixed in a cast iron frame and remains stationary.
- Rotation of nut inside the frame is prevented by pressing a set screw against it.
- Screw is rotated in the nut by means of a handle, which passes through a hole in the head of a screw.
- Head carries a cup, which supports the load and remains stationary, while the screw is being rotated.
- There is a collar friction due to the annular contacting surfaces between the cup and head of the screw.
- A washer is fixed to the other end of screw inside the frame, which prevents the screw to be completely turned out of the nut.

**Design Procedure for Screw Jack :**

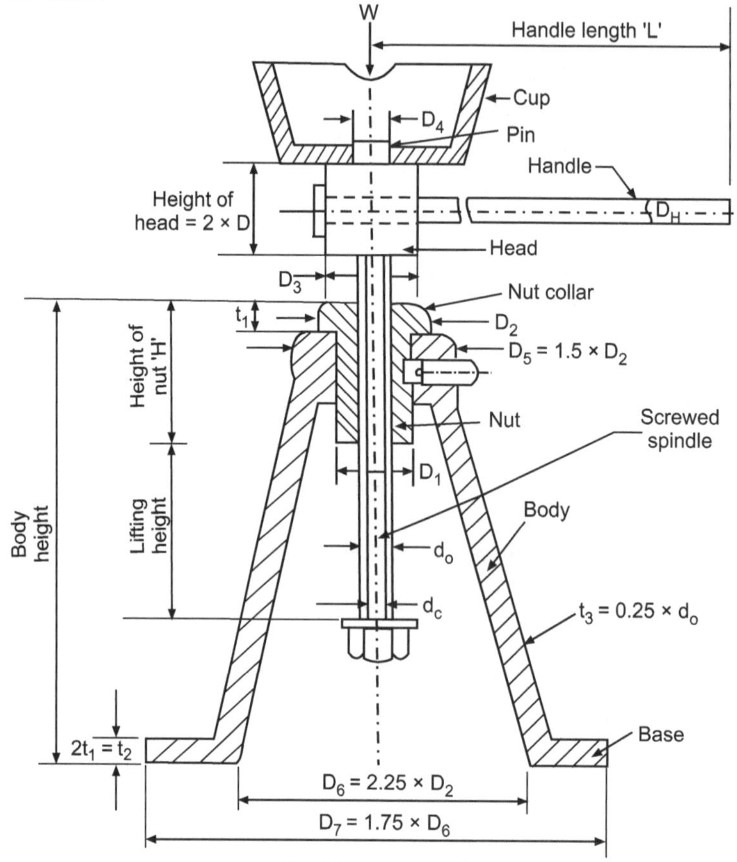


**Fig. 4.8 : Screw jack**

***(A) Design of Screw:***

**(1) Find core diameter ($d_c$):**

Considering the screw under pure compression, use the formula of direct compressive stress being induced due to axial compressive load (W) load to calculate value of $d_c$.

$$\sigma_c = \frac{W}{\frac{\pi}{4} \times d_c^2}$$

In addition to compressive stress, screw is also subjected to torsional shear stress. To account for this, increase the value of $d_c$ considerably by referring the standard table 4.1 of square threads, where, we get the values of $d_o$ and p.

Now, calculate new '$d_c$' using the formula, $d_c = d_o - p$.

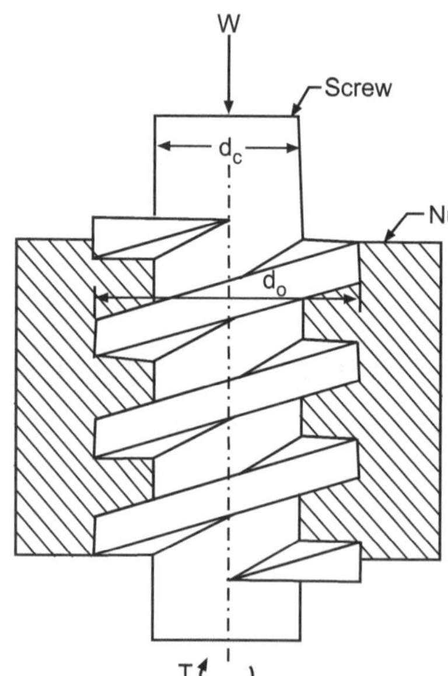


**Fig. 4.9 : Screw and nut assembly**

**2. Find torque $T_1$ required to rotate the screw and hence, find the induced shear stress ($\tau$) due to this torque :**

$$T_1 = P \times \frac{d}{2} = W \tan (\alpha + \phi) \times \frac{d}{2}$$

where, P = Effort required at circumference of screw

and d = Mean diameter of screw

We know that, torque transmitted at the circumference of screw,

$$T_1 = \frac{\pi}{16} \times \tau \times d_c^3$$

Find out the torsional shear stress due to torque $T_1$ in the screw designed in step (1).

∴ Induced torsional shear stress in screw,

$$\tau_{ind} = \frac{16\, T_1}{\pi \cdot d_c^3}$$

Also, find the induced direct compressive stress ($\sigma_c$) due to axial compressive load in the screw designed in step (1).

∴ Induced compressive stress in screw,

$$\sigma_{c\ ind} = \frac{W}{\frac{\pi}{4} \times d_c^2}$$

**3. Check maximum principal stresses :** When screw is simultaneously subjected to two types of stresses, i.e. direct stress (compressive) and torsional shear stress, then design must be based on,

(a) Maximum shear stress theory, according to which, maximum shear stress induced on minimum cross-section is,

$$\text{Induced } \tau_{max} = \frac{1}{2} \times \sqrt{\sigma_c^2 + 4\tau^2}$$

(b) Maximum normal stress theory, according to which, maximum normal stress induced on minimum cross-section is,

$$\text{Induced } \sigma_{c\ max} = \frac{1}{2}\sigma_c + \frac{1}{2} \times \sqrt{\sigma_c^2 + 4\tau^2}$$

**Note :** We can put the value of $\tau_{max}$ calculated in previous step 3 (a) in the above expression, for easy calculation of $\sigma_{c\ max}$ using the following formula.

$$\therefore \quad \sigma_{c\ max} = \frac{1}{2}\sigma_c + \tau_{max}$$

Check whether the principal stresses induced ($\tau_{max}$ and $\sigma_{c\ max}$) are less than corresponding permissible stress values or not ? If yes, then design of screw is safe.

***(B) Design of Nut :***

**4. Find the height of nut by considering the bearing pressure acting on it :**

By considering the bearing pressure on nut, we have,

$$P_b = \frac{W}{\frac{\pi}{4}(d_o^2 - d_c^2) \times n}$$

where, $n$ = Number of threads of screw in contact (engagement) with nut

(a) Calculate 'n' from the above equation.

(b) Calculate, Height of nut, $H = n \times p$

(c) Calculate thickness or width of screw thread, $t = \frac{p}{2}$.

**5. Check transverse shear stresses induced in screw and nut as follows :**

(a) Induced Transverse Shear stress for screw, $\tau_{screw} = \frac{W}{\pi \cdot d_c \cdot n \cdot t}$

(b) Induced Transverse Shear stress for nut, $\tau_{nut} = \frac{W}{\pi \cdot d_o \cdot n \cdot t}$

By using the above formulae, calculate $\tau_{screw}$ and $\tau_{nut}$ and verify that, these induced stresses are less than their permissible values.

6. **Find inner diameter ($D_1$), outer diameter ($D_2$) and thickness ($t_1$) of collar :**

(a) **Inner diameter ($D_1$) :** It is found by considering tensile failure of nut.

$$\sigma_t = \frac{W}{\frac{\pi}{4}(D_1^2 - d_o^2)}$$

(b) **Outer diameter ($D_2$) :** It is found by considering crushing failure of nut collar.

$$\sigma_{ck} = \frac{W}{\frac{\pi}{4}(D_2^2 - D_1^2)}$$

(c) **Thickness of nut collar ($t_1$) :** It is found by considering shear failure of nut collar.

$$\tau = \frac{W}{\pi D_1 t_1}$$

7. **Fix the diameter of head ($D_3$) and diameter of pin ($D_4$) fitted on top of screw :**

By using empirical relation, we can find $R_3$ and $R_4$ as under.

Diameter of head = $D_3 = 1.75\ d_o$

The cup is fitted with a pin.

Diameter of pin = $D_4 = \frac{D_3}{4}$ [Put the value of calculated $D_3$ and find out $D_4$]

Due to known values of $D_3$ and $D_4$, we can find, $R_3 = \frac{D_3}{2}$ and $R_4 = \frac{D_4}{2}$.

8. **Find torque ($T_2$) required to overcome friction at top of screw (collar friction) :**

$$T_2 = \frac{2}{3} \times \frac{\mu_1 \times W \times (R_3^3 - R_4^3)}{(R_3^2 - R_4^2)} \quad \text{(uniform pressure condition)}$$

OR $T_2 = \mu_1 \times W \times R$ (uniform wear condition)

where, $R_3$ = Radius of head, $R_4$ = Radius of pin and

Mean radius, $R = \frac{R_3 + R_4}{2}$

9. **Find the total torque, which is applied at the end of handle/lever :**

$$T = T_1 + T_2$$

Assuming that, a person can apply a force or effort ($P_1$) of 300 N intermittently,

Find the length of handle as, $L = \frac{T}{P_1} = \frac{T}{300}$ ($\because T = P_1 \times L$)

10. **Diameter of handle ($D_H$) may be obtained by considering bending moment :**

Maximum bending moment = $M = \frac{\pi}{32} \times \sigma_b \times D_H^3$

Put the value of torque (T) as maximum bending moment (M) and calculate diameter of handle ($D_H$).

11. **Height of head = H = 2D**

12. **Fix the diameter of body of screw jack as follows :**

(i) Diameter of body at top = $D_5 = 1.5 \times D_2$.

(ii) Thickness of body = $t_3 = 0.25 \times d_o$.

(iii) Inside diameter at bottom = $D_6 = 2.25 \times D_2$.

(iv) Outside diameter at bottom = $D_7 = 1.75 \times D_6$.

(v) Thickness of base, $t_2 = 2 \times t_1$.

(vi) Height of body = Maximum lift + Height of nut + 100 mm.

**Note for students :** If design of screw and nut is only asked in the examination instead of complete screw jack, then step (1) to step (5) are essential to write.

**Table 4.1 : Standard table of square threads for nominal diameter and corresponding pitch**

| Nominal diameter ($d_o$) in mm | Pitch (p) in mm | Nominal diameter ($d_o$) in mm | Pitch (p) in mm |
|---|---|---|---|
| 22 | | 55 | |
| 24 | 5 | 58 | 9 |
| 26 | | 60 | |
| 28 | | 62 | |
| 30 | | 65 | |
| 32 | 6 | 68 | |
| 34 | | 70 | |
| 36 | | 72 | 10 |
| 38 | | 75 | |
| 40 | 7 | 78 | |
| 42 | | 80 | |
| 44 | | 82 | |
| 46 | | 85 | |
| 48 | 8 | 88 | 12 |
| 50 | | 90 | |
| 52 | | 92 | |

**Numerical Type No. 3 : "Design of Screw Jack"**

**Problem 4.10 :** *A screw jack is to lift a load of 80 kN through a height of 400 mm. The elastic strength of screw material in tension and compression is 300 MPa and in shear is 120 MPa. The material for nut is phosphor bronze, for which, elastic limits are 150 MPa in tension, 135 MPa in compression and 160 MPa in shear. The bearing pressure between the nut and screw is not to be exceeding 18 N/mm². Design and draw the screw jack. Take $\mu = 0.15$, $\mu_1 = 0.2$.*

**Solution : Given data : W** = 80 kN = $\mathbf{80 \times 10^3}$ **N**, $\mu = 0.15$, $\mu_1 = 0.2$, $P_b = 18$ N/mm², Maximum lift = 400 mm

**Procedure :** Let us assume factor of safety for screw and nut material as 5.

∴ For screw,
$$\sigma_c = \frac{300}{5} = \mathbf{60\ MPa}$$
$$\tau = \frac{120}{5} = \mathbf{24\ MPa}$$

∴ For nut,
$$\sigma_t = \frac{150}{5} = \mathbf{30\ MPa}$$
$$\sigma_{ck} = \frac{135}{5} = \mathbf{27\ MPa}$$
(Assuming elastic limit for nut in crushing as 135 MPa)
$$\tau = \frac{160}{5} = \mathbf{32\ MPa}$$

**Design of Screw Jack :**

***(A) Design of Screw:***

**(1) Find core diameter ($d_c$) :**

Considering the screw under pure compression, use the formula of direct compressive stress due to axial load.

$$\sigma_c = \frac{W}{\frac{\pi}{4} \times d_c^2}$$

$$\therefore \quad 60 = \frac{80 \times 10^3}{\frac{\pi}{4} \times d_c^2}$$

i.e. $\mathbf{d_c = 41.20\ mm}$

In addition to compressive stress, screw is also subjected to torsional shear stress. Therefore, to account for this, the value of $d_c$ is increased considerably for safe design and is taken from the standard table for square threaded screw.

Let us take $\mathbf{d_o = 68^*}$ **mm** and **p = 10 mm**

*[**Note :** *For this, multiply the calculated value of $d_c$ by an arbitrarily chosen factor say 1.6. We will get, $d_c = 41.20 \times 1.6 = 65.92$ mm. Now, refer the standard table (given on previous page) for square threads and choose the value of $d_o$ higher than 65.92. Referring to the table, we can take values of $d_o$ and p as, $d_o = 68$ mm and p = 10 mm.*]

$$\therefore \quad \mathbf{d_c} = d_o - p = 68 - 10 = \mathbf{58\ mm}$$

$$\therefore \quad \mathbf{d} = \frac{d_o + d_c}{2} = \frac{68 + 58}{2} = \mathbf{63\ mm}$$

**(2) Find torque $T_1$ required to rotate the screw and find the induced shear stress ($\tau$) due to this torque :**

We have,
$$\tan\alpha = \frac{\text{Lead}}{\pi d} = \frac{p}{\pi d} = \frac{10}{\pi \times 63} = 0.05052 \qquad \text{[For single start]}$$

$$\therefore \quad \alpha = \tan^{-1}(0.05052) = \mathbf{2.8924^\circ}$$

Also,
$$\mu = \tan\phi$$

$$\therefore \quad \phi = \tan^{-1}(\mu) = \tan^{-1}(0.15) = \mathbf{8.5307^\circ}$$

Torque ($T_1$) required to rotate the screw,

$$T_1 = P \times \frac{d}{2} = W \times \tan(\alpha + \phi) \times \frac{d}{2}$$

$$\therefore \quad \mathbf{T_1} = 80 \times 10^3 \times \tan(2.8924 + 8.5307) \times \frac{63}{2} = \mathbf{509.178 \times 10^3\ N.mm}$$

Torsional shear stress ($\tau$) induced due to torque $T_1$ is,

$$\tau = \frac{16T_1}{\pi \times d_c^3} = \frac{16 \times 509.178 \times 10^3}{\pi \times (58)^3} = \mathbf{13.29\ MPa < 24\ MPa\ (Permissible\ stress)}$$

Direct compressive stress induced due to axial load,

$$\sigma_c = \frac{W}{\frac{\pi}{4} \times d_c^2} = \frac{80000}{\frac{\pi}{4} \times (58)^2} = \mathbf{30.279\ MPa < 60\ MPa\ (Permissible\ stress)}$$

Thus, induced torsional shear and compressive stresses are less than permissible limits (i.e. permissible stress values). So, design is safe.

**(3) Check maximum principal stresses :** When screw is simultaneously subjected to two types of stresses, like direct stress (compressive) and torsional shear stress, then design must be based on,

(a) Maximum shear stress theory, according to which, induced maximum shear stress on minimum cross-section is,

$$\tau_{max} = \frac{1}{2}\sqrt{\sigma_c^2 + 4\tau^2} = \frac{1}{2}\sqrt{(30.279)^2 + 4\,(13.29)^2} = \mathbf{20.145\ MPa < 24\ MPa\ (Permissible\ stress)}$$

(b) Maximum normal stress theory, according to which, induced maximum normal stress on minimum cross-section is,

$$\sigma_{c\,max} = \frac{1}{2}\sigma_c + \frac{1}{2}\sqrt{\sigma_c^2 + 4\tau^2} = \frac{1}{2} \times (30.279) + \frac{1}{2}\sqrt{(30.279)^2 + 4\,(13.29)^2} = \mathbf{35.2845\ MPa < 60\ MPa\ (P.S.)}$$

***(B) Design of Nut :***

**(4) Find the height of nut, considering the bearing pressure on nut :**

$$P_b = \frac{W}{\frac{\pi}{4}(d_o^2 - d_c^2) \times n}$$

$$\therefore \quad n = \frac{W}{\frac{\pi}{4}(d_o^2 - d_c^2) \times P_b}$$

$\therefore$ Number of threads of nut in contact with the screw,

$$\mathbf{n} = \frac{80 \times 10^3}{\frac{\pi}{4}[(68)^2 - (58)^2] \times 18} = \mathbf{4.491 \cong 5\ (say)}$$

$\therefore$ $\mathbf{H}$ = Height of nut = $n \times p = 5 \times 10 = \mathbf{50\ mm}$

and $\mathbf{t}$ = thickness or width of screw = $\frac{p}{2} = \frac{10}{2} = \mathbf{5\ mm}$

**(5) Check the transverse shear stresses induced in screw and nut as follows :**

Transverse shear stress for screw,

$$\tau_{screw} = \frac{W}{\pi\, d_c\, n\, t} = \frac{80 \times 10^3}{\pi \times 58 \times 5 \times 5} = \mathbf{17.56\ MPa < 24\ MPa\ (Permissible\ stress)}$$

Transverse shear stress for nut,

$$\tau_{nut} = \frac{W}{\pi\, d_o\, n\, t} = \frac{80 \times 10^3}{\pi \times 68 \times 5 \times 5} = \mathbf{14.98\ MPa < 32\ MPa\ \ (Permissible\ stress)}$$

Thus, induced stresses are less than permissible limits. Hence, design of screw and nut is safe.

**(6) Find inner diameter ($D_1$) and outer diameter ($D_2$) and thickness ($t_1$) of collar :**

**(a) Inner diameter ($D_1$) :** It is found by considering tensile failure of nut.

$$\therefore \quad \sigma_t = \frac{W}{\frac{\pi}{4}(D_1^2 - d_o^2)}$$

$$\therefore \quad 30 = \frac{80 \times 10^3}{\frac{\pi}{4}[D_1^2 - (68)^2]}$$

$$\therefore \quad D_1 = \mathbf{89.55\ mm \cong 90\ mm\ (say)}$$

**(b) Outer diameter ($D_2$) :** It is found by considering crushing of nut collar.

$$\sigma_{ck} = \frac{W}{\frac{\pi}{4}(D_2^2 - D_1^2)}$$

$$\therefore \quad 27 = \frac{80 \times 10^3}{\frac{\pi}{4}[D_2^2 - (90)^2]}$$

$$\therefore \quad D_2 = \mathbf{108.96\ mm \cong 110\ mm\ (say)}$$

**(c) Thickness of nut collar ($t_1$) :** It is found by considering shear failure of nut collar,

$$\tau = \frac{W}{\pi_1 \cdot D_1 \cdot t_1}$$

$$\therefore \quad 32 = \frac{80000}{\pi \times 90 \times t_1}$$

$\therefore$ Thickness of nut collar, $t_1$=**8.842 mm ≅ 10 mm (say)**

**(7) Using empirical relations, calculate diameter of head ($D_3$) and diameter of pin ($D_4$) fitted on top of screw :**

Head diameter, $D_3 = 1.75\, d_o = 1.75 \times 68 = \mathbf{119\ mm}$

$$\therefore \quad \mathbf{R_3} = \frac{D_3}{2} = \frac{119}{2} = \mathbf{59.5\ mm}$$

The cup is fitted with a pin of diameter $D_4$, which is calculated by,

$$D_4 = \frac{\text{Head diameter}}{4} = \frac{D_3}{4} = \frac{119}{4} = \mathbf{29.75\ mm}$$

$$\therefore \quad \mathbf{R_4} = \frac{D_4}{2} = \frac{29.75}{2} = \mathbf{14.875\ mm}$$

**(8) Find torque required ($T_2$) to overcome friction at the top of screw :**

We have, $$\mathbf{R} = \frac{R_3 + R_4}{2} = \frac{59.5 + 14.875}{2} = \mathbf{37.1875\ mm}$$

Also, we know, $T_2 = \mu_1 \cdot W \cdot R$ (Assuming uniform wear condition)

$\therefore$ $\mathbf{T_2} = 0.2 \times 80000 \times 37.1875 = \mathbf{595 \times 10^3\ N.mm}$

**(9) Find the total torque, to which, handle will be subjected :**

$$\mathbf{T} = T_1 + T_2 = (509.178 \times 10^3) + (595 \times 10^3) = \mathbf{1104.178 \times 10^3\ N.mm}$$

**(10) Find out length of handle by considering torque :**

Assuming that a person can apply a force of 300 N intermittently, length of handle required is calculated as,

$$T = P_1 \times L = 300 \times L \qquad [\because \text{ Effort applied, } P_1 = 300 \text{ N}]$$

$$\therefore \quad \mathbf{L} = \frac{T}{300} = \frac{(1104.178 \times 10^3)}{300} = \mathbf{3680.59\ mm \cong 3682\ mm = 3.682\ m\ (say)}$$

**(11) Find out diameter of handle ($D_H$) by considering bending moment :**

$$\text{Maximum bending moment} = M = T = \frac{\pi}{32} \times \sigma_b \times D_H^3$$

$$\therefore \quad 1104.178 \times 10^3 = \frac{\pi}{32} \times 30 \times D_H^3 \qquad (\text{Assume } \sigma_b = 30 \text{ MPa})$$

$$\therefore \quad D_H = \mathbf{72.10\ mm \cong 74\ mm\ (say)}$$

**(12) Using empirical relations, we can fix the dimensions of body of screw jack as follows :**

Diameter of body at top = $D_5 = 1.5 \times D_2 = 1.5 \times 110 = \mathbf{165\ mm}$

Thickness of body = $t_3 = 0.25 \times d_o = 0.25 \times 68 = \mathbf{17\ mm}$

Inside diameter of body at bottom = $D_6 = 2.25 \times D_2 = 2.25 \times 110 = \mathbf{247.5\ mm}$

Outside diameter of body at bottom = $D_7 = 1.75 \times D_6 = 1.75 \times 247.5 = \mathbf{433.125\ mm}$

Thickness of base = $t_2 = 2 \times t_1 = 2 \times 12 = \mathbf{24\ mm}$

Height of body = Maximum lift + Height of nut + 100 mm

= 400 + 50 +100 = **550 mm**

---

**Problem 4.11 :** *A screw jack is used to lift a load of 50 kN through a maximum lift of 200 mm. The material used for a screw is steel of allowable stresses in tension and compression as 100 N/mm² and 50 N/mm² respectively. The pitch of screw is 8 mm. The nut is made of phosphor bronze with allowable stresses as 50 N/mm² and 45 N/mm² in tension and crushing. The allowable shear stress is not to exceed 20 N/mm². If the coefficient of friction between screw and nut is 0.14, design the screw and nut.*

**Solution : Given data : W** = 50 kN = $\mathbf{50 \times 10^3\ N}$, p = 8 mm

For screw material (steel), $\sigma_t = 100$ N/mm² and $\sigma_c = 50$ N/mm²

For nut material (phosphor bronze), $\sigma_t = 50$ N/mm², $\sigma_{ck} = 45$ N/mm², $\tau = 40$ N/mm², $P_b = 20$ N/mm², $\mu = 0.14$.

**Procedure :**

**(A) Design of screw :**

**(1) Find core diameter of screw :**

Initially, we will consider that, the screw is under pure compression. Therefore, considering its failure in compression, we have,

$$\sigma_c = \frac{W}{\frac{\pi}{4} \times d_c^2}$$

$$\therefore \quad 50 = \frac{50 \times 10^3 \times 4}{\pi \times d_c^2}$$

$$\therefore \quad d_c = \mathbf{35.68\ mm}$$

But, in addition to direct compressive load, the screw is also subjected to torque for rotation. Therefore, to design the diameter of screw, both compressive stress and torsional shear stress with their combined effect should be considered. Therefore, in order to consider the additional torsional shear stress effect, the calculated value of $d_c$ (35.68 mm) is increased considerably.

From standard table of square thread, let us take,

$$\mathbf{d_o} = \mathbf{58\ mm} \text{ with } \mathbf{p = 9\ mm}$$

$$\therefore \quad \mathbf{d_c} = d_o - p = 58 - 9 = \mathbf{49\ mm}$$

$$\therefore \quad \mathbf{d} = \frac{d_o + d_c}{2} = \frac{58 + 49}{2} = \mathbf{53.5\ mm}$$

**(2) Check induced stresses in the screw designed as above :**

**(i) Induced compressive stress :**

$$\sigma_{c\ ind} = \frac{W}{\frac{\pi}{4} \times d_c^2} = \frac{50 \times 10^3 \times 4}{\pi \times (49)^2} = \mathbf{26.51\ N/mm^2 < 50\ MPa\ (Permissible\ stress)}$$

**(ii) Induced torsional shear stress :**

$$\alpha = \tan^{-1}\left[\frac{\text{Lead}}{\pi d}\right] = \tan^{-1}\left[\frac{p}{\pi \cdot d}\right] = \tan^{-1}\left[\frac{9}{\pi \times 53.5}\right] = \mathbf{3.0651°}$$

$$\phi = \tan^{-1}(\mu) = \tan^{-1}(0.14) = \mathbf{7.9696°}$$

We have, $T_1 = W \tan(\alpha + \phi) \times \frac{d}{2}$

$$\therefore \quad T_1 = 50 \times 10^3 \times \tan(3.0651 + 7.9656) \times \frac{53.5}{2} = \mathbf{260824.39\ N.mm}$$

$$\therefore \quad \text{Induced torsional shear stress, } \tau_{ind} = \frac{16T_1}{\pi \cdot d_c^3} = \frac{16 \times 260824.39}{\pi \times (49)^3} = \mathbf{11.29\ N/mm^2}$$

**(3) Check induced principal stresses in the screw :**

(i) $$\sigma_{c\ max} = \frac{1}{2}\sigma_c + \left[\frac{1}{2}\sqrt{\sigma_c^2 + 4\tau^2}\right] = \left[\frac{1}{2} \times 26.51\right] + \left[\frac{1}{2} \times \sqrt{(26.51)^2 + 4 \times (11.29)^2}\right]$$

$$\therefore \quad \sigma_{c\ max} = 13.255 + [17.41] = \mathbf{30.665\ N/mm^2 < 50\ N/mm^2\ (Permissible\ stress)}$$

(ii) $$\tau_{max} = \frac{1}{2}\sqrt{\sigma_c^2 + 4\tau^2} = \mathbf{17.41\ N/mm^2}$$

As all the induced stresses are less than permissible stresses, design of screw is safe.

**(B) Design of nut :**

If n is taken as number of threads of screw in engagement with nut or number of threads present inside nut, then due to revolution of screw inside the nut, bearing pressure will be induced between nut and screw. It is given by,

$$P_b = \frac{W}{\frac{\pi}{4} \times (d_o^2 - d_c^2) \times n}$$

$$\therefore \quad 20 = \frac{50 \times 10^3}{\frac{\pi}{4} \times [(58)^2 - (49)^2] \times n}$$

$$\therefore \quad n = 3.3053 \cong \mathbf{4\ (say)}$$

Height of nut, $\mathbf{H} = n \times p = 4 \times 9 = \mathbf{36\ mm}$

Thickness of thread, $\mathbf{t} = \frac{p}{2} = \frac{9}{2} = \mathbf{4.5\ mm}$

To check induced transverse shear stresses in screw and nut, we use

$$\tau_{screw} = \frac{W}{\pi\, d_c\, n \cdot t} = \frac{50 \times 10^3}{\pi \times 49 \times 4 \times 4.5} = \mathbf{18.044\ N/mm^2}$$

$$\tau_{nut} = \frac{W}{\pi\, d_o\, n\, t} = \frac{50 \times 10^3}{\pi \times 58 \times 4 \times 4.5} = \mathbf{15.24\ N/mm^2 < Permissible\ stress\ (40\ N/mm^2)}$$

Since the induced stresses are less than permissible stresses, design of nut is safe.

**Problem 4.12 :** *A power screw on a machine has single start square thread with non-rotating bronze nut. Axial force on the screw is 15 kN. Allowable stresses for screw material in compression and shear are 85 MPa and 37 MPa respectively. Allowable bearing pressure for the screw nut pair is 5 MPa. Find :*

*(i) Core diameter of screw. (ii) Length of the nut.*

*(iii) Efficiency of power screw, if coefficient of friction between screw and nut is 0.12.*

*(iv) Shear stresses in the threads of screw and nut.*

**Solution : Given data : W** = 15 kN = $\mathbf{15 \times 10^3}$ **N**

**For screw :** $\sigma_c$ = 85 MPa and $\tau$ = 37 MPa, $P_b$ = 5 MPa or N/mm$^2$, $\mu$ = 0.12

**Procedure :**

**(i) Core diameter of screw :** Considering the screw under pure compression, we take

$$\sigma_c = \frac{W}{\frac{\pi}{4} \times d_c^2}$$

$$\therefore \quad d_c^2 = \frac{W \times 4}{\sigma_c \times \pi} = \frac{15 \times 10^3 \times 4}{85 \times \pi} = 224.689$$

$$\therefore \quad d_c = \mathbf{14.98\ mm}$$

In addition to compressive load, the screw is also subjected to twisting moment or torque for its rotation. Therefore, to account for torsional shear stress, value of $d_c$ is considerably increased.

From standard table of square threads, let us take

$\mathbf{d_o = 24\ mm}$ with pitch, $\mathbf{p = 5\ mm}$ and thickness of thread, $\mathbf{t} = \frac{p}{2} = \frac{5}{2} = \mathbf{2.5\ mm}$

$$\therefore \quad \mathbf{d_c} = d_o - p = 24 - 5 = \mathbf{19\ mm} \quad \text{And} \quad \mathbf{d} = \frac{d_o + d_c}{2} = \frac{24 + 19}{2} = \mathbf{20.5\ mm}$$

**(ii) Length of nut :** Considering bearing pressure between screw and nut, we have,

$$p_b = \frac{W}{\frac{\pi}{4} \times (d_o^2 - d_c^2) \times n}$$

$$\therefore \quad 5 = \frac{15 \times 10^3 \times 4}{\pi \times (24^2 - 19^2) \times n}$$

$$\therefore \quad n = \mathbf{17.76 \cong 18\ (say)}$$

$\therefore$ Length or Height of nut, H = n × p = 18 × 5 = **90 mm**

**(iii) Efficiency of screw :**

We have, $\tan \alpha = \frac{\text{Lead}}{\pi d} = \frac{p}{\pi d}$ [For single start square thread, Pitch = Lead]

$$\therefore \quad \alpha = \tan^{-1}\left(\frac{p}{\pi d}\right) = \tan^{-1}\left(\frac{5}{\pi \times 20.5}\right) = \mathbf{4.4393°}$$

Also, $\phi = \tan^{-1}(\mu) = \tan^{-1}(0.12) = \mathbf{6.8427°}$

We know that, $\eta_{screw} = \frac{\tan \alpha}{\tan(\alpha + \phi)} = \frac{\tan(4.4393)}{\tan(4.4393 + 6.8427)}$

$$\therefore \quad \eta_{screw} = 0.3891 = \mathbf{38.91\%}$$

**(iv)** (a) Transverse shear stress in screw, $\tau = \frac{W}{\pi d_c n t} = \frac{15 \times 10^3}{\pi \times 19 \times 18 \times 2.5} = \mathbf{5.583\ N/mm^2}$

(b) Transverse shear stress in nut, $\tau = \frac{W}{\pi d_o n t} = \frac{15 \times 10^3}{\pi \times 24 \times 18 \times 2.5} = \mathbf{4.42\ N/mm^2}$

(c) Torsional shear stress in screw, $\tau = \frac{16 T_1}{\pi d_c^3}$

$$\therefore \quad \tau = \frac{16 \times W \tan(\alpha + \phi) \cdot \frac{d}{2}}{\pi \times d_c^3} = \frac{16 \times 15 \times 10^3 \times \tan(4.4393 + 6.8427) \times 20.5}{\pi \times (19)^3 \times 2} = \mathbf{22.77\ N/mm^2}$$

**Problem 4.13 :** *Construction of a shaft straightener used on the shop floor is shown in Fig. 4.10. The screw has single start square threads of 80 mm nominal diameter and 10 mm pitch. The screw is required to exert a maximum axial force of 10 kN. Mean radius of friction collar is 30 mm. Axial length of nut is 40 mm. Coefficient of friction at the threads and the collar is 0.12. Mean diameter of the rim of handwheel is 500 mm.*

*Calculate : (i) Force exerted at the rim to drive the screw.*

*(ii) Efficiency of the straightener and (iii) Bearing pressure on the threads in the nut.*

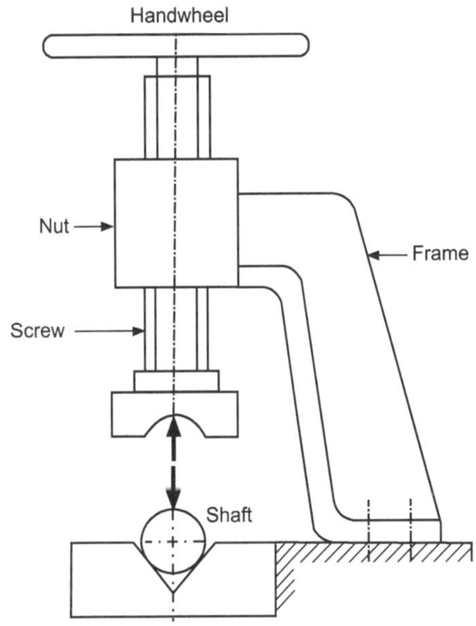


**Fig. 4.10 : Shaft straightener**

**Solution : Given data :** $d_o = 80$ mm, $p = 10$ mm, $\mathbf{W} = 10 \text{ kN} = \mathbf{10 \times 10^3 \ N}$, $\mu = \mu_1 = 0.12$,

$$D_H = 500 \text{ mm}; \therefore \mathbf{R_H} = \frac{D_H}{2} = \mathbf{250 \ mm}, R = 30 \text{ mm}, H = 40 \text{ mm}$$

**Procedure :** We have,

$$\mathbf{d_c} = d_o - p = 80 - 10 = \mathbf{70 \ mm}$$

$$\therefore \quad \text{Mean diameter, } \mathbf{d} = \frac{d_o + d_c}{2} = \frac{80 + 70}{2} = \mathbf{75 \ mm}$$

Also,

$$H = n \cdot p$$

$$\therefore \quad \mathbf{n} = \frac{H}{p} = \frac{40}{10} = \mathbf{4}$$

We have,

$$\alpha = \tan^{-1}\left(\frac{\text{Lead}}{\pi d}\right) = \tan^{-1}\left(\frac{p}{\pi d}\right) = \tan^{-1}\left(\frac{10}{\pi \times 75}\right) = \mathbf{2.4302^\circ}$$

Also,

$$\phi = \tan^{-1}(\mu) = \tan^{-1}(0.12) = \mathbf{6.8427^\circ}$$

Torque ($T_1$) required to overcome the thread friction,

$$T_1 = P \times \frac{d}{2} = W \tan(\alpha + \phi) \times \frac{d}{2}$$

$$\therefore \quad \mathbf{T_1} = 10 \times 10^3 \times \tan(2.4302 + 6.8427) \times \frac{75}{2} = \mathbf{61226.49 \ N.mm}$$

Torque ($T_2$) required to overcome the collar friction,

$$T_2 = \mu_1 \cdot W \cdot R = 0.12 \times 10 \times 10^3 \times 30 = \mathbf{36000 \ N.mm}$$

$$\therefore \quad \text{Total torque required} = T_1 + T_2$$

$$\therefore \quad \mathbf{T} = 61226.49 + 36000 = \mathbf{97226.49 \ N.mm}$$

**(i) Force exerted at the rim to drive the screw :**

We know, Torque exerted or applied at the rim (circumference) of hand wheel to rotate the screw is,

$$T = P_1 \times R_H$$

$$\therefore \quad 97226.49 = P_1 \times 250$$

$$\therefore \quad P_1 = \frac{97226.49}{250} = \mathbf{388.9 \ N}$$

**(ii) Efficiency of straightener :**

$$\text{Overall efficiency of screw, } \eta_o = \frac{T_o}{T} = \frac{W \cdot \tan\alpha \cdot \frac{d}{2}}{T} = \frac{10 \times 10^3 \times \tan(2.4302) \times \frac{75}{2}}{97226.49} = \mathbf{0.1637 = 16.37\%}$$

**(iii) Bearing pressure on threads in the nut :** $P_b = \dfrac{W}{\frac{\pi}{4}(d_o^2 - d_c^2)\, n} = \dfrac{10 \times 10^3}{\frac{\pi}{4}(80^2 - 70^2) \times 4} = \mathbf{2.122 \ N/mm^2}$

## 4.9 DESIGN OF SCREW HAVING ACME OR TRAPEZOIDAL THREADS

- In case of square threads, normal reaction is given by,

$$R_N = W \cos \alpha$$

where, $\alpha$ = Helix angle

- But, in case of Acme or trapezoidal threads, the normal reaction between screw and nut is increased, because, axial component of this normal reaction is equal to axial load (W).
- Let $2\beta$ = Angle of Acme thread = **29°**

$\therefore$ $\beta$ = Semi-angle of Acme thread

Referring the Fig. 4.11,

$$W = R_N \cdot \cos \beta \text{ or } R_N = \frac{W}{\cos \beta} \quad \text{... (4.22)}$$

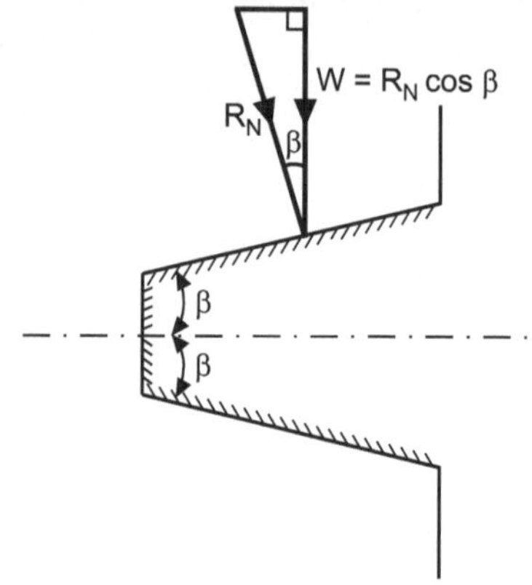


**Fig. 4.11 : Design of Acme thread or Trapezoidal thread**

We know that, frictional force is given by ,

$$F = \mu \cdot R_N$$

$$\therefore \quad F = \mu \times \frac{W}{\cos \beta} \quad \text{[From equation (4.22)]}$$

$$\therefore \quad F = \mu' \times W$$

where $\mu'$ = Virtual coefficient of friction = $\frac{\mu}{\cos \beta} = \tan \phi'$

- When virtual coefficient of friction $\mu'$ is considered, then acme thread becomes ***equivalent*** to a square thread and thus in case of acme threads, $\phi'$ is substituted in place of $\phi$ in square threads.

Therefore, in Acme threads,

$$T_1 = W \tan (\alpha + \phi') \times \frac{d}{2}$$

**Note :** *For Trapezoidal threads, 2β = 30°.*

---

**Numerical Type No. 4 : "Design of Acme Threads and Trapezoidal Threads"**

**Problem 4.14 :** *The lead screw of a lathe has Acme threads of 50 mm outside diameter and 8 mm pitch. The screw must exert an axial pressure of 2500 N in order to drive the tool carriage. The thrust is carried on a collar having 110 mm outside diameter and 55 mm inside diameter and lead screw rotates at 30 r.p.m. Determine :*

*(i) Power required to drive the screw,*

*(ii) Efficiency of lead screw.*

*Assume* $\mu$ = *0.15 for screw*

$\mu_1$ = *0.12 for collar*

**Solution : Given data :** $d_o$ = 50 mm, p = 8 mm, W = 2500 N, $D_1$ = 110 mm, $D_2$ = 55 mm, N = 30 r.p.m., $\mu$ = 0.15, $\mu_1$ = 0.12.

**Procedure :**

$$\mathbf{R_1} = \frac{D_1}{2} = \frac{110}{2} = \mathbf{55\ mm} \; ; \; \mathbf{R_2} = \frac{D_2}{2} = \frac{55}{2} = \mathbf{27.5\ mm}$$

We have,

$$\mathbf{R} = \frac{R_1 + R_2}{2} = \frac{55 + 27.5}{2} = \mathbf{41.25\ mm}$$

And

$$\mathbf{d} = d_o - \frac{p}{2} = 50 - \frac{8}{2} = \mathbf{46\ mm}$$

We have,

$$\tan \alpha = \frac{p}{\pi d}$$

$$\therefore \quad \alpha = \tan^{-1}\left(\frac{8}{\pi \times 46}\right) = \mathbf{3.1685°}$$

For acme threads,

$$2\beta = 29°$$

$$\therefore \quad \beta = \mathbf{14.5°}$$

We have, $\mu' = \frac{\mu}{\cos\beta} = \frac{0.15}{\cos(14.5°)} = \mathbf{0.1549}$

But, $\mu' = \tan\phi'$

$\therefore$ $\phi' = \tan^{-1}(\mu') = \tan^{-1}(0.1549) = \mathbf{8.8051°}$

Therefore, torque required to overcome friction between screw and nut is,

$$T_1 = W\tan(\alpha + \phi')\cdot\frac{d}{2} = 2500 \times \tan(3.1685 + 8.8051) \times \frac{46}{2} = \mathbf{12194.31\ N.mm}$$

Assuming uniform wear condition, torque required to overcome collar friction is,

$$T_2 = \mu_1 \cdot W \cdot R = 0.12 \times 2500 \times 41.25 = \mathbf{12375\ N.mm}$$

$\therefore$ Total torque required, $T = T_1 + T_2$

$\therefore$ $T = 12194.31 + 12375 = 24569.31\text{ N.mm} = \mathbf{24.569\ N.m}$

**(i) Power required to drive the screw :**

$$P = \frac{2\pi NT}{60} = \frac{2\pi \times 30 \times 24.569}{60} = 77.18\text{ W} = \mathbf{0.07718\ kW}$$

**(ii) Efficiency of lead screw :**

$$\eta = \frac{T_o}{T} = \frac{W\cdot\tan\alpha\cdot\frac{d}{2}}{T_1 + T_2} = \frac{2500 \times \tan(3.1685°) \times \frac{46}{2}}{24569.31} = 0.1295 = \mathbf{12.95\%}$$

---

**Problem 4.15 :** *The C-clamp as shown in Fig. 4.12, has trapezoidal threads of 12 mm outside diameter and 2 mm pitch. The coefficient of friction for screw thread is 0.15 and for the collar is 0.28. The mean radius of the collar is 8 mm. If the force exerted by the operator at the end of handle is 80 N, find :*

*(i) The length of handle.*

*(ii) The bearing pressure on the threads.*

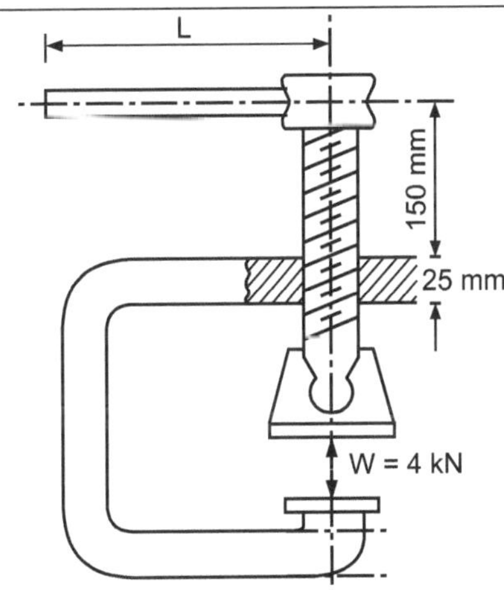


**Fig. 4.12**

**Solution : Given data :** $d_o = 12\text{ mm}, p = 2\text{ mm}, \mu = 0.15, \mu_1 = 0.28, R = 8\text{ mm}, P_1 = 80\text{ N}, H = 25\text{ mm}$

$\mathbf{W} = 4\text{ kN} = \mathbf{4 \times 10^3\ N}$

**Procedure :** We have, $d = d_o - \frac{p}{2}$

$\therefore$ $\mathbf{d} = 12 - \frac{2}{2} = \mathbf{11\ mm}$

Also, $\tan\alpha = \frac{p}{\pi d} = \frac{2}{\pi \times 11} = 0.05787$

$\therefore$ $\alpha = \tan^{-1}(0.0578) = \mathbf{3.3122°}$

For trapezoidal threads, $2\beta = 30°$

$\therefore$ $\beta = \mathbf{15°C}$

We have, $\mu' = \frac{\mu}{\cos\beta} = \frac{0.15}{\cos(15°)} = \mathbf{0.1553}$

But, $\mu' = \tan\phi'$

$\therefore$ $\phi' = \tan^{-1}(\mu') = \tan^{-1}(0.1553) = \mathbf{8.8275°}$

Torque required to overcome friction between screw and nut is given by,

$$T_1 = W \cdot \tan(\alpha + \phi') \times \frac{d}{2}$$

$$\therefore \quad \mathbf{T_1} = 4 \times 10^3 \times \tan(3.3122 + 8.8275) \times \frac{11}{2} = \mathbf{4732.33\ N.mm}$$

Torque required to overcome collar friction,

$$\mathbf{T_2} = \mu_1 W R = 0.28 \times 4 \times 10^3 \times 8 = \mathbf{8960\ N.mm}$$

∴ Total torque required for clamping is given by,

$$\mathbf{T} = T_1 + T_2 = 4732.33 + 8960 = \mathbf{13692.33\ N.mm}$$

**1. Length of handle :**

We have, $T = P_1 \times L$

$$\therefore \quad 13692.33 = 80 \times L$$

$$\therefore \quad L = 171.15 \text{ mm} \cong \mathbf{172\ mm\ (say)}$$

**2. Bearing pressure on threads :**

We have, Height of nut, $H = n \times p$

$$\therefore \quad \mathbf{n} = \frac{H}{p} = \frac{25}{2} = \mathbf{12.5}$$

Also, $d_o - d_c = p$

$$\therefore \quad 12 - d_c = 2$$

$$\therefore \quad \mathbf{d_o} = 10 \text{ mm}$$

We have, $$P_b = \frac{W}{\frac{\pi}{4} \times (d_o^2 - d_c^2) \times n} = \frac{4 \times 10^3}{\frac{\pi}{4} \times [(12)^2 - (10)^2] \times 12.5} = \mathbf{9.26\ N/mm^2}$$

## 4.10 DESIGN OF TOGGLE JACK

- Toggle jack is a type of screw jack, where maximum load is lifted with minimum effort.
- It consists of eight symmetrical links, two nuts and a screw having right hand-left hand threads.

**Application :** Used for four wheelers to lift the punctured wheel.

**Design Procedure :**

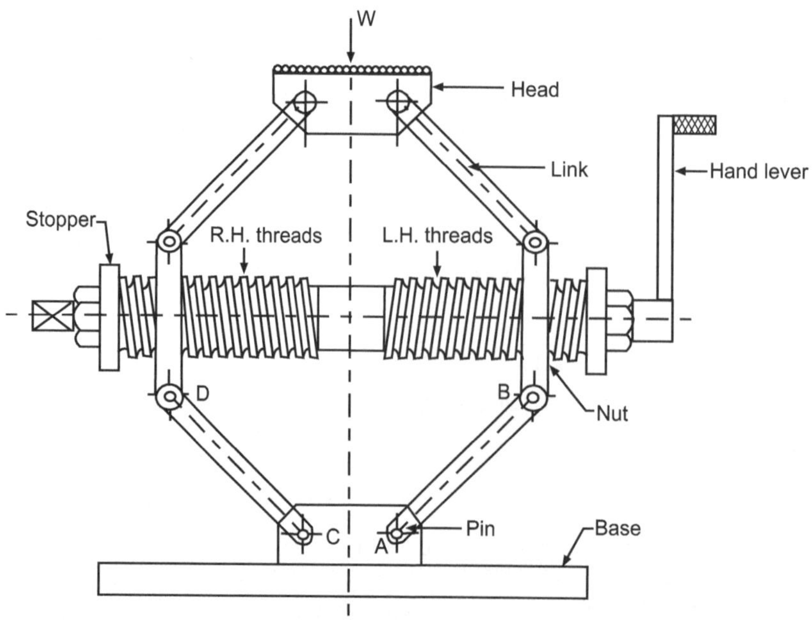


**Fig. 4.13 : Toggle jack**

***(A) Design of Screw :***

**1. Find the value of $W_1$ :**

- We know that, load on the square threaded screw will be maximum, when the toggle jack is in its bottom position. The position of link CD is shown in Fig. 4.14.
- Suppose a pull 'F' acts on the left nut, then another pull 'F' of similar value will act on the other nut.

∴ Total tensile pull on the square threaded screw,

$$W_1 = 2F$$

From right angle triangle DOC, we can write,

$$\tan\theta = \frac{\frac{W}{2}}{F}$$

$$\therefore \quad F = \frac{W}{2\tan\theta}$$

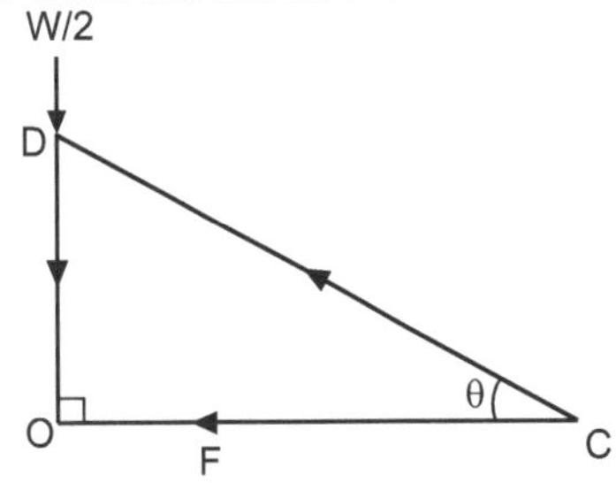


**Fig. 4.14: Position of link CD**

**2. Find the value of core diameter and mean diameter :**

Let $d_c$ = Core diameter of screw.

- Direct tensile or compressive stress due to axial load is given by,

$$\sigma_c = \frac{W_1}{\frac{\pi}{4}\times d_c^2}$$

- As the screw is also subjected to torsional shear stress, therefore to account for this, the value of $d_c$ is increased considerably. For this purpose, select the values of nominal diameter '$d_o$' and pitch 'p' from standard table of square threads. Then, find the increased value of $d_c$ using the following relation.

$$d_c = d_o - p$$

Also, Mean diameter, $d = \frac{d_o + d_c}{2}$

**3. Find torsional shear stress induced due to torque 'T' required to rotate screw :**

Calculate the value of torque 'T' by following relation.

$$T = P\times\frac{d}{2} = W_1\cdot\tan(\alpha+\phi)\times\frac{d}{2}$$

where, P = Effort required at the circumference of screw,

d = Mean diameter of screw.

Find the torsional shear stress induced in the screw by using the following equation.

Torsional shear stress, $\tau = \frac{16\,T}{\pi\times d_c^3}$ $\left(\because T = \frac{\pi}{16}\times\tau\times d_c^3\right)$

**4. Find the direct compressive stress induced in the screw due to axial load :**

We have, $$\sigma_c = \frac{W_1}{\frac{\pi}{4}\times d_c^2}$$

**5. Check maximum principal stresses :**

When screw is simultaneously subjected to two types of stresses i.e. direct stress (compressive) and torsional shear stress, then design is based on -

(a) Maximum shear stress theory, according to which, maximum shear stress induced on minimum cross-section is,

$$\tau_{max} = \frac{1}{2}\times\sqrt{\sigma_c^2 + 4\tau^2}$$

(b) Maximum normal stress theory, according to which, maximum normal stress induced on minimum cross-section is,

$$\sigma_{c\,max} = \frac{1}{2}\sigma_c + \frac{1}{2}\times\sqrt{\sigma_c^2 + 4\tau^2}$$

Check whether the induced principal stresses are less than permissible limits.

***(B) Design of Nut :***

Let n = Number of threads of screw in contact with the nut.

To calculate the value of n, consider the bearing pressure on nut.

$$P_b = \frac{W_1}{\frac{\pi}{4}(d_o^2 - d_c^2) \times n}$$

$$\therefore \quad n = \frac{W_1}{\frac{\pi}{4}(d_o^2 - d_c^2) \times P_b}$$

To calculate the Height of nut, use following relation,

$$H = n \times p$$

Also, find thickness or width of screw thread, $t = \frac{p}{2}$

***(C) Design of Pins in the Nuts :***

If $d_1$ is the diameter of pins in the nuts, then considering the pins in double shear, we have,

$$\text{Shear stress,} \quad \tau = \frac{F}{2 \times \left[\frac{\pi}{4} \times d_1^2\right]}$$

From this equation, diameter of pins ($d_1$) can be determined.

**Numerical Type No. 5 : "Design of Toggle Jack"**

**Problem 4.16 :** *A toggle jack as shown in Fig. 4.15 is to be designed for lifting a load of 4 kN. When the jack is in its top position, the distance between the centerlines of nuts is 50 mm and in bottom position is 210 mm. The eight links of the jack are symmetrical and 110 mm long. The link pins in the base are set 30 mm apart. The links and pins are made up of mild steel, for which, the permissible stresses are 100 N/mm² in tension and 50 N/mm² in shear. The bearing pressure on the pins is limited to 20 N/mm². Assume, pitch of threads as 6 mm and coefficient of friction between threads as 0.2.*

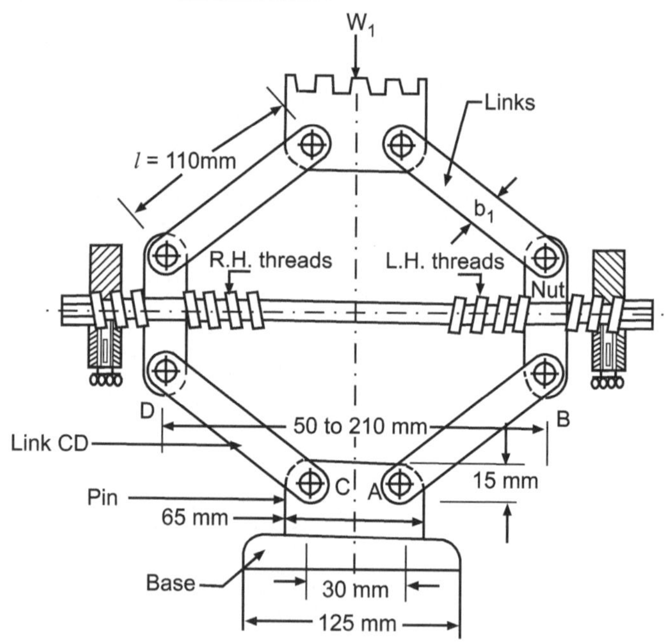


**Fig. 4.15**

**Solution : Given data : W** = 4 kN = **4000 N**, $l$ = 110 mm, p = 6 mm, $\mu$ = 0.2, $\sigma_c$ = 100 N/mm², $\tau$ = 50 N/mm², $P_b$ = 20 N/mm²

**Procedure :**

**(1) Design of screw :**

We know that, load on the square threaded screw will be maximum, when the toggle jack is in its bottom position. The position of link CD is shown in Fig. 4.16 (a).

Let, $\theta$ = Angle made by link CD with the horizontal (X-axis),

$$\therefore \quad \cos\theta = \frac{105 - 15}{110} = \frac{90}{110} = 0.8182$$

$$\therefore \quad \theta = \cos^{-1}(0.8182) = \mathbf{35.094°}$$

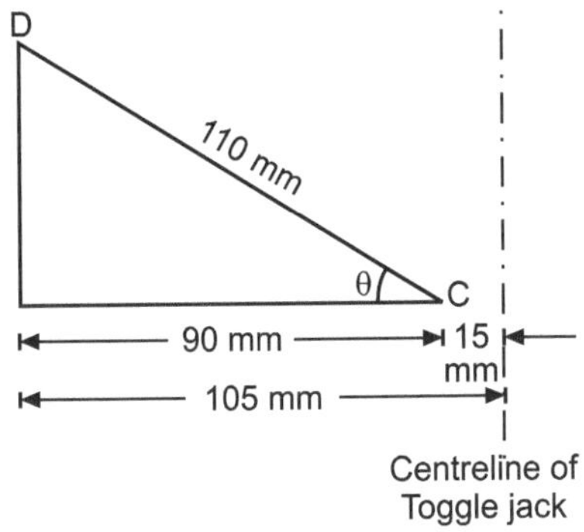


Fig. 4.16 (a)

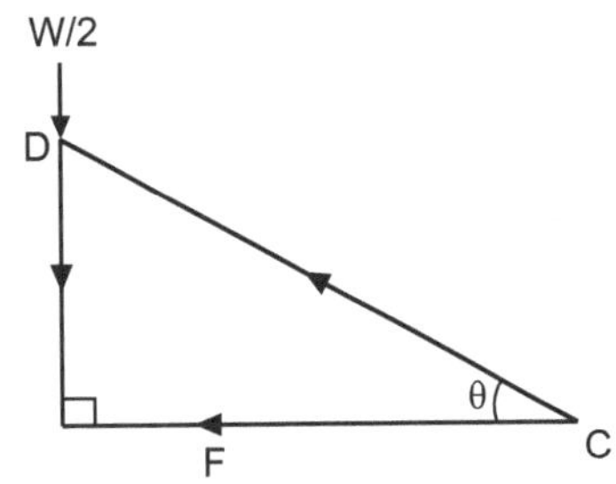


Fig. 4.16 (b)

Suppose a pull force 'F' acts on the left nut, due to which, link CD is in tension.

From Fig. 4.16 (b), we have,

$$\tan\theta = \frac{\left(\frac{W}{2}\right)}{F}$$

$$\therefore \quad F = \frac{W}{2\tan\theta} = \frac{4000}{2\times\tan(35.094°)} = \mathbf{2846.3\ N}$$

Then another pull F of similar value will act on the other nut.

∴ Total tensile pull on the square threaded screw is given by,

$$W_1 = 2F = 2\times 2846.3 = \mathbf{5692.6\ N}$$

Direct tensile or compressive stress due to axial load is given by,

$$\sigma_c = \frac{W_1}{\frac{\pi}{4}\times d_c^2}$$

$$\therefore \quad d_c^2 = \frac{W_1}{\frac{\pi}{4}\times\sigma_c} = \frac{5692.6}{\frac{\pi}{4}\times 100}$$

Thus, $d_c = \mathbf{8.51\ mm}$

As the screw is also subjected to torsional shear stress, therefore to account for this, the value of $d_c$ is increased considerably.

∴ From standard table of square threads, let us take $\mathbf{d_o = 22\ mm}$ **and** $\mathbf{p = 5\ mm}$

$$\therefore \quad \mathbf{d_c} = d_o - p = 22 - 5 = \mathbf{17\ mm}$$

Mean diameter, $\mathbf{d} = \frac{d_c + d_o}{2} = \frac{17+22}{2} = \mathbf{19.5\ mm}$

Also, $\tan\alpha = \frac{p}{\pi d}$; Therefore, $\boldsymbol{\alpha} = \tan^{-1}\left(\frac{p}{\pi d}\right) = \tan^{-1}\left(\frac{5}{\pi\times 19.5}\right) = \mathbf{4.666°}$

And, $\mu = \tan\phi$

$$\therefore \quad \phi = \tan^{-1}(\mu) = \tan^{-1}(0.2) = \mathbf{11.3099°}$$

Torque required to rotate the screw,

$$\mathbf{T} = P\times\frac{d}{2} = W_1\tan(\alpha+\phi)\times\frac{d}{2} = 5692.6\times\tan(4.666+11.3099)\times\frac{19.5}{2} = \mathbf{15.89\times 10^3\ N\text{-}mm}$$

We have,

Induced shear stress,

$$\tau = \frac{16\,T}{\pi\times d_c^3} = \frac{16\times 15.89\times 10^3}{\pi\times(17)^3} = \mathbf{16.47\ N/mm^2 < 50\ N/mm^2\ (Permissible\ stress)}$$

Induced compressive stress, $\sigma_c = \frac{W_1}{\frac{\pi}{4}\times d_c^2} = \frac{5692.6}{\frac{\pi}{4}\times(17)^2} = \mathbf{25.08\ N/mm^2 < 100\ N/mm^2\ (Permissible\ stress)}$

Maximum principal stresses induced are,

$$\tau_{max} = \frac{1}{2}\sqrt{\sigma_c^2 + 4\tau^2} = \frac{1}{2}\sqrt{(25.08)^2 + 4 \times (16.47)^2} = \textbf{20.70 N/mm}^2 < \textbf{50 N/mm}^2 \textbf{ (Permissible stress)}$$

$$\sigma_{c\,max} = \frac{1}{2}\sigma_c + \frac{1}{2}\sqrt{\sigma_c^2 + 4\tau^2} = \frac{1}{2} \times 25.08 + \frac{1}{2}\sqrt{(25.08)^2 + 4 \times (16.47)^2}$$

$$= \textbf{33.24 N/mm}^2 < \textbf{100 N/mm}^2 \textbf{ (Permissible stress)}$$

As the induced stresses are less than permissible values, design of screw is safe.

**(2) Design of nut :**

We have, $$P_b = \frac{W_1}{\frac{\pi}{4}(d_o^2 - d_c^2) \times n}$$

$$\therefore \quad 20 = \frac{5692.6}{\frac{\pi}{4} \times (22^2 - 17^2) \times n}$$

$$\therefore \quad n = 1.85 \cong \textbf{4 (say)}$$

$\therefore$ Height of nut, H = n × p = 4 × 5 = **20 mm**

**(3) Design of pins :** Considering failure of pins under double shear, we can write transverse shear stress as,

$$\tau = \frac{F}{2 \times \frac{\pi}{4} d_1^2}$$

$$\therefore \quad d_1^2 = \frac{F}{2 \times \frac{\pi}{4} \times \tau} = \frac{2846.3}{2 \times \frac{\pi}{4} \times 50} = 36.24$$

$$\therefore \quad d_1 = \textbf{6.019 mm} \cong \textbf{7 mm (say)}$$

## 4.11 INTRODUCTION TO GEARS

- Gear drive is used, when centre to centre distance between driver and driven shafts is very small. Fig. 4.17 shows a spur gear drive.
- Gear drive consists of two wheels. The smaller wheel is called as **pinion** and the larger wheel is called as **gear**.
- In case of belt or rope drives, slipping of belt or rope is a common phenomenon, which leads to loss of power transmission between two shafts and decrease in velocity ratio of the system.
- In gear drive, **slip** is absent. Therefore, it gives **exact and uniform velocity ratio**. Due to its ability of maximum power transmission and exact velocity ratio, gear drive is called as **perfect positive drive.**
- **Gears** are defined as, *"toothed wheels, which can transmit power and motion from one shaft to another shaft by means of successive engagement of teeth."*
- Gears are also called as *toothed wheels.*

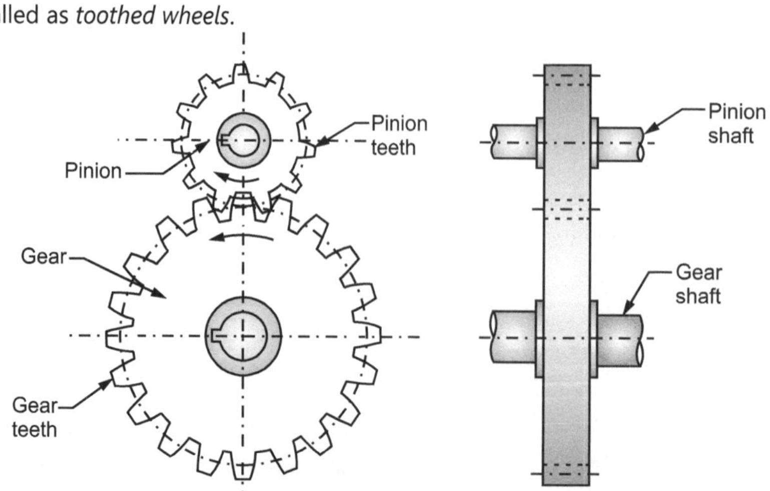


**Fig. 4.17 : Spur gear drive**

### 4.11.1 Advantages of Gear Drive

(1) Positive drive giving exact and constant velocity ratio.
(2) Higher speed ratio than belt or chain drives.
(3) Compact layout, therefore, occupies less space.
(4) Ability of transmitting large power as compared to belt or chain drives.
(5) Ability of transmitting large power at much higher speed than belt or chain drives.
(6) Gear drives can be used for changing the speed ratios over a wide range.
(7) More efficient.
(8) Low maintenance.
(9) Reliable service.

### 4.11.2 Limitations of Gear Drive

(1) Manufacturing of gears requires special tools and equipment. This leads to high production cost.
(2) Complicated manufacturing process.
(3) Gear drive requires precise alignment of shafts.
(4) Costly lubrication system is required for smooth operation and longer life of gear teeth.
(5) Any error induced in gear during the process of teeth cutting leads to noisy operation with severe vibrations.

### 4.11.3 Applications of Gear Drive

(1) Gear box of vehicle.
(2) Gear mechanism of wrist watches.
(3) Machine tools.
(4) Differential mechanism of automobile.
(5) Dial indicator.
(6) Cement mixing unit.

## 4.12 CLASSIFICATION OF GEARS

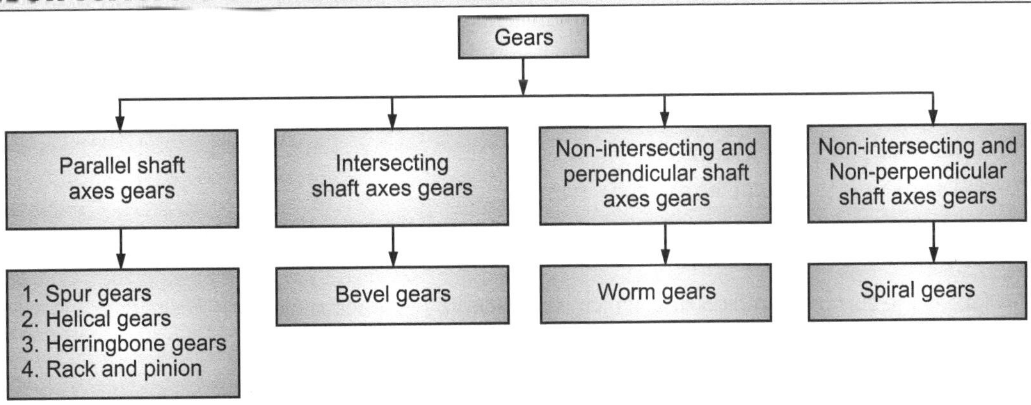


**Chart 4: Classification of gears**

## 4.13 MATERIALS USED FOR GEARS

- Gears are manufactured from metallic and non-metallic materials.

1. **Metallic Materials:**

**(a) Cast iron:**

- Cast iron is commonly used material for manufacturing gears due to its good properties, such as, good machinability, high wear resistance etc.

**(b) Steel:**

- Steel is used, where high strength is required. Steel is heat treated to get good toughness and tooth hardness.
- There are various types of steels used for gears ranging from plain carbon steel to high alloy steels and low carbon steel to high carbon steel.
- Plain carbon steels are used for medium duty applications, whereas, alloy steels are used for heavy duty applications.

**(c) Non-ferrous metal:**

- Non-ferrous metals like copper, zinc, tin, aluminum, manganese etc. are used in various combinations as gear materials.
- Most commonly used material is bronze due to its ability to withstand heavy loads.
- Non-ferrous metal is preferably used in case of worm gears due to its ability of reducing the effect of wear, which is a common problem found in other metallic materials.

**2. Non-metallic materials:**

- In case of light duty applications, gears may be made up of non-metallic materials like, wood, nylon etc.

### 4.13.1 Criteria for Selection of Gear Materials

(1) Strength. (2) Wear resistance. (3) Service condition.
(4) Shock resistance. (5) Speed. (6) Degree of accuracy.

### 4.13.2 Need of Selecting Involute Curves for Gear Tooth Profile

- For a gear tooth profile, involute curves are adopted and preferred due to following advantages:

(1) Involute tooth profile has a single curve. Therefore, it is easy to manufacture.

(2) In any gear drive, it is desired that, speed ratio must be always constant. By using involute tooth profile, it is possible to obtain a constant speed ratio of gears in mesh.

(3) Speed ratio of gears in mesh does not remain constant, if there is variation in centre to centre distance between two mating gears. This difficulty can be overcome by using involute tooth profile, in which, small variations in centre to centre distance does not affect the speed ratio.

(4) In case of involute tooth profile gears, pressure angle is constant throughout the engagement, i.e. from the point of commencement of engagement to the point of disengagement.

## 4.14 GEAR TERMINOLOGY

**Base Circle:**

- Base Circle is the circle, from which, involute form is generated.

**Pitch Circle:**

- Pitch Circle is an imaginary circle on gear, which by pure rolling action produce the same motion as the actual gear.

**Pitch Circle Diameter (PCD) or Pitch Diameter:**

- Pitch Circle Diameter is the diameter of pitch circle.
- Generally, the size of gear is specified by Pitch Circle Diameter. It is denoted by D.

**Pitch Surface :**

- Pitch Surface is the surface of rolling disc, which is replaced by toothed gear at pitch circle.

**Pitch Point :**

- Pitch Point is the point of contact between pitch circles of two gears in mesh.

**Centre Distance :**

- Centre Distance is the distance between the axes of two mating gears.
- It is also referred as centre to centre distance between two gears in mesh.

**Top land:**

- Top land is the surface of top of the tooth.

**Face of tooth:**

- Face of tooth is the part of tooth surface, which is above the pitch surface.

**Face width:**

- Face width is the width of gear tooth measured along a line parallel to gear axis.

**Flank:**

- Flank is the part of tooth surface, which is below the pitch surface.

**Tooth Profile:**

- Tooth Profile is the curve formed by the face and flank of tooth.

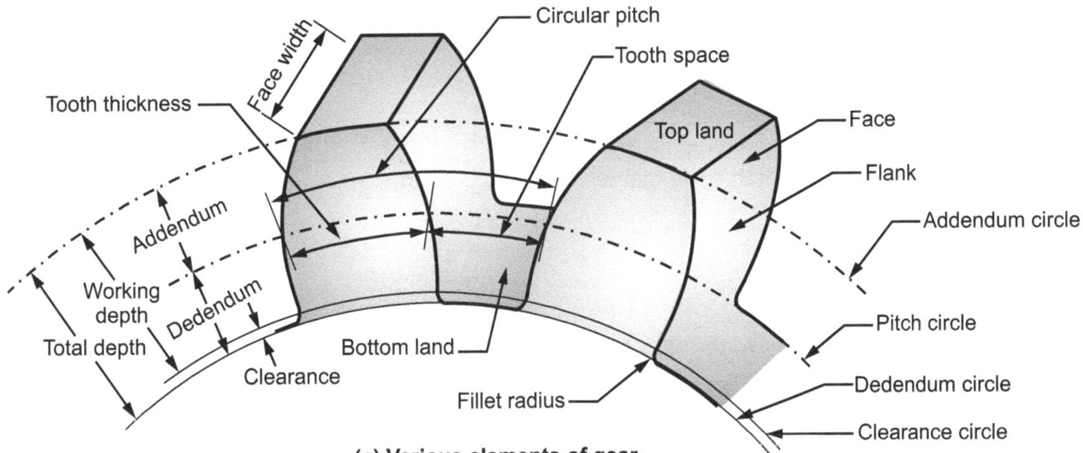


(a) Various elements of gear

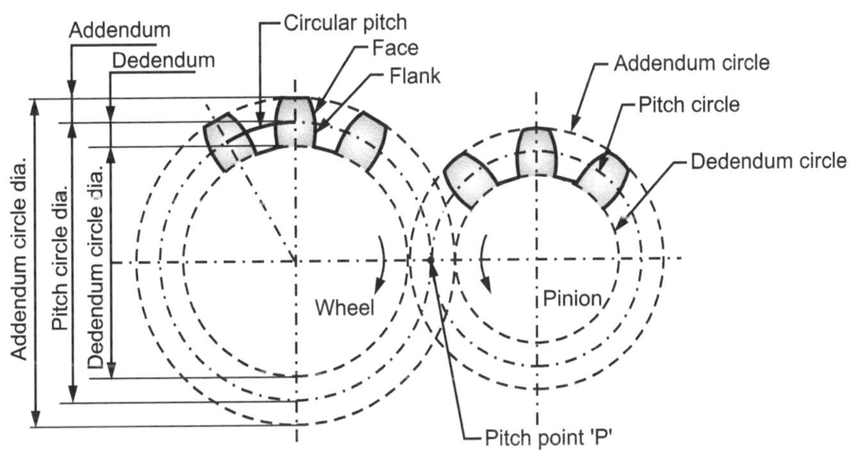


(b) Two external gears in mesh

**Fig. 4.18 : Spur Gear Terminology**

**Module :**

- Module is defined as, *"the ratio of pitch circle diameter to number of teeth".*
- It is denoted by 'm'.
- Mathematically, $m = \frac{D}{T}$ in 'mm'

  where, D = Pitch circle diameter in 'mm'

  T = Number of teeth on wheel (Gear)
- Module specifies the size of gear tooth. Size of gear increases with increase in value of module.

**Circular Pitch:**

- Circular pitch is defined as, *"the distance measured along the circumference of pitch circle from a point on one tooth to corresponding point on the next or adjacent tooth."*
- Circular pitch ($p_c$) for a spur gear of pitch circle diameter 'D' having number of teeth 'T' is given by,

$$p_c = \frac{\pi D}{T} = \pi \cdot m \qquad \left(\text{where, module } m = \frac{D}{T}\right)$$

- It is measured in mm.

**Diametral Pitch:**

- Diametral pitch is defined as, *"the ratio of number of teeth of a gear to its pitch circle diameter."*
- Mathematically, Diametral pitch ($p_d$) is given by,

$$p_d = \frac{\text{Number of teeth}}{\text{Pitch circle diameter}} = \frac{T}{D} = \frac{1}{\left(\frac{D}{T}\right)} = \frac{1}{m} \qquad \left[\because m = \frac{D}{T}\right]$$

- SI unit of diametral pitch is $mm^{-1}$.

**Addendum Circle:**

- Addendum circle is the circle drawn through top of gear teeth.
- It is concentric with the pitch circle.

**Addendum:**

- Addendum is the radial distance from pitch circle to top of tooth.
- Its value is equal to module 'm'. Therefore, Addendum = m.

**Dedendum Circle:**

- Dedendum circle is the circle drawn through bottom of gear teeth.
- It is concentric with the pitch circle.

**Dedendum:**

- Dedendum is the radial distance from pitch circle to bottom of tooth.
- Mathematically, Dedendum = 1.157 m

**Clearance:**

- Clearance is the radial distance from top of tooth of a gear to bottom of tooth space of mating gear, when teeth are symmetrically engaged.
- Mathematically, Clearance = Dedendum − Addendum = 1.157 m − m = 0.157 m.

**Total Depth:**

- Total depth is the radial distance between addendum circle and dedendum circle.
- It is equal to sum of addendum and dedendum.
- Mathematically, Total depth = Addendum + Dedendum = m + 1.157 m = 2.157 m

**Working Depth:**

- Working depth is the greatest depth, to which, a tooth of one gear can extend into tooth space of a mating gear.
- Working depth can be defined as, *"radial distance between addendum circle and clearance circle"*.
- It is given by,

  Working depth = Addendum + Dedendum − Clearance = m + 1.157 m − 0.157 m = 2m

**Tooth Thickness:**

- Tooth Thickness is the actual distance measured along the pitch circle from its intercept with one flank to its intercept with other flank of same tooth. In short, it is the width of tooth measured along pitch circle.

**Tooth Space:**

- Tooth space is the gap or width of space between the two adjacent teeth, which is measured along the pitch circle.

**Pressure Angle:**

- Pressure angle is the angle between the common normal (perpendicular) and the common tangent, both drawn at the pitch point.
- It is denoted by $\phi$.
- Standard values of pressure angle are $14\frac{1}{2}°$ and 20°.

**Blank Diameter:**

- Blank Diameter is equal to sum of pitch circle diameter (D) and value equal to twice the addendum.
- ∴ Mathematically, Blank diameter = D + 2m [Where, m = Module = Addendum]

**Backlash :**

- Backlash is the difference between tooth space and tooth thickness measured along pitch circle.
- Theoretically, it should be zero. But, in actual practice, some backlash must be allowed to prevent jamming of teeth due to tooth errors and thermal expansion during manufacturing.

## 4.15 DESIGN CONSIDERATIONS FOR A GEAR DRIVE

- Following requirements must be fully satisfied in the design of a gear drive:

(1) Gear teeth should have sufficient strength, so that, they will not fail during normal running condition.
(2) Gear teeth should have wear resistant property to work satisfactorily for longer period.
(3) Use of space and material should be economical.
(4) Alignment of gears and deflection of shafts must be considered to improve the performance of the gears.
(5) Gears should get satisfactory lubrication.
(6) High efficiency and speed ratio.
(7) Low cost.

### 4.15.1 Objectives in Design of Gear Drives

(1) High power transmission.
(2) Compact in size.
(3) High efficiency.
(4) Less initial cost.
(5) Less maintenance and running cost.
(6) Noiseless and vibration free operation.

### 4.15.2 Data Usually given, While Designing a Gear Drive

(1) Amount of power to be transmitted.
(2) Speed of driving gear.
(3) Speed of driven gear or velocity ratio.
(4) Centre to centre distance between two shafts.

## 4.16 MODES OF GEAR TOOTH FAILURE

Different modes of gear tooth failure and their possible remedies to avoid the failure are :

1. **Bending failure:**
   - Every gear tooth acts as a cantilever beam. If the total repetitive dynamic load acting on gear tooth is greater than the beam strength of gear tooth, then the gear tooth with fail in bending, i.e., the gear tooth will break.
   - In order to avoid such failure, module and face width of the gear are adjusted in such a way that, beam strength will be always greater than dynamic load.
2. **Pitting:**
   - It is a surface fatigue failure, which occurs due to many repetitions of surface contact stresses. Failure of tooth occurs, when the surface contact stresses are higher than endurance limit of material. Failure of tooth starts with formation of pits, which continue to grow resulting in rupture of tooth surface.
   - In order to avoid the pitting, the dynamic load should be less than wear strength of the gear tooth.
3. **Scoring :**
   - Excessive heat is generated, when there is an excessive surface pressure, high speed or failure in supply of lubricants.
   - It is a stick-slip phenomenon, in which, alternate shearing and welding takes place rapidly at high spots.
   - This type of failure can be avoided by properly designing the parameters, such as, speed, pressure and proper flow of lubricant, so that, temperature at the rubbing faces lies within the permissible limits.
4. **Abrasive wear:**
   - Foreign particles in the lubricants such as dirt, dust etc. enter between the teeth and damage the form of tooth.
   - This type of failure can be avoided by providing filters for the lubricating oil or by using high viscosity lubricant oil, which enables the formation of thicker oil film and hence permits easy passage of such particles without damaging the gear surface.
5. **Corrosive wear:**
   - Corrosion of tooth surfaces is mainly caused due to the presence of corrosive elements, such as additives present in the lubricating oils.
   - In order to avoid this type of wear, proper anti-corrosive additives should be added in lubricants.

## 4.17 BEAM STRENGTH OF SPUR GEAR TOOTH – LEWIS EQUATION

- **'Lewis'** derived an equation for determining the approximate stress in a gear tooth by considering it as a **cantilever beam of uniform strength.**
- Consider each tooth as a cantilever beam loaded by a normal load ($W_N$) as shown in Fig. 4.19.
- This normal load is resolved in two mutually perpendicular components.
  1. **Tangential component ($W_T$) acting perpendicular to centerline of tooth:**
     - It induces a bending stress, which tends to break the tooth.
     - It is also called as **beam strength of tooth**.
  2. **Radial component ($W_R$) acting parallel to centerline of tooth:**
     - It induces a compressive stress of relatively smaller magnitude; therefore, its effect on tooth may be neglected.
- Hence, bending stress is used as a basis for design calculations.
- Critical section subjected to maximum bending stress may be obtained by drawing a parabola passing through 'A' and tangential tooth curves at 'B' and 'C'. Refer Fig. 4.19.
- Parabola outlines a beam of uniform strength. It means that, there will be same stress at all sections of tooth.
- But, tooth is larger than parabola at every section except BC.
- Therefore, we say that, section BC is critical section subjected to maximum stress.

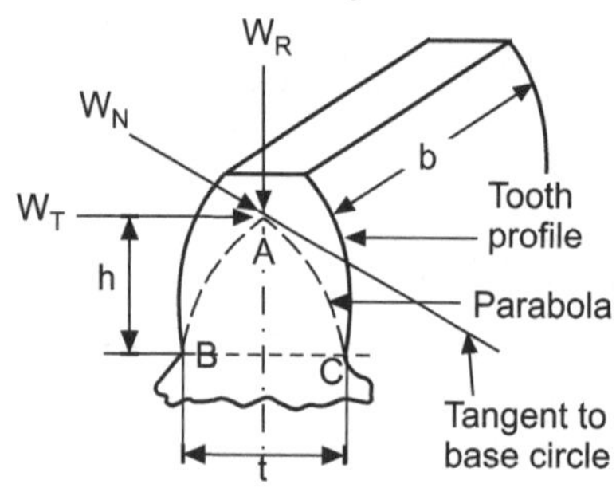


**Fig. 4.19: Tooth considered as cantilever beam**

Maximum value of bending stress at the section 'BC' is given by,

$$\sigma_w = \frac{M}{I} \cdot y \qquad \text{... (4.23)}$$

where, $M$ = Maximum bending moment at critical section 'BC' in 'N.mm'

$y$ = Half of tooth thickness $= \frac{t}{2}$ in 'mm'

$I$ = Moment of inertia about centerline of tooth in 'mm$^4$'

Maximum bending moment at critical section 'BC' is calculated by,

$$M = W_T \times h \quad \text{in 'N.mm'} \qquad \text{... (4.24)}$$

where, $W_T$ = Tangential load acting on tooth in 'N'

$h$ = Length of tooth in 'mm'

Moment of inertia about centerline of tooth is given by,

$$I = \frac{1}{12} bt^3 \quad \text{in 'mm}^4\text{'} \qquad \text{... (4.25)}$$

where, $b$ = Facewidth of gear in 'mm'

$t$ = Tooth thickness in 'mm'

Also, $$y = \frac{t}{2} \quad \text{in 'mm'} \qquad \text{... (4.26)}$$

Substituting the values of M, I and y from equations (4.2), (4.3) and (4.4) in equation (4.1), we get,

$$\sigma_w = \frac{(W_T \times h) \cdot \frac{t}{2}}{\frac{1}{12} \cdot b \cdot t^3} = \frac{6 \cdot W_T \cdot h}{b \cdot t^2}$$

$$\therefore \quad W_T = \frac{\sigma_w \cdot b \cdot t^2}{6 \cdot h} \quad \ldots (4.27)$$

Here, 't' and 'h' are the variables depending upon size of tooth and its profile.

Let, $t = x \cdot p_c$ and $h = k \cdot p_c$; where, x and k are constants.

Substituting the above terms of 't' and 'h' in equation (4.27), we get,

$$W_T = \sigma_w \cdot b \cdot \frac{x^2 \cdot p_c^2}{6 \cdot k \cdot p_c}$$

$$\therefore \quad W_T = \sigma_w \cdot b \cdot p_c \cdot \frac{x^2}{6 \cdot k}$$

Let, $\frac{x^2}{6 \cdot k} = y$ = An another constant

Therefore, equation (4.27) becomes,

$$\therefore \quad \mathbf{W_T = \sigma_w \cdot b \cdot p_c \cdot y} \quad \ldots (4.28)$$

This is called as *Lewis equation*. It is applied to the gear or pinion, whichever is weaker element.

Equation (4.28) can also be written as,

$$\mathbf{W_T = \sigma_w \cdot b \cdot \pi \cdot m \cdot y} \quad (\because p_c = \pi \cdot m) \ldots (4.29)$$

where,

$W_T$ = Tangential load on tooth in 'N'

$\sigma_w$ = Permissible working stress in 'N/mm$^2$'

m = module in 'mm'

y = Lewis form factor

**Note :**

(a) When both pinion and gear are made of same material, then pinion is weaker.

(b) When pinion and gear are made of different material, then the product of $(\sigma_w \cdot y)$ or $(\sigma_o \cdot y)$ is deciding factor. Lewis equation is applied (used) to that wheel (either pinion or gear), for which, $(\sigma_w \cdot y)$ or $(\sigma_o \cdot y)$ is less.

## 4.15.1 Systems of Gear Teeth

- Following four systems of gear teeth are commonly used in practice.

1. $14^1/_2°$ Composite system
2. $14^1/_2°$ Full depth involute system
3. 20° Full depth involute system
4. 20° Stub involute system

## 4.17.2 Lewis Form Factor (y)

- Lewis form factor in the Lewis equation is an unitless quantity.
- It is dented by 'y'.
- Value of Lewis form factor is different for different **system of gear teeth**.
- If 'T' is number of teeth, then Lewis form factor for corresponding system of gear teeth is given in the following table.

**Table 4.2: Lewis form factor**

| Sr. No. | System of gear teeth | Lewis form factor (y) |
|---|---|---|
| 1 | $14\frac{1°}{2}$ composite and full depth involute system | $0.124 - \left(\frac{0.684}{T}\right)$ |
| 2 | 20° full depth involute system | $0.154 - \left(\frac{0.912}{T}\right)$ |
| 3 | 20° stub system | $0.175 - \left(\frac{0.861}{T}\right)$ |

### 4.17.3 Permissible Working Stress for Gear Teeth in the Lewis Equation ($\sigma_w$)

- Permissible Working Stress ($\sigma_w$) in the Lewis equation depends upon the material, for which, the allowable static stress ($\sigma_o$) may be determined.
- Allowable static stress is the stress at elastic limit of the material.
- While designing gears, value of allowable static stress is reduced to working permissible stress in order to account for the dynamic effects, which become more severe with increase in pitch line velocity.
- Permissible Working Stress is given by,

$$\sigma_w = \sigma_o \times C_v \quad \text{... (4.30)}$$

where, $\sigma_o$ = Allowable static stress in 'N/mm$^2$'

$C_v$ = Velocity factor

- Please note that, value of $C_v$ is always less than one. Refer article 4.17.4 for more details.

### 4.17.4 Velocity Factor ($C_v$)

- Velocity factor is used to make approximate allowance for the effect of dynamic loading. It is denoted by $C_v$.
- Value of velocity factor mainly depends upon following parameters.
  1. Quality of manufacturing process used for cutting teeth on gears, and
  2. Pitch line velocity (V)
- Following table 4.3 gives the value of velocity factor for various conditions of gears stated on the basis of above parameters.

**Table 4.3: Value of velocity factor**

| Sr. No. | Condition of gears | Value of velocity factor ($C_v$) |
|---|---|---|
| 1. | For **ordinarily cut gears** operating at velocities up to **12.5 m/s.** | $C_v = \frac{3}{3 + V}$ |
| 2. | For **carefully cut gears** operating at velocities up to **12.5 m/s.** | $C_v = \frac{4.5}{4.5 + V}$ |
| 3. | For **accurately cut and ground metallic gears** operating at velocities up to 20 m/s. | $C_v = \frac{6}{6 + V}$ |
| 4. | For **precision gears** cut with high accuracy operating at velocities up to 12.5 m/s. | $C_v = \frac{0.75}{0.75 + \sqrt{V}}$ |
| 5. | For **non-metallic** gears | $C_v = \left(\frac{0.75}{1 + V}\right) + 0.25$ |

### 4.17.5 Static Tooth Load ($W_s$)

- From Lewis equation, **tangential tooth load** is given by,

$$W_T = \sigma_w \cdot b \cdot pc \cdot y$$

- **Static tooth load** is obtained by substituting **flexural endurance limit or elastic limit stress ($\sigma_e$)** in place of permissible working stress ($\sigma_w$).

- Therefore, **Static tooth load** is given by,

$$W_S = \sigma_e \cdot b \cdot pc \cdot y$$

- Static tooth load is also called as **beam strength or endurance strength of tooth**.

### 4.17.6 Dynamic Tooth Load by Buckingham's Equation ($W_D$)

- Gear teeth are subjected to dynamic loads due to following reasons.
  1. Inaccuracies of tooth spacing, 2. Irregularities in tooth profiles, and 3. Deflections of teeth under load.
- **Dynamic tooth load** is determined by using the following **Buckingham equation**.

$$W_D = W_T + W_I \quad \text{... (4.31)}$$

where, $W_T$ = Tangential load on tooth = $\frac{P}{V}$

$W_I$ = Increment load due to dynamic action = $\frac{21V\,(b \cdot C + W_T)}{21V + \sqrt{b \cdot C + W_T}}$

- Equation (4.31) can also be written as,

$$W_D = W_T + \frac{21V\,(b \cdot C + W_T)}{21V + \sqrt{b \cdot C + W_T}} \quad \text{... (4.32)}$$

where, $W_D$ = Dynamic tooth load in 'N'

$W_T$ = Tangential load on tooth in 'N'

$V$ = Pitch line velocity in 'm/s'

$b$ = Facewidth of gear in 'mm'

$C$ = Deformation factor or Dynamic factor.

### 4.17.7 Wear Tooth Load ($W_W$)

- Wear tooth load is determined by using the following Buckingham equation.
- Mathematically, wear tooth load is given by,

$$W_W = D \cdot b \cdot Q \cdot k \quad \text{... (4.33)}$$

where, $Q$ = Ratio factor

$$= \frac{2 \times V.R.}{V.R. + 1} = \frac{2T_G}{T_G + T_P} \quad \text{[For external gearing]}$$

$$= \frac{2 \times V.R.}{V.R. - 1} = \frac{2T_G}{T_G - T_P} \quad \text{[For internal gearing]}$$

where, $V.R. = \frac{T_G}{T_P}$

$D$ = Pitch circle diameter in 'mm'

$k$ = Load stress factor or material combination factor in 'N/mm$^2$

$$= \frac{(\sigma_{es})^2 \sin\phi}{1.4} \cdot \left(\frac{1}{E_P} + \frac{1}{E_G}\right)$$

where, $\sigma_{es}$ = Surface endurance limit in 'N/mm$^2$'

$\phi$ = Pressure angle in 'degrees'

$E_P$ = Young's modulus for pinion material in 'N/mm$^2$'

$E_G$ = Young's modulus for gear material in 'N/mm$^2$'

### 4.17.8 Power Transmission Capacity of Spur Gear in Bending

- Power transmission capacity can be calculated by the relation,

$$W_T = \frac{P}{V} \cdot C_s$$

$$\therefore \quad P = \frac{W_T \cdot V}{C_s} \quad \text{... (4.34)}$$

where, $C_s$ is the service factor.

- For steady load conditions with 8 to 10 hours of service per day, its value can be taken as unity.

$\therefore$ $$C_s = 1$$

- Substituting the value of $C_s = 1$ in equation (4.34), power transmission capacity can be calculated by,

$$P = W_T \cdot V \quad \text{... (4.35)}$$

where, $P$ = Power transmitted in 'Watts'

$W_T$ = Permissible tangential tooth load in 'N'

$V$ = Pitch line velocity in 'm/s'

## 4.18 DESIGN PROCEDURE FOR SPUR GEAR

**Step I : Calculate design value of tangential tooth load :**

- Design value of tangential tooth load is obtained from power transmitted and the pitch line velocity by using the following relation :

$$W_T = \frac{P}{V} \times C_s \quad \text{... (4.36)}$$

where, $C_s$ = Service factor = 1

$W_T$ = Permissible tangential tooth load in 'N'

$P$ = Power transmitted in 'Watts'

$V$ = Pitch line velocity = $\frac{\pi DN}{60}$ in 'm/s'

where, $D$ = Pitch circle diameter in 'm'

$N$ = Speed in 'r.p.m.'

**Step II : Apply Lewis equation to the weaker element (either pinion or gear):**

- From Lewis equation, tangential tooth load or tangential load on tooth is given by,

$$W_T = \sigma_w \cdot b \cdot p_c \cdot y \quad \text{in 'N'}$$

$$\therefore \quad W_T = \sigma_w \times b \times \pi m \times y \quad [\because p_c = \pi m] \quad \text{... (4.37)}$$

where, $\sigma_w$ = Permissible working stress in 'N/mm²'

$b$ = Facewidth of gear in 'mm'

$m$ = Module in 'mm'

$y$ = Lewis form factor

- Equation (4.37) can also be written as,

$$\mathbf{W_T = \sigma_o \times C_v \times b \times \pi m \times y} \quad [\because \sigma_w = \sigma_o \cdot C_v] \quad \text{... (4.38)}$$

- **Apply Lewis equation** to the weaker element of gear drive, which may be either pinion or gear. **To find out weaker element** in gear drive, whether pinion or gear**,** follow the guidelines given below.

a) When both pinion and gear are made up of same material, then pinion is weaker.

b) When both, pinion and gear are made up of different materials;

   (i) The deciding factor is $(\sigma_o \times y)$ or $(\sigma_w \times y)$. Let us consider $(\sigma_o \times y)$.

   (ii) Find $(\sigma_o \times y)_{gear}$ and $(\sigma_o \times y)_{pinion}$. If $(\sigma_o \times y)_{pinion}$ is less, pinion is considered as weaker element. But, if $(\sigma_o \times y)_{gear}$ is less, then gear will be the weaker element.

- If pinion is weaker element, then equation (4.37) becomes,

$$W_T = \sigma_{oP} \times C_v \times b \times \pi m \times y_P \quad \text{... (4.39)}$$

- If gear is weaker element, then equation (4.37) becomes,

$$W_T = \sigma_{oG} \times C_v \times b \times \pi m \times y_G \quad \text{... (4.40)}$$

- Find out **dimensions of gear** by using the **appropriate Lewis equation**, either equation (4.39) or (4.40).

**Step III: Calculate dynamic load ($W_D$) acting on tooth:**

- Using the following **Buckingham equation**, calculate the value of dynamic load ($W_D$) acting on tooth.

$$W_D = W_T + W_I \quad \dots (4.41)$$

where,

$W_T$ = Steady load due to transmission of torque = $\frac{P}{V}$

$W_I$ = Increment load due to dynamic action

$$= \frac{21V\,(b \cdot C + W_T)}{21V + \sqrt{b \cdot C + W_T}}$$

Substituting the value of $W_I$ in equation (4.41), we can write,

$$\mathbf{W_D = W_T + \frac{21V\,(b \cdot C + W_T)}{21V + \sqrt{b \cdot C + W_T}}} \quad \dots (4.42)$$

**Step IV: Check whether static tooth load (i.e. beam strength or endurance strength of tooth) is more than dynamic load ($W_D$) or not?**

- Calculate the value of static tooth load by using the following equation,

$$\mathbf{W_S = \sigma_e \cdot b \cdot p_c \cdot y = \sigma_e \cdot b \cdot \pi m \cdot y} \quad \dots (4.43)$$

- **Check** whether the value of **static tooth load ($W_S$)** calculated using equation (4.43) is greater than value of dynamic load **($W_D$)** calculated in step III or not? If yes, then it is concluded that, **design is safe.**

**Step V: Check whether maximum wear tooth load (i.e. limiting load for wear) is more than dynamic load ($W_D$) or not?**

- Calculate the value of maximum wear tooth load ($W_W$) by using the following equation,

$$\mathbf{W_W = D \cdot b \cdot Q \cdot k} \quad \dots (4.44)$$

where,

D = Pitch circle diameter in 'mm'

b = Facewidth of gear in 'mm'

Q = Ratio factor

$$= \frac{2 \times V.R.}{V.R. + 1} = \frac{2T_G}{T_G + T_P} \quad \text{[For external gearing]}$$

$$= \frac{2 \times V.R.}{V.R. - 1} = \frac{2T_G}{T_G - T_P} \quad \text{[For internal gearing]}$$

where,

$$V.R. = \frac{T_G}{T_P}$$

k = Load stress factor or material combination factor in 'N/mm$^2$'

$$= \frac{(\sigma_{es})^2 \sin\phi}{1.4} \cdot \left(\frac{1}{E_P} + \frac{1}{E_G}\right)$$

$\sigma_{es}$ = Surface endurance limit in 'N/mm$^2$'

$\phi$ = Pressure angle in 'degrees'

$E_P$ = Young's modulus for pinion material in 'N/mm$^2$'

$E_G$ = Young's modulus for gear material in 'N/mm$^2$'

- **Check** whether the value of **maximum wear tooth load ($W_W$)** calculated using equation (4.44) is greater than value of **dynamic load ($W_D$)** calculated in step III or not? If yes, then it is concluded that, **design is safe.**

**Additional Formulae to Solve 'Numerical Problems '**

**(1) Module:** $m = \frac{D}{T}$ in 'mm'

where,

D = Pitch circle diameter in 'mm'

T = Number of teeth on wheel (Gear)

**(2) Pitch line velocity:** $V = \frac{\pi DN}{60}$

where, D = Pitch circle diameter in 'mm'

N = Speed in 'r.p.m.'

**(3) Velocity ratio :** $V.R. = \frac{T_G}{T_P} = \frac{D_G}{D_P} = \frac{N_P}{N_G}$

**(4) Power transmitted :** $P = \frac{W_T \cdot V}{C_s}$ OR $P = W_T \cdot V$

where, $W_T$ = Permissible tangential tooth load in 'N'

$C_s$ = Service factor = 1

For steady load conditions with 8 to 10 hours of service per day, value of $C_s$ can be taken as unity. Therefore, $C_s = 1$

## Numerical Type No. 6 : "Design of Spur Gears"

**Problem 4.17 :** *A bronze spur pinion rotating at 600 r.p.m. drives a cast iron gear at a transmission ratio of 4 : 1. The allowable static stresses for the bronze pinion and cast iron gear are 84 N/mm² and 105 N/mm² respectively. The pinion has 16 standard 20° full depth involute teeth of module 8 mm. Face width of both pinion and gear is 90 mm. Find the power, which can be transmitted from standpoint of strength.*

**Solution : Given data :** $N_P = 600$ r.p.m., $T_P = 16$, Velocity ratio = V.R. = $\frac{T_G}{T_P} = 4$; $\mathbf{T_G} = 4 \times T_P = 4 \times 16 = \mathbf{64}$

$\sigma_{oP} = 84\ N/mm^2$, $\sigma_{oG} = 105\ N/mm^2$, m = 8 mm, b = 90 mm

**Procedure :** We have, Module, $m = \frac{\text{Pitch circle diameter of pinion}}{\text{Number of teeth on pinion}}$

$\therefore \quad m = \frac{D_P}{T_P}$

$\therefore \quad \mathbf{D_p} = m \times T_p = 8 \times 16 = 128\ mm = \mathbf{0.128\ m}$

We have, Pitch line velocity, $V = \frac{\pi D_P \cdot N_P}{60}$

$\therefore \quad \mathbf{V} = \frac{\pi \times 0.128 \times 600}{60} = \mathbf{4.02\ m/s}$

Since pitch line velocity is less than 12.5 m/s, therefore,

Velocity factor, $\mathbf{C_v} = \frac{3}{3+V} = \frac{3}{3+4.02} = \mathbf{0.427}$

We know that, for 20° full depth involute teeth, tooth form factor for the pinion is given by,

$$\mathbf{y_P} = 0.154 - \left(\frac{0.912}{T_P}\right) = 0.154 - \left(\frac{0.912}{16}\right) = \mathbf{0.097}$$

Similarly, tooth form factor for the gear is given by,

$$\mathbf{y_G} = 0.154 - \left(\frac{0.312}{T_G}\right) = 0.154 - \left(\frac{0.912}{64}\right) = \mathbf{0.14}$$

Now, let us calculate $(\sigma_{oP} \times y_P)$ and $(\sigma_{oG} \times y_G)$

$\boldsymbol{\sigma_{oP} \times y_P} = 84 \times 0.097 = \mathbf{8.148}$ ... (1)

$\boldsymbol{\sigma_{oG} \times y_G} = 105 \times 0.14 = \mathbf{14.7}$ ... (2)

On comparing equation (1) and (2), we find that, $(\sigma_{oP} \times y_P)$ is less than $(\sigma_{oG} \times y_G)$. Therefore, we conclude that, pinion is weaker.

Therefore, applying Lewis equation to pinion, we can write,

Tangential load on tooth, $W_T = \sigma_{wP} \cdot b \cdot p_c \cdot y_P$

$\therefore \quad W_T = \sigma_{wP} \times b \times \pi \cdot m \times y_P = \sigma_{oP} \times C_v \times \pi \cdot m \times y_P$ $(\because \sigma_{wP} = \sigma_{oP} \times C_v)$

$\therefore \quad \mathbf{W_T} = 84 \times 0.427 \times \pi \times 8 \times 0.097 = \mathbf{7870\ N}$

**Power Transmitted (P) :**

We have, $\mathbf{P} = W_T \times V = 7870 \times 4.02 = 31640$ Watts = **31.64 kW**

**Problem 4.18 :** *A pair of straight teeth gear is to transmit 20 kW, when the pinion rotates at 300 r.p.m. Velocity ratio is 1 : 3. The allowable static stresses for pinion and gear materials are 120 N/mm² and 100 N/mm² respectively. The pinion has 15 teeth and its face width is 14 times the module. Determine : 1. Module, 2. Face width, 3. Pitch circle diameters of both pinion and gear from standpoint of strength, taking into consideration the effect of dynamic loading.*

*Tooth form factor 'y' can be taken as* $y = 0.154 - \frac{0.912}{\text{Number of teeth}}$ *and Velocity factor as,* $C_v = \frac{3}{3 + V}$

**Solution : Given data :** $P = 20\text{ kW} = \mathbf{20 \times 10^3\ W}$, $N_P = 300$ r.p.m., $T_P = 15$,

$\text{V.R.} = \frac{T_G}{T_P} = 3$; $\therefore$ $\mathbf{T_G} = 3T_P = 3 \times 15 = \mathbf{45}$, $\sigma_{oP} = 120\text{ N/mm}^2$, $\sigma_{oG} = 100\text{ N/mm}^2$, $b = 14 \times m$.

**Procedure :**

**1. Module (m) :**

We know, $$m = \frac{D_P}{T_P}; \quad \mathbf{D_P = m \times T_P}$$

We have, Pitch line velocity, $$V = \frac{\pi D_P N_P}{60} = \frac{\pi \cdot m \cdot T_P \cdot N_P}{60}$$

$$\therefore \quad \mathbf{V} = \frac{\pi \cdot m \times 15 \times 300}{60} = \mathbf{236 \times m} \quad \text{in 'mm/s'}$$

where, module m is in mm.

$$\therefore \quad \mathbf{V} = 236 \times m \times 10^{-3} = \mathbf{0.236\ m} \quad \text{in 'm/s'} \quad \ldots (1)$$

We know that, tangential tooth load is given by,

$$P = W_T \times V$$

$$\therefore \quad \mathbf{W_T} = \frac{P}{V} = \frac{20 \times 10^3}{0.236\ m} = \mathbf{\frac{84745.76}{m}} \quad \ldots (2)$$

We have, Velocity factor, $$\mathbf{C_v} = \frac{3}{3 + V} = \frac{3}{3 + 0.236\ m} \quad \ldots (3)$$

Tooth form factor for pinion is given as,

$$\mathbf{y_P} = 0.154 - \frac{0.912}{T_P} = 0.154 - \frac{0.912}{15} = \mathbf{0.0932} \quad \ldots (4)$$

Similarly, tooth form factor for gear is given as,

$$\mathbf{y_G} = 0.154 - \frac{0.912}{T_G} = 0.154 - \frac{0.912}{45} = \mathbf{0.1337}$$

Now, let us calculate $(\sigma_{oP} \times y_P)$ and $(\sigma_{oG} \times y_G)$.

We get, $$\mathbf{\sigma_{oP} \times y_P} = 120 \times 0.0932 = \mathbf{11.184}$$

Similarly, $$\mathbf{\sigma_{oG} \times y_G} = 100 \times 0.1337 = \mathbf{13.37}$$

From the above calculations, we find that, $(\sigma_{oP} \times y_P)$ is less than $(\sigma_{oG} \times y_G)$. Therefore, we conclude that, pinion is weaker. So, applying Lewis equation to pinion, we can write,

Tangential load on tooth, $$W_T = \sigma_{wP} \cdot b \cdot p_c \cdot y_P$$

$$\therefore \quad W_T = \sigma_{oP} \times C_v \times b \times \pi \cdot m \times y_P \quad [\because p_C = \pi \cdot m \text{ and } \sigma_{wP} = \sigma_{oP} \times C_v]$$

$$\therefore \quad \frac{84745.76}{m} = 120 \times \left(\frac{3}{3 + 0.236\ m}\right) \times 14 \times m \times \pi \times m \times 0.0932$$

[From equation (1), (2), (3) and (4)]

$$\therefore \quad 0.0174\ m^3 - 0.236\ m - 3 = 0$$

$$\therefore \quad m = 6.37\text{ mm} \cong \mathbf{6.8\ mm\ (say)}$$

**2. Face width (b) :** We know, $b = 14 \times m = 14 \times 8 = \mathbf{112\ mm}$

**3. Pitch circle diameters of pinion and gear ($D_P$ and $D_G$) :**

We know, $$D_P = m \times T_P = 8 \times 15 = \mathbf{120\ mm}$$

Similarly, $$T_P = m \times T_G = 8 \times 45 = \mathbf{360\ mm}$$

**Problem 4.19 :** *A gear drive is required to transmit a maximum power of 22.5 kW. The velocity ratio is 1 : 2 and r.p.m. of the pinion is 200. The approximate centre distance between the shafts may be taken as 600 mm. Teeth has 20° stub involute profile. The static stress for the gear material (which is cast iron) may be taken as 60 N/mm² and face width as 10 times the module. Find the module, face width and number of teeth on each gear. Check the design for dynamic and wear loads. The deformation or dynamic factor in the Buckingham equation may be taken as 80 and material combination factor for the wear as 1.4. Take flexural endurance limit for cast iron as 84 N/mm².*

**Solution : Given data :** $\mathbf{P} = 22.5 \text{ kW} = \mathbf{22.5 \times 10^3\ W}$, V.R. $= \frac{T_G}{T_P} = \frac{D_G}{D_P} = 2$; $\therefore \mathbf{D_G = 2D_P}$

$N_P = 200$ r.p.m., $L = 600$ mm, $\boldsymbol{\sigma_{oP} = \sigma_{oG} = 60\ N/mm^2}$, $b = 10 \times m$; $C = 80$, $k = 1.4$, $\sigma_e = 84\ N/mm^2$.

**Procedure : 1. Module (m) :**

Centre to centre distance between two shafts, on which, gear and pinion are mounted, is given by

$$L = \frac{D_P}{2} + \frac{D_G}{2}$$

$$\therefore \quad 600 = \frac{D_P}{2} + \frac{2D_P}{2} = 1.5D_P \qquad [\because D_G = 2D_P]$$

$$\therefore \quad \mathbf{D_P} = \frac{600}{1.5} = 400 \text{ mm} = \mathbf{0.4\ m}$$

and

$$\mathbf{D_G} = 2D_p = 2 \times 400 = 800 \text{ mm} = \mathbf{0.8\ m}$$

It is given that, $\sigma_{oP} = \sigma_{oG} = 60\ N/mm^2$. It means that, both pinion and gear are made up of same material. Therefore, we can observe that, $\sigma_{oP} \times y_P < \sigma_{oG} \times y_G$, because $y_P$ is always less than $y_G$ due to lesser number of teeth. So, we conclude that, pinion is weaker element and therefore, design will be based upon the pinion.

Pitch line velocity of pinion is given by,

$$V = \frac{\pi\ D_P \cdot N_P}{60} = \frac{\pi \times 0.4 \times 200}{60} = \mathbf{4.2\ m/s}$$

Since V is less than 12 m/s, therefore, velocity factor is calculated as,

$$\mathbf{C_v} = \frac{3}{3 + V} = \frac{3}{3 + 4.2} = \mathbf{0.417}$$

We know that,

$$m = \frac{D_P}{T_P}$$

$$\therefore \quad \mathbf{T_P} = \frac{D_P}{m} = \frac{\mathbf{400}}{\mathbf{m}} \qquad \ldots (1)$$

$\therefore$ Tooth form factor for pinion,

$$y_P = 0.175 - \frac{0.841}{T_P} \qquad \text{(For given 20° stub system)}$$

$$\therefore \quad y_P = 0.175 - \frac{0.841}{\left(\frac{400}{m}\right)} = 0.175 - 0.0021\ m \qquad \ldots (2)$$

We know that, tangential teeth load is given by,

$$W_T = \frac{P}{V} \times C_s$$

where, $C_s$ is the service factor and value of $C_s$ can be taken as 1 for steady load conditions with 8 to 10 hours of service per day.

Therefore, above equation becomes,

$$\mathbf{W_T} = \frac{22500}{4.2} \times 1 = \mathbf{5357\ N}$$

Also, we have,

$$W_T = \sigma_{wP} \cdot b \cdot p_c \cdot y_P$$

$$\therefore \quad 5357 = (\sigma_{oP} \times C_v) \cdot b \cdot \pi \cdot m \cdot y_P \qquad [\because \sigma_{wP} = \sigma_{oP} \times C_v \text{ and } p_c = \pi \cdot m]$$

$$\therefore \quad 5357 = (60 \times 0.417) \times 10\ m \times \pi m \times (0.175 - 0.0021\ m)$$

On solving, we get,

$$1.65\ m^3 - 137.6\ m^2 + 5357 = 0$$

$$\therefore \quad m = 6.5 \text{ mm} \cong \mathbf{8\ mm\ (say)}$$

**2. Face width (b) :**

$$\text{Face width, } b = 10\ m = 10 \times 8 = \mathbf{80\ mm}$$

**3. Number of teeth on the gears :**

We know that, $$m = \frac{D_P}{T_P} = \frac{D_G}{T_G}$$

$$\therefore \quad T_P = \frac{D_P}{m} \quad \text{and} \quad T_G = \frac{D_G}{m}$$

$$\therefore \quad T_P = \frac{400}{8} = \mathbf{50} \quad \text{and} \quad T_G = \frac{800}{8} = \mathbf{100}$$

**4. Checking the gears for dynamic and wear load :**

By Buckingham equation, dynamic load is given by,

$$W_D = W_T + W_I$$

$$\therefore \quad W_D = W_T + \frac{21 \times V \times (b \cdot C + W_T)}{21 \times V + \sqrt{b \cdot C + W_T}}$$

$$\therefore \quad W_D = 5357 + \frac{21 \times 4.2 \times (80 \times 80 + 5357)}{21 \times 4.2 + \sqrt{(80 \times 80) + 5357}} = \mathbf{10630\ N}$$

From equation (2), $$\mathbf{y_P} = 0.175 - 0.0021\ m = 0.175 - (0.0021 \times 8) = \mathbf{0.1582}$$

Static tooth load or beam strength or endurance strength of tooth is calculated as,

$$W_S = \sigma_e \times b \times p_c \times y_P = \sigma_e \times b \times \pi \cdot m \times y_P \qquad [\because p_c = \pi \cdot m]$$

$$\therefore \quad W_S = 84 \times 80 \times \pi \times 8 \times 0.1582 = \mathbf{26722\ N} \qquad \dots (4)$$

Maximum or limiting load for wear or wear tooth load is calculated as,

$$W_W = D_p \cdot b \cdot Q \cdot k$$

$$\therefore \quad W_W = 400 \times 80 \times Q \times 1.4 \qquad \dots (5)$$

Ratio factor (Q) is given by, $$\mathbf{Q} = \frac{2 \times V.R.}{V.R. + 1} = \frac{2 \times 2}{2 + 1} = \mathbf{1.33}$$

By substituting value of Q in equation (5), we get,

$$W_W = 400 \times 80 \times 1.33 \times 1.4 = \mathbf{59584\ N} \qquad \dots (6)$$

After comparing equations (3), (4) and (6), it can be observed that, values of $W_S$ and $W_W$ are more than $W_D$. Therefore, we conclude that, design is safe.

---

**Problem 4.20 :** *A pair of straight teeth spur gears, having 20° involute full depth is to transmit 12 kW at 300 r.p.m. of the pinion. The speed ratio is 3 : 1. The allowable static stresses for gear of cast iron and pinion of steel are 60 N/mm² and 105 N/mm² respectively. Assume the following : $T_P$ = 16, Face width = 14 times module, Velocity ratio ($C_v$) = $\frac{4.5}{4.5 + V}$ and tooth form factor, $y = 0.154 - \frac{0.912}{T}$. Determine the module, face width and pitch circle diameters of gears. Check the gears for wear. Take, $\sigma_{es}$ = 600 N/mm², $E_P = 200 \times 10^3$ N/mm², $E_G = 100 \times 10^3$ N/mm².*

**Solution : Given data :** $\phi = 20°$, $P = 12\ kW = 12 \times 10^3\ W$, $N_P = 300$ r.p.m., $V.R. = \frac{T_G}{T_P} = \frac{D_G}{T_P} = 3$, $\sigma_{oG} = 60\ N/mm^2$, $\sigma_{oP} = 105\ N/mm^2$, $T_P = 16$, $b = 14 \times m$, $\sigma_{es} = 600\ N/mm^2$, $E_P = 200 \times 10^3\ N/mm^2$, $E_G = 100 \times 10^3\ N/mm^2$.

**Procedure : 1. Module :**

We have, $$\text{Pitch line velocity, } V = \frac{\pi \cdot D_P \cdot N_P}{60} = \frac{\pi \cdot m \cdot T_P \cdot N_P}{60} \qquad \left(\because m = \frac{D_P}{T_P}\right)$$

$$\therefore \quad V = \frac{\pi \cdot m \times 16 \times 300}{60} = 251 \times m \qquad \text{in 'mm/sec'}$$

$$\therefore \quad \mathbf{V = 0.251\ m} \qquad \text{in 'm/sec'}$$

Assuming steady load conditions and 8-10 hours of service per day, the service factor can be taken as 1.

$$\therefore \quad \mathbf{C_s = 1}$$

We have, design tangential tooth load as,

$$W_T = \frac{P}{V} \times C_s = \frac{12 \times 10^3}{0.251\ m} \times 1 = \mathbf{\frac{47.8 \times 10^3}{m}} \quad \ldots (2)$$

Also, Velocity factor, $C_v = \dfrac{4.5}{4.5 + V} = \dfrac{4.5}{4.5 + 0.251\ m}$ [From equation (1)] ... (3)

Tooth form factor for pinion is calculated as,

$$\mathbf{y_P} = 0.154 - \frac{0.912}{T_P} = 0.154 - \frac{0.912}{16} = \mathbf{0.097}$$

Similarly, tooth form factor for gear is calculated as,

$$y_G = 0.154 - \frac{0.912}{T_G}$$

$$\therefore \quad y_G = 0.154 - \frac{0.912}{3 \times T_p} \qquad \left(\because \frac{T_G}{T_P} = 3\right)$$

$$\therefore \quad \mathbf{y_G} = 0.154 - \frac{0.912}{3 \times 16} = \mathbf{0.135} \quad \ldots (4)$$

We know, $\sigma_{oP} \times y_P = 105 \times 0.097 = \mathbf{10.185}$

Similarly, $\sigma_{oG} \times y_G = 60 \times 0.097 = \mathbf{8.1}$

Since $(\sigma_{oG} \times y_G)$ is less than $(\sigma_{oP} \times y_P)$, therefore, gear is weaker. Therefore, applying Lewis equation to gear, we write,

$$W_T = \sigma_{wG} \times b \times p_c \times y_G$$

$$\therefore \quad W_T = (\sigma_{oG} \times C_v) \times b \times \pi \cdot m \times y_G \qquad [\because \sigma_{wG} = \sigma_{oG} \times C_v \text{ and } p_c = \pi \cdot m]$$

$$\therefore \quad \frac{47.8 \times 10^3}{m} = \left(60 \times \frac{4.5}{4.5 + 0.251\ m}\right) \times 14\ m \times \pi \cdot m \times 0.135 \quad \text{[From eq. (2), (3) and (4)]}$$

$$\therefore \quad \frac{47.8 \times 10^3}{m} = \frac{1603.4\ m^2}{4.5 + 0.251\ m}$$

On solving, we get,

$$0.0335\ m^3 - 0.251\ m - 4.5 = 0$$

$$\therefore \quad m = 5.6 \text{ mm} \cong \mathbf{6 \text{ mm (say)}}$$

**2. Face width (b) :** $b = 14 \times m = 14 \times 6 = \mathbf{84 \text{ mm}}$

**3. Pitch circle diameters of gear and pinion :**

$$D_G = m \times T_G = m \times (3 \times T_P) = 6 \times 3 \times 16 = \mathbf{288 \text{ mm}}$$

$$D_P = m \times T_P = 6 \times 16 = \mathbf{96 \text{ mm}}$$

**4. Checking the gears for wear :**

We know, Ratio factor, $\mathbf{Q} = \dfrac{2 \times V.R.}{V.R. + 1} = \dfrac{2 \times 3}{3 + 1} = \mathbf{1.5}$

Load stress factor, $k = \dfrac{(\sigma_{es})^2 \sin \phi}{1.4} \times \left(\dfrac{1}{E_p} + \dfrac{1}{E_G}\right)$

$$\therefore \quad \mathbf{k} = \frac{(600)^2 \times \sin (20°)}{1.4} \times \left[\frac{1}{200 \times 10^3} + \frac{1}{100 \times 10^3}\right] = \mathbf{1.32 \text{ N/mm}^2}$$

Tangential load on the tooth or beam strength of the tooth is,

$$W_T = \frac{47.8 \times 10^3}{m} \qquad \text{[From equation (2)]}$$

$$\therefore \quad \mathbf{W_T} = \frac{47.8 \times 10^3}{6} = \mathbf{7967\ N}$$

Maximum or limiting load for wear (i.e. wear tooth load) is,

$$\mathbf{W_W} = D_P \cdot b \cdot Q \cdot k = 96 \times 84 \times 1.5 \times 1.32 = \mathbf{15967\ N}$$

Since, maximum wear tooth load ($W_T$) is much greater than tangential load on the tooth, therefore, the design is safe from the standpoint of wear.

## Theory Questions for Practice

1. Enlist the various types of power threads. Give at least one practical application for each type. Discuss their relative advantages and disadvantages.
2. Why square threads are preferable over V-threads for power transmission ?
3. What do you understand by overhauling of screws ?
4. What is self locking property of threads and where it is necessary ?
5. Explain "Efficiency of power screw".
6. Show that, "Efficiency of self locking screw is less than 50 %".
7. Explain "How to improve efficiency of power screw ?"
8. Obtain the expression for efficiency and maximum efficiency for square threaded screw.
9. Explain design procedure of screw jack with neat sketch.
10. Explain design procedure of toggle jack with neat sketch.
11. Enlist the applications of gear drive.
12. State any two materials used for manufacturing gears. Justify their use for making gears.
11. Give the criteria considered, while selecting materials for producing gears.
13. State any six design considerations, while designing the spur gear.
14. State different modes of failure of gear teeth and suggest possible remedies to avoid the failure.
15. Write Lewis equation for strength of gear tooth. Give the meaning and SI unit of each term.
16. Write down the equation for calculating Static Tooth Load. Give the meaning and SI unit of each term.
17. Write down the equation for calculating Dynamic Tooth Load. Give the meaning and SI unit of each term.
18. Write down the equation for calculating Wear Tooth Load. Give the meaning and SI unit of each term.
19. Write down the equation for calculating power transmission capacity of spur gear in bending? Give the meaning and SI unit of each term.
20. Write down the design procedure for spur gear.

## Numerical Problems for Practice

1. The lead screw of a lathe machine, having square thread is required to exert an axial force of 15 kN to the carriage. The coefficient of friction at the screw and collar are 0.1 and 0.15 respectively. Assume compressive stress and shear stress for screw material as 60 MPa and 33.33 MPa respectively, whereas bearing pressure on thread = 10 MPa. Design the screw and nut. If mean diameter of collar is 100 mm and screw rotates at 32 r.p.m., find power required to drive it. (**Ans.** $d_o$ = 30 mm, p = 6 mm, n = 6, H = 36 mm, P = 0.493 kW)

2. A screw jack carries a load of 25 kN. If the co-efficient of friction between screw and nut is 0.15, design the screw and nut. Neglect collar friction and column action. Take $\sigma_c$ = 42 N/mm$^2$ and $\tau$ = 30 N/mm$^2$ for screw and take $\tau_{nut}$ = 20 N/mm$^2$. The permissible bearing pressure on the nut is 14 N/mm$^2$. (Use single start thread).

   (**Ans :** $d_o$ = 46 mm, p = 8 mm, d = 42 mm, n = 4, t = 4 mm, H = 32 mm)

3. A lead screw of a lathe has square threads of 24 mm outside diameter and 5 mm pitch. It supplies drive to a tool carriage, which needs an axial force of 2500 N. A collar bearing with inner and outer radii as 15 mm and 30 mm respectively is provided. The coefficient of friction for screw threads is 0.12 and for collar is 0.1. Find the torque required to drive the screw and the efficiency of the screw.

   If the lead screw rotates at 30 r.p.m., find the power required to drive the screw.

   (**Ans.** T = 10.886 N.m, P = 34.199 kW, $\eta$ = 18.27%)

4. In a machine tool application, the tool holder is pulled by means of an operating nut mounted on a screw. The tool holder travels at a speed of 5 m/min. The screw has single start square threads of 48 mm nominal diameter and 8 mm pitch. The operating nut exerts a force of 500 N to drive the tool holder. The mean radius of the friction collar is 40 mm. The coefficient of friction at thread and collar surfaces is 0.15. Calculate :

   (i) Power required to drive the screw. (ii) The efficiency of the mechanism. (**Ans.** (i) 0.35 kW, (ii) 12%)

5. A 50 kN capacity screw jack consists of a square threaded steel screw meshing with a bronze nut. The nominal diameter is 60 mm and the pitch is 9 mm. The permissible bearing pressure is 10 N/mm$^2$. Calculate :

   (i) The length of the nut. (ii) The transverse shear stress in the nut. (**Ans.** (i) 63 mm, (ii) 8.42 N/mm$^2$)

6. A vertical double start square threaded screw of 80 mm outside diameter and 14 mm pitch, supports a vertical load of 1.5 kN. The axial thrust on screw is taken up by collar bearing having outside diameter as 200 mm and inside diameter as 73 mm. Assuming uniform pressure condition, find the force required at the end of lever 500 mm long to raise and lower the load. Take coefficient of friction as 0.15 for screw and 0.20 for collar.

❑❑❑

*Chapter* **5**

# DESIGN OF SPRINGS

## 5.1 INTRODUCTION TO SPRING

- A *spring* is an elastic body, whose primary function is to deflect or distort under the application of load. When the load is removed, the spring restores its original shape.
- **Spring** is defined as, *"an elastic member, which offers a resisting force to externally applied load causing change in its shape and size (deflection or distortion) and it can regain its original shape and size, when externally applied load is removed"*.

### 5.1.1 Functions and Applications of Spring

Following table illustrates different functions performed by springs along with practical applications.

| Sr. No. | Function performed by springs | Practical applications and examples |
|---|---|---|
| 1. | **To cushion, absorb or control energy due to shock or vibration.** | (i) Car springs<br>(ii) Shock absorbers<br>(iii) Vibration damper<br>(iv) Buffer springs of railways |
| 2. | **To apply forces.** | (i) Brakes of an automobile<br>(ii) Clutch of an automobile<br>(iii) Spring loaded valves of I.C. engines, steam engines, reciprocating compressors etc. |
| 3. | **To measure forces.** | (i) Spring balances<br>(ii) Rope brake dynamometer used to measure engine power. |
| 4. | **To store energy.** | (i) Watches<br>(ii) Toys |
| 5. | **To control motion by maintaining contact between two elements.** | (i) Cam and follower |
| 6. | **To change vibration characteristics of a member.** | (i) Flexible mounting of motors |
| 7. | **To provide clamping force.** | (i) Jigs and fixtures |

### 5.1.2 Classification of Spring

**Classification of Springs**

- Helical springs
  - (a) Helical compression
  - (b) Helical tension
- Spiral spring
- Leaf springs
  - (a) Elliptical
  - (b) Semi-elliptical
  - (c) Cantilever
- Belleville spring
- Volute spring

**Chart 1 : Classification of spring**

- Various types of springs are classified according to their shape in following way.

## 1. Helical Springs :

- Helical springs are made of a wire wound in the form of helix.
- Cross-section of spring wire may be circular, square and rectangular depending upon the type of loading and deflection required.
- A compressive or tensile load is applied on helical spring. Axis of load applied on spring is same as the axis of spring or helix.
- Therefore, we say, helical spring is designed to withstand this axial load. Major stress induced in helical spring is torsional shear stress due to twisting moment or torque, when axial load (W) is applied.
- Depending upon type of axial load, helical spring is further classified as:

### (i) Helical Compression Spring :

- It is used to withstand an axial compressive load.
- Coils of the spring are separate from each other.

**Applications :**

(a) Clutches.

(b) Brakes.

(c) Valve mechanisms of I.C. engines.

(d) Telescopic shock absorbers.

### (ii) Helical Tension Spring :

- It is used to withstand an axial tensile load.
- Coils of spring are held closed to each other. Here, hooks are provided at the ends of spring for applying loads.

**Applications :**

(a) Spring loaded safety valve.

(b) Spring balances.

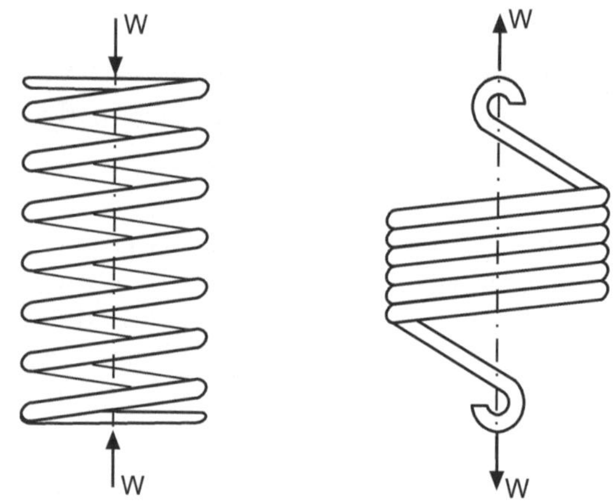


**(a) Helical compression spring** **(b) Helical tension spring**

**Fig. 5.1 : Helical springs**

- Depending upon the value of helix angle, helical springs are also classified as :

### (i) Close Coiled Helical Spring :

- Here, helix angle (α) is very small. Usually its value is less than 10°. Plane containing each coil is nearly perpendicular to the axis of helix.
- Spring wire is so coiled that, there exists a small gap between the two adjacent coils.

### (ii) Open Coiled Helical Spring :

- Here, helix angle (α) is more than 10°.
- Spring wire is so coiled that, there is a large gap between two adjacent coils.

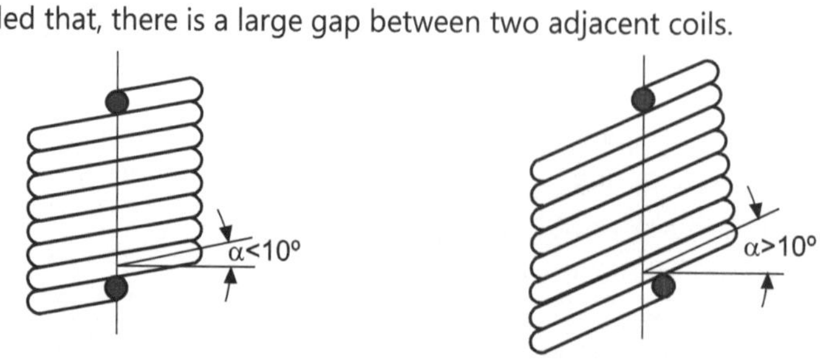


**(a) Close coiled helical spring** **(b) Open coiled helical spring**

**Fig. 5.2**

## 2. Conical and Volute Springs :

- In conical as well as volute spring, coils are arranged in the shape of frustum of cone.
- Wire used in both springs may have either rectangular or circular cross-section.
- When conical or volute spring is loaded axially, each element of spring behaves like a curved bar in tension.
- They are used, when load is increasing. Therefore, to withstand against increasing load, higher stiffness is required.

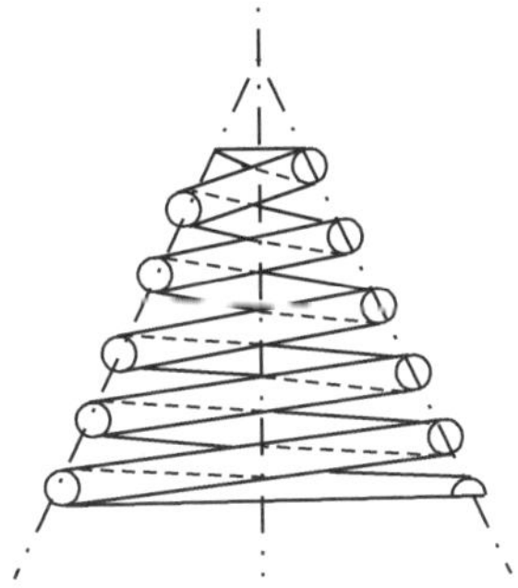

**Fig. 5.3 : Conical spring**

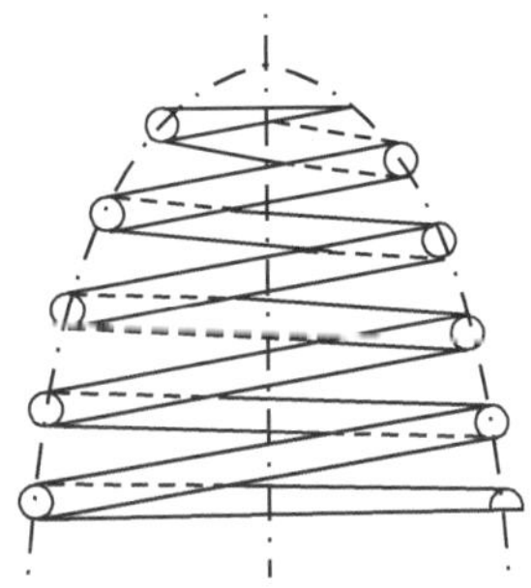

**Fig. 5.4 : Volute spring**

- Conical spring is wound with uniform pitch, whereas, volute spring is wound in the form of parabolid (parabolic shape) with uniform pitch and lead angle.
- Both springs may be either partially or completely telescoping. In both cases, number of active coils gradually decreases, which results in an increasing spring rate (also called as stiffness).

## 3. Torsional Springs :

- Torsional springs are classified in to two types:

**(i) Spiral torsional spring :** It is made up of a flat spring, which is wounded in the form of spiral and loaded in torsion.

**Applications of Spiral Torsional Springs :**

(a) Mechanical watches.

(b) Clocks.

**(ii) Helical torsional spring :** It is manufactured by wounding a wire like compression spring, where load is applied tangentially. It is subjected to a twisting moment or torque acting about centroidal axis of spring coil. This induces bending stress in the spring wire.

**Applications of Helical Torsional Springs :**

(a) Door hinges. (b) Automobile starters. (c) Levers.

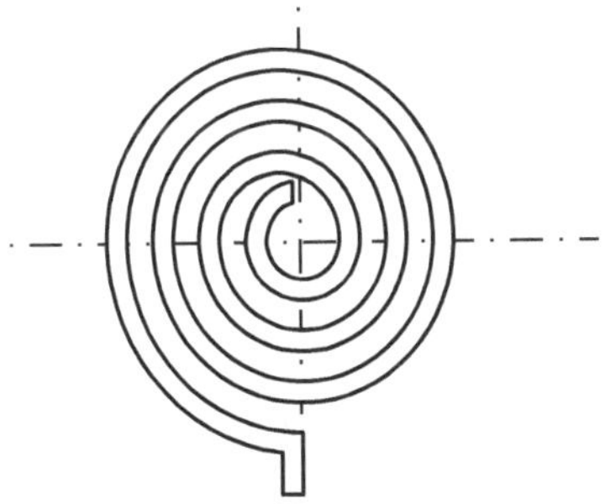

**(a) Spiral torsional spring**

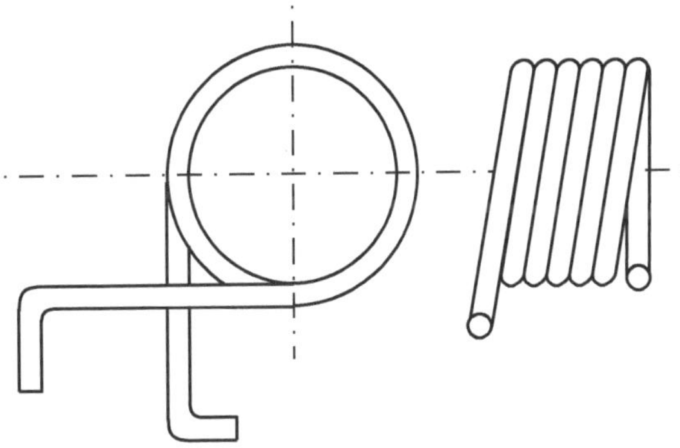

**(b) Helical torsional spring**

**Fig. 5.5 : Torsional springs**

**4. Leaf Spring :**

- A leaf spring has semi-elliptical form. Refer Fig. 5.6.
- It consists of number of leaves held together by means of clamps and bolts.
- Major function of leaf spring is **'To damp out vibrations or shocks'**.
- Leaf spring is also able to carry lateral loads, brake torque, driving torque etc. in addition to vibrations and shocks.

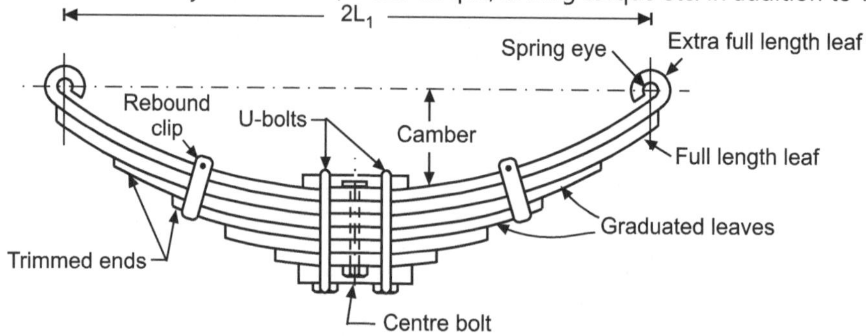


**Fig. 5.6 : Leaf spring**

- To understand constructional features of leaf spring in detail, Fig. 5.16 may be referred.

**Applications :**

(a) Automobile suspension system.
(b) Shock absorber in railways.

**5. Belleville Spring :**

- Belleville spring consists of several annular discs of conical shape.
- Desired 'Load Deflection' characteristic can be obtained by using these multiple number of conical discs.
- These conical discs are arranged as one on top of other, in order to increase the deflection.

**Applications :**

(a) Belleville springs are used, where high capacity compression springs must fit into small spaces.
(b) Belleville spring is used in the application, where spring having higher stiffness is desired or required.

**Fig. 5.7 : Belleville spring**

## 5.1.3 Advantages of Helical Spring 1.

1. Highly reliable, and hence, prediction regarding performance of helical spring is more accurate.
2. 'Load-Deflection' characteristics can be varied by changing the dimensions of spring.

## 5.1.4 Advantages of Helical Compression Spring Over Helical Tension Spring

- Helical tension (also called as extension) springs are rarely preferred to use as compared to helical compression springs due to following reasons.
  1. In case of helical tension springs, more stress concentration occurs at ends of coil than other portions due to presence of hook. This leads to uneven or non-uniform stress distribution. But, in case of helical compression springs, hooks are not used. Therefore, very less stress concentration occurs at ends of coil.
  2. Helical tension springs are likely to be stressed beyond elastic limit. Under such conditions, spring will not be restored to original shape after removal of load.
  3. Complex manufacturing.

## 5.1.5 Advantages of Torsional and Spiral Springs

1. They have ability to store more amount of strain energy.
2. They can be used in transmission of small torque.
3. Compact and occupy less space.

### 5.1.6 Practical Application of Each Type of Spring

| Sr. no. | Type of Spring | Applications |
|---|---|---|
| 1. | **Helical spring** | Engine valve mechanism. |
| 2. | **Torsion spring** | Clock and watches. |
| 3. | **Leaf spring** | Automobiles. |
| 4. | **Belleville spring** | Used, where high spring rates are needed. |
| 5. | **Conical and volute spring** | Used, where increasing spring rate is desired to withstand against increasing load. |

### 5.1.7 Difference between Close Coiled and Open Coiled Helical Spring

| Close coiled helical spring | Open coiled helical spring |
|---|---|
| 1. In this type of spring, the spring wire is coiled, such that, all turns or coils are very close to each other and gap between two adjacent or consecutive turns is very small. We observe that, plane containing each turn is nearly right angles to the axis of helix. | 1. In this type of spring, spring wire is coiled, such that, gap between two adjacent or consecutive turns is considerable. |
| 2. This type of spring is subjected to torsion only. | 2. This type of spring can take up compressive load in addition to torsion. |

### 5.1.8 Difference between Tension Spring and Torsion Spring

| Tension spring | Torsion spring |
|---|---|
| 1. This type of spring is provided with hooks at the ends. Hooks may be made by turning whole coil or half of the coil. | 1. This type of spring is wound in a similar way as tension spring, but its ends are so shaped that, it can transmit torque. |
| 2. This type of spring is primarily subjected to *torsional shear stresses.* | 2. This type of spring is primarily subjected to *bending stress.* |
| 3. This type of spring is used, where constant spring rate is required. | 3. This type of spring is used for transmitting torques in door hinges, automobile starters etc. |

## 5.2 TERMS USED IN HELICAL SPRING

**1. Solid height / Solid length :**

- When compression spring is compressed until the coils come in contact with each other, then spring is said to be solid.
- "*Solid length of a spring is the product of total number of coils and diameter of wire*".
- Mathematically, Solid length = Total number of coils × Diameter of spring wire

$$\therefore \quad \mathbf{L_s = n' \times d} \quad \text{in 'mm'}$$

where, n' = Total number of coils, and

d = Diameter of spring wire in 'mm'

**2. Free length :**

- "*Free length is the length of spring, when the spring is in free or unloaded condition*".
- $$\text{Free length} = \text{Solid length} + \begin{bmatrix}\text{Max. deflection}\\ \text{(Compression)}\end{bmatrix} + \begin{bmatrix}\text{Clearance between}\\ \text{adjacent coils}\end{bmatrix}$$

$$\therefore \quad \mathbf{L_f = n' \cdot d + \delta_{max} + 0.15\, \delta_{max}}$$

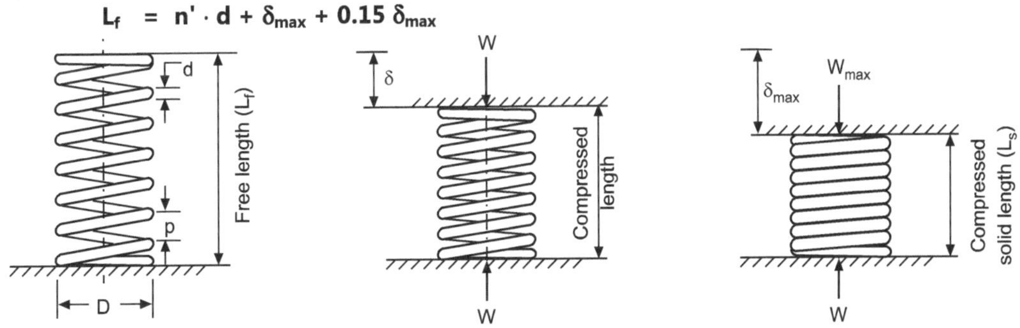


**(a) Free or unloaded condition** **(b) Loaded condition** **(c) Compressed solid (Fully loaded)**

**Fig. 5.8 : Nomenclature of helical compression spring**

**3. Spring index :**

- Spring index is defined as, "*ratio of mean diameter of spring coil to diameter of spring wire*".
- Mathematically, spring index is expressed as,

$$C = \frac{D}{d}$$

- Generally, value of spring index lies between 5 to 10. Spring index is unitless quantity.

**4. Stiffness / Spring rate / Spring constant :**

- Stiffness is also known as spring rate or spring constant.
- Mathematically, stiffness is expressed as, $S = \frac{W}{\delta}$ in 'N/mm'

**5. Pitch :**

- Pitch is defined as, "*the axial distance between the two adjacent coils of spring in uncompressed state*".
- In other words, when spring is in free or unloaded condition, then the axial distance measured in between two adjacent coils is **pitch.**
- Mathematically, pitch is expressed as, $p = \frac{\text{Free length}}{n' - 1}$ in 'mm'

## 5.2.1 Material with Specifications Used for Springs

- **Various materials used for spring are :**

1. Plain carbon steel having carbon 0.7 to 1%.
2. Medium carbon alloy steels.
3. Corrosion resistant steel.
4. Non-ferrous spring materials such as phosphor, bronze, brass, nickel alloys and beryllium copper.

- Selection of material for spring depends upon the type of application and load (static load or fluctuating load).
- **Some of the spring steels are :**

(a) Patented and cold drawn steel wires of Grade I, II, III and IV.

(b) Oil hardened and tempered spring steel wires :

   (i) Chromium – Vanadium spring steel with Cr-1.5%, V-0.2%, C-0.35% to 5%.

   (ii) Silicon – Manganese spring steel with Si-2%, Mn-1%, C-0.6 to 1.2%.

## 5.2.2 Desirable Properties of Spring Materials

(a) High resilience. (b) Highly ductile.

(c) High static strength. (d) High fatigue strength. (e) Non-corrosive.

## 5.2.3 Manufacturing of Springs

- Springs are manufactured either by hot or cold working processes, depending upon the size of wire, spring index and types of properties desired.
- Small size springs are cold wound, whereas large size springs are hot wound.
- During winding, unwanted residual stresses are induced in spring wires. A suitable heat treatment process like annealing is carried out to relieve these stresses. Pre-hardened wire should not be used for springs having spring index less than 4, i.e. $C < 4$.
- Springs are heat treated in different ways to obtain the desired mechanical properties. For example : Oil quenching followed by tempering to adjust the hardness of wire. Care must be taken to avoid formation of cracks due to quenching.
- Special toughing treatments are carried out to have high resilience and toughness in spring wire.

## 5.2.4 Types of Spring Ends for Helical Compression Springs

- End connections of helical compression springs are suitably formed in order to apply the load.
- In springs, the end coils produce an eccentric application of load, increasing stress on one side of spring. When the length of spring is small, this effect must be taken into consideration.

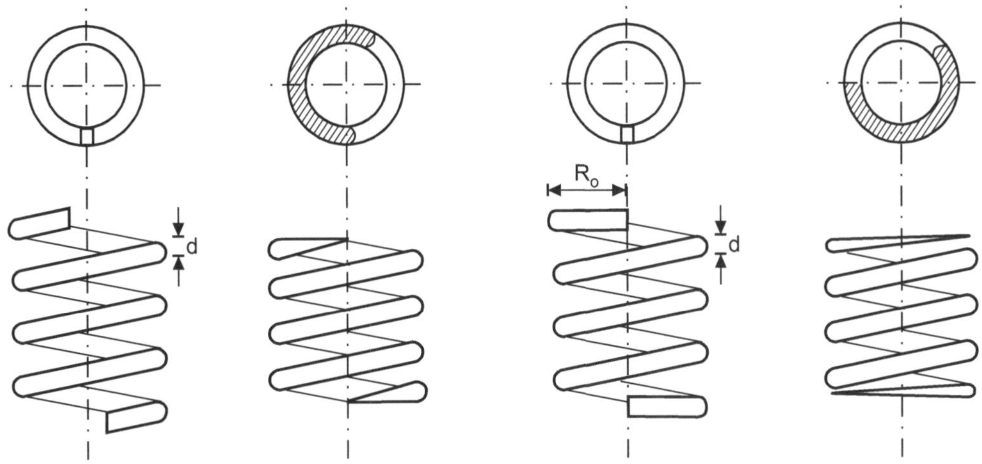


(a) Plain ends (b) Plain and grounded ends (c) Squared ends (d) Squared and grounded ends

**Fig. 5.9 : Types of ends**

- The commonly used spring ends are of four types.

**(a) Plain ends :** In this, the load axis does not coincide with the axis of spring coil. Hence, they are not much used in practice.

**(b) Plain and grounded ends :** In this, grounding the ends eliminates the drawback of plain ends. The total seat area is sufficient for providing spring support.

**(c) Squared ends :** In this, the end coils are pressed, such that, they become square to the axis of spring. Due to this, total seat area for spring support increases. But, as the ends are not grounded, proper seating is not available.

**(d) Squared and grounded ends :** Nearest approach to axial loading is secured by the use of squared and grounded ends. Thus, they are widely used in helical compression springs.

**Inactive turns :** The portion of coil, which is in contact with seat and do not contribute in spring action, is said to be inactive. The turns involved are called as *inactive turns*. Inactive turns may be 0 or 2, depending upon the type of spring ends.

**Active turns (n) :** The turns, which take part in spring action, are called as *active turns*.

| Sr. No. | Type of ends | Total number of turns (n') |
|---|---|---|
| 1. | Plain ends | n |
| 2. | Plain and grounded ends | n |
| 3. | Squared ends | n + 2 |
| 4. | Squared and Grounded ends | n + 2 |

where, n = number of active turns or coils.

## 5.2.5 Types of Ends for Helical Tension Spring

- Design of helical tension spring differs from design of helical compression due to,
  - (a) Types of spring ends.
  - (b) Bending stress in tension springs.
  - (c) Initial tension in the springs.
- In case of helical tension springs, the ends are formed like hooks for applying loads, so that, stress concentration effects caused due to presence of bends are minimized.
- Fig. 5.10 shows different types of ends for tension loading.

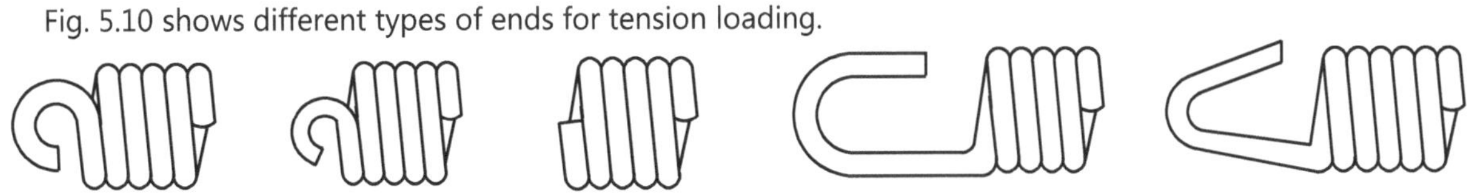

Full loop Small eye Plain square cut ends Long round end hook V hook

**Fig. 5.10 : Spring ends for helical tension spring**

- Here, all coils are active coils.

  Therefore, Total number of coils (n') = Number of active coils (n).

## 5.3 DESIGN OF HELICAL SPRING

### 5.3.1 Stresses in Helical Springs

- Consider a helical compression spring of circular cross-section subjected to an axial load, as shown in Fig. 5.11.

Let,

| | | |
|---|---|---|
| D | = Mean diameter of spring coil | in 'mm' |
| d | = Diameter of spring wire | in 'mm' |
| n | = Number of active coils | |
| G | = Modulus of rigidity for spring material | in 'N/mm²' |
| W | = Axial load on the spring | in 'N' |
| $\tau$ | = Maximum shear stress induced in the wire | in 'N/mm²' |
| C | = Spring index $= \frac{D}{d}$ | |
| p | = Pitch of spring coil | in 'mm' |
| $\delta$ | = Axial deflection of spring | in 'mm' |

- Axial load 'W' tends to produce twisting moment in the wire. To resist the twisting moment, torsional shear stress '$\tau_1$' gets induced in the wire. The torsional moment or torque is given by,

$$T = W \times \frac{D}{2}$$

Also,

$$T = \frac{\pi}{16} \tau_1 d^3$$

Equating the above two equations, we get,

$$\boldsymbol{\tau_1 = \frac{8\,WD}{\pi d^3}} \quad \text{... (Neglecting the stress due to curvature of wire) ... (5.1)}$$

- In addition to torsional shear stress ($\tau_1$), the following stresses are also induced in wire.
  (a) Direct shear stress ($\tau_2$) due to axial load 'W'.
  (b) Stress due to curvature of wire. (For details, Refer Article 5.2.2).
- Direct shear stress ($\tau_2$) due to load 'W' is given by,

$$\tau_2 = \frac{\text{Load}}{\text{Cross-sectional area of spring wire}} = \frac{W}{\frac{\pi}{4} \cdot d^2} = \boldsymbol{\frac{4W}{\pi \cdot d^2}} \quad \text{... (5.2)}$$

- The resultant shear stress ($\tau$) can be determined as, algebraic sum of two shear stresses, $\tau_1$ and $\tau_2$ from equations (5.1) and (5.2).

$$\tau = \tau_1 + \tau_2$$

$$\therefore \quad \tau = \frac{8\,WD}{\pi d^3} + \frac{4W}{\pi d^2} = \frac{8\,WD}{\pi d^3} \cdot \left(1 + \frac{d}{2D}\right) \quad \left(\text{Taking } \frac{8\,WD}{\pi d^3} \text{ out of bracket}\right)$$

$$\therefore \quad \tau = \frac{8\,WD}{\pi d^3} \cdot \left(1 + \frac{1}{2C}\right) \quad \left(\because C = \frac{D}{d}\right)$$

$$\therefore \quad \boldsymbol{\tau = K_s \cdot \frac{8\,WD}{\pi d^3} = K_s \times \frac{8\,WC}{\pi d^2}} \quad \text{... (5.3)}$$

where, $K_s$ = Shear stress factor $= 1 + \frac{1}{2C}$

and $C$ = Spring index $= \frac{D}{d}$

- When the stress due to curvature of wire is to be considered, equation (5.3) is modified as,

$$\boldsymbol{\tau = K \cdot \frac{8\,WD}{\pi d^3} \text{ or } K \cdot \frac{8\,WC}{\pi d^2}} \quad \text{... (5.4)} \left(\because C = \frac{D}{d}\right)$$

This equation (5.4) is the design equation used to determine various dimensions of spring. It can be termed as design equation for designing helical springs.

where, $K$ = Wahl's correction factor = $\frac{4C-1}{4C-4}+\frac{0.615}{C}$

$K$ is an unitless quantity.

$W$ = Load applied on spring in 'N'

$d$ = Diameter of spring wire in 'mm'

$D$ = Mean diameter of spring coil in 'mm'

$\tau$ = Maximum shear stress induced in spring wire in 'N/mm$^2$'

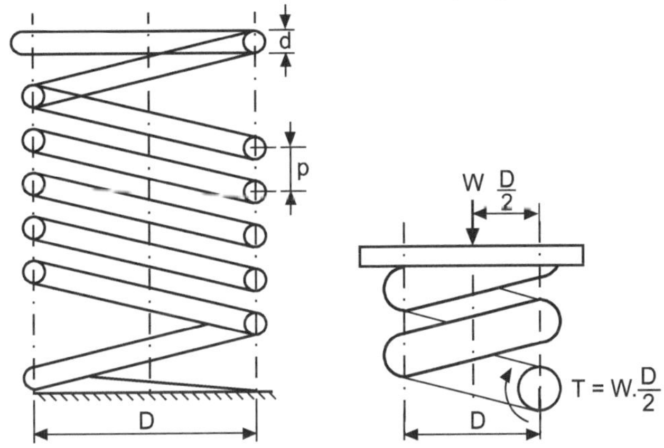


**Fig. 5.11 : Stresses in helical springs**

### 5.3.2 Wahl's Correction Factor

- In order to find the dimensions of spring, the maximum shear stress induced in the wire is considered, which is given by,

$$\tau = K_s \times \frac{8\,WD}{\pi d^3} \quad \text{... [Refer equation 5.3] ... (5.5)}$$

where, $K_s$ = Shear stress factor or Shear multiplication factor = $1+\frac{1}{2C}$

- In the above equation (5.5), we have considered two types of stresses, (i) Torsional shear stress and (ii) Direct shear stress due to direct load W.

**Effect of Curvature of Coil on Stress Distribution :**

- We have not considered stress induced in the spring due to curvature of wire in equation (5.5).
- When a bar is bent in the form of helical coil, the curvature of coil increases stress on inner side of the spring and at the same time, curvature of coil decreases stress on outer side of spring. Thus, the length of inner fiber is less than the length of outer fiber. Therefore, inner fiber gets subjected to stress concentration. The effect of stress concentration is severe and considerably larger for high values of spring index.
- Therefore, a *Wahl's stress factor* is introduced in the above equation (5.5), and the equation is modified as,

$$\tau = K \times \frac{8\,WD}{\pi d^3} \quad \text{(Refer equation 5.4)}$$

where, $K$ = Wahl's stress factor or Wahl's correction factor or Spring stress factor = $\frac{4C-1}{4C-4}+\frac{0.615}{C}$

$W$ = Load in 'N'

$D$ = Mean diameter of spring in 'mm'

$d$ = Diameter of spring wire in 'mm'

$C$ = Spring index = $\frac{D}{d}$

- Wahl's correction factor is also considered as composed of two sub-factors $K_s$ and $K_c$, such that,

$$K = K_s \cdot K_c$$

where, $K_s$ = Shear stress factor, and

$K_c$ = Stress concentration factor due to wire curvature effect

- To plot the graph of Wahl's stress factor versus spring index, we will put the values of C as 3, 4 ... upto 16 in the equation of Wahl's stress factor, i.e. $K = \frac{4C-1}{4C-4}+\frac{0.615}{C}$ to obtain the values of K.

**Table 5.1 : Values of Wahl's stress factor for varying spring index**

| Spring index 'C' | Wahl's stress factor |
|---|---|
| 3 | 1.58 |
| 4 | 1.4 |
| 5 | 1.31 |
| 6 | 1.25 |
| 7 | 1.21 |
| 8 | 1.18 |
| 9 | 1.16 |
| 10 | 1.14 |

And so on .........

- Taking spring index (C) on X-axis and Wahl's stress factor (K) on Y-axis, we can plot the graph with the help of values of C and K from the above table.
- Following graph shows the variation of Wahl's stress factor with varying value of spring index.

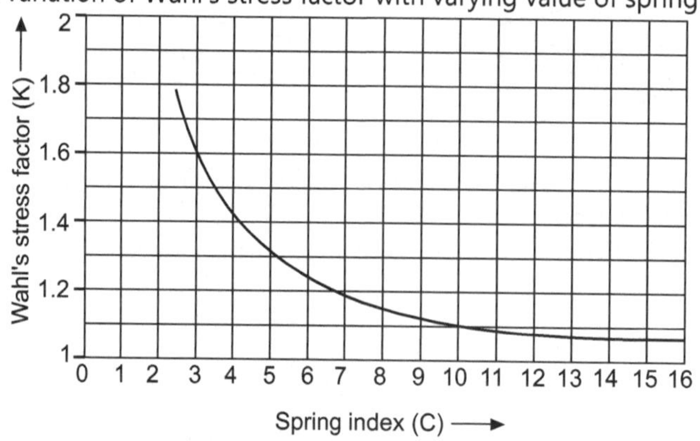


**Fig. 5.12 : Graph of Wahl's stress factor versus Spring index**

## 5.3.3 Deflection of a Helical Compression Spring

Let, $D$ = Mean diameter of spring coil in 'mm'

$n$ = Number of active coils

Therefore, Total active length of wire,

$$L = (\text{Length of one coil}) \times (\text{Number of active coils})$$

$$\therefore \quad L = \pi D \cdot n \quad \text{in 'mm'} \quad \dots (5.6)$$

- The twisting moment or torque developed due to axial load W is given by,

$$T = W \times \frac{D}{2} \quad \text{in 'N/mm}^2\text{'} \quad \dots (5.7)$$

We know that, $J$ = Polar M.I. of spring wire $= \frac{\pi}{32} d^4$ in 'mm$^4$' ... (5.8)

- If $\theta$ = Angular deflection of wire, when subjected to torque 'T' in 'radians'

Then, Axial deflection of spring,

$$\delta = \theta \times \frac{D}{2} \quad \text{in 'mm'} \quad \dots (5.9)$$

- From basic torsion equation, we have,

$$\frac{T}{J} = \frac{G\theta}{L}$$

$$\therefore \quad \theta = \frac{T \cdot L}{G \cdot J}$$

$$\therefore \quad \theta = \frac{\left(W \times \frac{D}{2}\right) \times (\pi D n)}{G \cdot \left(\frac{\pi}{32} d^4\right)} \quad \text{[From equations (5.6), (5.7) and (5.8)]}$$

$$\therefore \quad \theta = \frac{16\ WD^2 n}{Gd^4}$$

Put this value of θ in equation (5.9) to obtain axial deflection of spring,

$$\delta = \frac{16\ WD^2 n}{Gd^4} \times \frac{D}{2}$$

$$\therefore \quad \boldsymbol{\delta = \frac{8\ WD^3 n}{Gd^4} = \frac{8\ WC^3 n}{Gd}} \qquad \left(\because C = \frac{D}{d}\right)$$

**Note :** If deflection per active turn is to be calculated, then take, $\frac{\delta}{n} = \frac{8WC^3}{G \cdot d}$.

### 5.3.4 Energy Stored in Helical Spring of Circular Wire

- If the spring material is not stressed beyond the elastic limit, the graph of load (W) versus axial deflection (δ) will be a straight line as shown in Fig. 5.13.
- Area under the curve gives the amount of energy stored in the spring.
- Therefore, energy stored in a spring is given by considering the triangle (shaded) area,

$$U = \frac{1}{2} \times W \times \delta \qquad \dots (5.10)$$

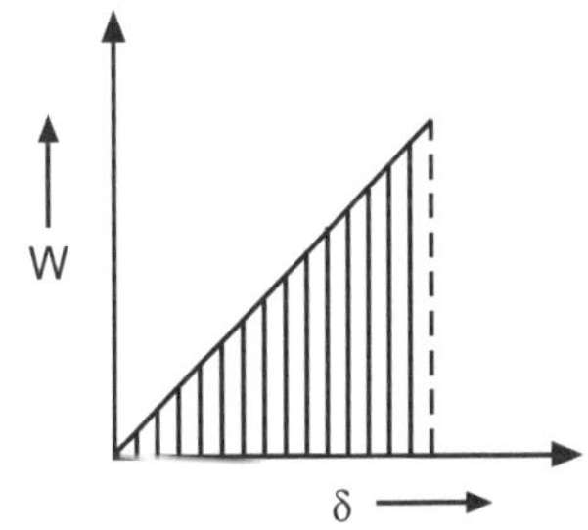


**Fig. 5.13 : Energy stored in spring shown with the help of 'Load versus Deflection' graph**

- We have, the maximum shear stress induced in the spring as,

$$\tau = K \times \frac{8\ WD}{\pi d^3}$$

$$\therefore \quad W = \frac{\pi \cdot \tau \cdot d^3}{8 \cdot K \cdot D} \qquad \dots (5.11)$$

And deflection as,

$$\delta = \frac{8\ WD^3 n}{Gd^4} \qquad \dots (5.12)$$

Put the values of W and δ in equation (5.10), we have,

$$U = \frac{1}{2} \times \frac{\pi \cdot \tau \cdot d^3}{8 \cdot K \cdot D} \times \frac{8\ WD^3 n}{Gd^4}$$

Again, put the value of W from equation (5.11) in the above expression, we get,

$$U = \frac{1}{2} \times \frac{\pi \cdot \tau \cdot d^3}{8 \cdot K \cdot D} \times \frac{8 \cdot \pi \cdot \tau \cdot d^3}{8 \cdot K \cdot D} \cdot \frac{D^3 n}{Gd^4}$$

$$\therefore \quad U = \frac{\tau^2}{4K^2 G} \cdot (\pi D n) \cdot \left(\frac{\pi}{4} d^2\right)$$

$$\therefore \quad \boldsymbol{U = \frac{\tau^2}{4K^2 G} \cdot V} \qquad \dots (5.13)$$

where, V = Volume of spring wire

∴ V = Length of spring × Cross-sectional area of spring wire

$$\therefore \quad V = (\pi D n) \cdot \left(\frac{\pi}{4} d^2\right) \quad \text{in 'mm}^3\text{'} \qquad \dots (5.14)$$

- When a load (say P) drops on a spring through a height h (case of impact loading), then the energy absorbed in a spring is given by,

$$\boldsymbol{U = P\,(h + \delta) = \frac{1}{2} \times W \times \delta} \qquad \dots (5.15)$$

where, W = Equivalent static load i.e. gradually applied load, which shall produce the same effect as by the falling load 'P',

δ = Axial deflection produced in the spring.

**Numerical Type No. 1 : "Design of Helical Spring Neglecting the Stress Concentration, Neglecting the Stress due to Curvature of Wire"**

**Problem 5.1 :** *A helical compression coil spring made up of alloy steel has following specifications : Mean diameter = 50 mm, Wire diameter = 5 mm, Number of active coils = 20.*

*If the spring is subjected to an axial load of 500 N, calculate the maximum shear stress, to which, the spring material is subjected. Neglect the curvature effect.*

**Solution : Given data :** D = 50 mm, d = 5 mm, n = 20, W = 500 N

**Procedure :** We have, Spring index, $\mathbf{C} = \frac{D}{d} = \frac{50}{5} = \mathbf{10}$

The axial load 'W' tends to produce twisting moment in the spring wire, due to which, torsional shear stress is induced in spring wire.

Torsional shear stress in the spring wire (neglecting the curvature effect) is given by,

$$\tau = K_s \times \frac{8\,WD}{\pi d^3} = \left(1 + \frac{1}{2C}\right) \times \frac{8\,WD}{\pi d^3} = \left(1 + \frac{1}{2 \times 10}\right) \times \frac{8 \times 500 \times 50}{\pi \times (5)^3} = \mathbf{534.76\ N/mm^2}$$

**Problem 5.2 :** *A closed coil helical spring is made of 10 mm diameter steel wire. Spring consists of 10 complete turns with a mean diameter of 120 mm. The spring carries an axial pull of 200 N. Determine the shear stress induced in the spring, neglecting the effect of stress concentration. Determine the deflection in spring, its stiffness and strain energy stored by it, if modulus of rigidity of spring material is 60 kN/mm$^2$.*

**Solution : Given data :** D = 120 mm, d = 10 mm, n = 10, W = 200 N, **G** = 60 kN/mm$^2$ = $\mathbf{60 \times 10^3\ N/mm^2}$.

**Procedure : (i) Shear stress induced :**

Shear stress induced in the spring (neglecting the stress concentration) is,

$$\tau = \frac{8\,WD}{\pi d^3} = \frac{8 \times 200 \times 120}{\pi \times 10^3} = \mathbf{61.115\ N/mm^2}$$

**(ii) Deflection in spring :**

$$\text{Deflection, } \delta = \frac{8\,WD^3 n}{Gd^4} = \frac{8 \times 200 \times (120)^3 \times 10}{60 \times 10^3 \times 10^4} = \mathbf{46.08\ mm}$$

**(iii) Stiffness :** Stiffness, $S = \frac{W}{\delta} = \frac{200}{46.08} = \mathbf{4.3402\ N/mm}$

**(iv) Strain energy stored :**

$$U = \frac{1}{2} \times W \times \delta = \frac{1}{2} \times 200 \times 46.08 = \mathbf{4608\ N.mm = 4.608\ N.m\ or\ J}$$

**Nunerical Type No. 2: "Design of Helical Spring Considering Stress due to Curvature of Wire"**

**Problem 5.3 :** *A helical spring is made from a wire of 8 mm diameter and has outside diameter 90 mm. If the permissible shear stress is 350 N/mm$^2$ and modulus of rigidity is 84 kN/mm$^2$, find the axial load, which the spring can carry and the deflection per active turn.*

*(i) Neglecting the effect of curvature. (ii) Considering the effect of curvature.*

**Solution : Given data :** d = 8 mm, $D_o$ = 90 mm, $\tau$ = 350 N/mm$^2$, **G** = 84 kN/mm$^2$ = $\mathbf{84 \times 10^3\ N/mm^2}$.

**Procedure :** We have, $\mathbf{D} = D_o - d = 90 - 8 = \mathbf{82\ mm}$

$$\mathbf{C} = \frac{D}{d} = \frac{82}{8} = \mathbf{10.25}$$

**Case (I) : Neglecting the effect of curvature :**

We have, $\mathbf{K_s} = 1 + \frac{1}{2C} = 1 + \frac{1}{2 \times 10.25} = \mathbf{1.0488}$

(i) Torsional shear stress induced in the spring (neglecting curvature effect) is given by,

$$\tau = K_s \times \frac{8\,WD}{\pi d^3}$$

$$\therefore \quad 350 = 1.0488 \times \frac{8 \times W \times 82}{\pi \times 8^3}$$

$$\therefore \quad W = \frac{350 \times \pi \times 8^3}{1.0488 \times 8 \times 82} = \mathbf{818.26\ N}$$

(ii) Deflection per active turn is given by,

$$\frac{\delta}{n} = \frac{8\ WD^3}{Gd^4} = \frac{8 \times 818.26 \times 82^3}{84 \times 10^3 \times 8^4} = \mathbf{10.49\ mm}$$

**Case (II) : Considering the effect of curvature :**

We know that, Wahl's stress concentration factor is,

$$\mathbf{K} = \frac{4C-1}{4C-4} + \frac{0.615}{C} = \frac{4 \times 10.25 - 1}{4 \times 10.25 - 4} + \frac{0.615}{10.25} = \mathbf{1.1411}$$

(i) Shear stress induced in the spring is given by,

$$\tau = K \times \frac{8\ WD}{\pi d^3}$$

$$\therefore \quad 350 = 1.1411 \times \frac{8 \times W \times 82}{\pi \times 8^3}$$

$$\therefore \quad W = \frac{350 \times \pi \times 8^3}{1.1411 \times 8 \times 82} = \mathbf{752.073\ N}$$

(ii) Deflection per active turn is given by,

$$\frac{\delta}{n} = \frac{8\ WD^3}{Gd^4} = \frac{8 \times 752.073 \times 82^3}{84 \times 10^3 \times 8^4} = \mathbf{9.6417\ mm}$$

---

**Problem 5.4 :** *A helical compression spring carries a load of 500 N with a deflection of 25 mm. The spring index may be taken as 8. Assume permissible $\tau$ = 350 MPa. Modulus of rigidity = 84 kN/mm², Wahl's factor as $\frac{4C-1}{4C-4} + \frac{0.615}{C}$, where C is spring index. Find number of active turns of spring.*

**Solution : Given data :** W = 500 N, $\delta$ = 25 mm, C = 8,

$\tau$ = 350 MPa = **350 N/mm²**, **G** = 84 kN/mm² = **84 × 10³ N/mm²**

**Procedure :** Wahl's stress concentration factor is,

$$\mathbf{K} = \frac{4C-1}{4C-4} + \frac{0.615}{C} = \frac{4 \times 8 - 1}{4 \times 8 - 4} + \frac{0.615}{8} = \mathbf{1.184}$$

Maximum shear stress induced is given by,

$$\tau = K \times \frac{8\ WD}{\pi d^3} = K \times \frac{8\ WC}{\pi d^2} \qquad \ldots \left(\because C = \frac{D}{d}\right)$$

$$\therefore \quad d^2 = K \times \frac{8\ WC}{\pi \cdot \tau} = 1.184 \times \frac{8 \times 500 \times 8}{\pi \times 350}$$

$$\therefore \quad \mathbf{d} = \mathbf{5.87\ mm}$$

and $\mathbf{D} = C \times d = 8 \times 5.87 = \mathbf{46.96\ mm}$

**Number of active turns :** It is calculated from the deflection formula.

We have, $$\delta = \frac{8\ WD^3 n}{Gd^4}$$

$$\therefore \quad 25 = \frac{8 \times 500 \times (46.96)^3 \times n}{84 \times 10^3 \times (5.87)^4}$$

$$\therefore \quad n = \mathbf{6.0190 \cong 7\ (say)}$$

Total number of active turns = n = **7**.

---

**Problem 5.5 :** *Design a helical compression spring with grounded ends of bronze material for valve used in an hydraulic circuit. Take following values.*

*(a) Spring index, C = 12, (b) Safe shear stress = 100 N/mm², (c) Maximum load = 100 N,*

*(d) Modulus of rigidity = 4 × 10⁴ N/mm², (e) Deflection = 15 mm.*

*Find out (i) Diameter of spring wire, (ii) Diameter of spring coil, (iii) Number of spring coils, (iv) Stiffness of spring.*

**Solution : Given data :** $\delta$ = 15 mm, W = 100 N, C = 12, $\tau$ = 100 N/mm$^2$, G = 4 × 10$^4$ N/mm$^2$

**Procedure :** Wahl's stress concentration factor is,

$$\mathbf{K} = \frac{4C-1}{4C-4} + \frac{0.615}{C} = \frac{4 \times 12 - 1}{4 \times 12 - 4} + \frac{0.615}{12} = \mathbf{1.1194}$$

**(i) Diameter of spring wire :** Maximum shear stress induced in spring material is given by,

$$\tau = K \times \frac{8\,WD}{\pi d^3} = K \times \frac{8\,WC}{\pi d^2} \qquad \ldots \left(\because C = \frac{D}{d}\right)$$

$$\therefore \quad 100 = 1.1194 \times \frac{8 \times 100 \times 12}{\pi \times d^2}$$

$$\therefore \quad \text{Wire diameter} = d = 5.8486 \text{ mm} \cong \mathbf{5.85\ mm}$$

**(ii) Mean diameter of spring coil :**

$$D = C \times d = 12 \times 5.85 = \mathbf{70.2\ mm}$$

**(iii) Number of spring coils (Turns) :** It can be calculated from the deflection formula

$$\delta = \frac{8\,WD^3 n}{Gd^4}$$

$$\therefore \quad n = \frac{\delta \times G \times d^4}{8 \times W \times D^3} = \frac{15 \times 4 \times 10^4 \times (5.85)^4}{8 \times 100 \times (70.2)^3} = \mathbf{2.539}$$

$\therefore$ Number of spring coils or turns = n = **2.539 ≅ 4 (say)**

**Note :** Minimum number of spring coils (active turns) should be taken as 4 for safe design of spring and its proper functioning.

**(iv) Stiffness of spring :**

We have, Spring stiffness, $S = \frac{W}{\delta} = \frac{100}{15} = \mathbf{6.667\ N/mm}$

---

**Problem 5.6 :** *A closed coil helical spring is used for an automobile suspension system. The spring has stiffness 85 N/mm with squared and grounded ends. Load on the spring causes a total deflection of 9 mm. By taking permissible shear stress of 400 MPa, find : (i) Wire diameter of spring. (ii) Length of spring.*

*Take spring index = 6 and G = 80 × 10$^3$ N/mm$^2$.*

**Solution : Given data :** S = 85 N/mm, $\delta$ = 9 mm, $\tau$ = 400 MPa = **400 N/mm$^2$**, C = 6, G = 80 × 10$^3$ N/mm$^2$

**Procedure : (i) Wire diameter of spring :**

We have, Spring stiffness, $S = \frac{W}{\delta}$

$$\therefore \quad \mathbf{W} = S \times \delta = 85 \times 9 = \mathbf{765\ N}$$

Wahl's stress concentration factor is,

$$\mathbf{K} = \frac{4C-1}{4C-4} + \frac{0.615}{C} = \frac{4 \times 6 - 1}{4 \times 6 - 4} + \frac{0.615}{6} = \mathbf{1.2525}$$

Maximum shear stress induced is given by,

$$\tau = K \times \frac{8\,WD}{\pi d^3} = K \times \frac{8\,WC}{\pi d^2} \qquad \ldots \left(\because C = \frac{D}{d}\right)$$

$$\therefore \quad d^2 = K \times \frac{8\,WC}{\pi \times \tau} = 1.2525 \times \frac{8 \times 765 \times 6}{\pi \times 400} = 36.5843$$

$$\therefore \quad \mathbf{d} = \mathbf{6.0497\ mm}$$

$$\text{and} \quad \mathbf{D} = C \times d = 6 \times 6.0497 = \mathbf{36.29\ mm}$$

$\therefore$ Diameter of spring wire = 6.0497 mm and Mean diameter of spring coil = 36.29 mm

**(ii) Length of spring :** Deflection is given by,

$$\delta = \frac{8\,WD^3 n}{G \times d^4}$$

$$\therefore \quad \mathbf{n} = \frac{\delta \times G \times d^4}{8\,WD^3} = \frac{9 \times 80 \times 10^3 \times (6.0497)^4}{8 \times 765 \times (36.29)^3} = \mathbf{3.29 \cong 4\ (say)}$$

$\therefore$ Total number of active turns (n) = 4.

Given squared and grounded ends, therefore, **total number of turns** is given by,

$$\mathbf{n'} = n + 2 = 4 + 2 = \mathbf{6}$$

Solid length, $\mathbf{L_s} = n' \times d = 6 \times 6.0497 = \mathbf{36.2982\ mm}$

Free length, $\mathbf{L_f} = n' \times d + \delta_{max} + 0.15 \times \delta_{max} = (6 \times 6.0497) + 9 + (0.15 \times 9) = \mathbf{46.648\ mm}$

**Problem 5.7 :** *Find the maximum shear stress and deflection induced in a helical spring, if it has to absorb 1000 N-m energy. Use the data as, D = 100 mm, d = 20 mm, n = 30, G = 84 kN/mm².*

**Solution : Given data : U** = 1000 N.m = **1000 × 10³ N-mm**, D = 100 mm,

d = 20 mm, n = 30, **G** = 84 kN/mm² = **84 × 10³ N/mm²**

**Procedure : (i) Maximum shear stress induced :**

We have, Spring index = $\mathbf{C} = \frac{D}{d} = \frac{100}{20} = \mathbf{5}$

∴ Wahl's stress concentration factor is,

$$\mathbf{K} = \frac{4C-1}{4C-4} + \frac{0.615}{C} = \frac{4 \times 5 - 1}{4 \times 5 - 4} + \frac{0.615}{5} = \mathbf{1.3105}$$

From the expression of strain energy stored in a spring, we have,

$$U = \frac{\tau^2}{4K^2 G} \times V$$

where, V = Volume of wire = Cross-sectional area of spring wire × Length of wire

$$\therefore \quad U = \frac{\tau^2}{4K^2 G} \times \left(\frac{\pi}{4} \times d^2\right) \times (\pi D n)$$

$$\therefore \quad \tau^2 = \frac{U \times 4 \times K^2 \times G}{\left(\frac{\pi}{4} \times d^2\right) \times (\pi D n)} = \frac{1000 \times 10^3 \times 4 \times (1.3105)^2 \times 84 \times 10^3}{\left(\frac{\pi}{4} \times (20)^2\right) \times (\pi \times 100 \times 30)} = 1{,}94{,}886{,}93$$

$$\therefore \quad \tau = \mathbf{441.46\ N/mm^2}$$

**(ii) Deflection :**

We have, $\tau = \frac{8 \times K \times W \times D}{\pi d^3}$

$$\therefore \quad \mathbf{W} = \frac{\tau \times \pi \times d^3}{8 \times K \times D} = \frac{441.46 \times \pi \times (20)^3}{8 \times 1.3105 \times 100} = \mathbf{10582.88\ N}$$

Deflection is given by,

$$\delta = \frac{8WD^3 n}{G \times d^4} = \frac{8 \times 10582.88 \times 100^3 \times 30}{84 \times 10^3 \times (20)^4} = \mathbf{188.98\ mm}$$

**Problem 5.8 :** *A load of 2 kN is dropped axially on a close coil helical spring from a height of 250 mm. The spring has 20 effective turns and it is made of 25 mm wire diameter. The spring index is 8. Find the maximum shear stress induced in spring and amount of compression produced. Modulus of rigidity for material of the spring wire is 84 kN/mm².*

**Solution : Given data : P** = 2 kN = **2 × 10³ N**, h = 250 mm, C = 8, d = 25 mm, n = 20, G = 84 × 10³ N/mm²

**Procedure :** We have, spring index = $C = \frac{D}{d}$

$$\therefore \quad \mathbf{D} = C \times d = 8 \times 25 = \mathbf{200\ mm}$$

∴ Wahl's stress concentration factor is,

$$\mathbf{K} = \frac{4C-1}{4C-4} + \frac{0.615}{C} = \frac{4 \times 8 - 1}{4 \times 8 - 4} + \frac{0.615}{8} = \mathbf{1.184}$$

Deflection is given by,

$$\delta = \frac{8WD^3 n}{G \times d^4} = \frac{8 \times W \times (200)^3 \times 20}{84 \times 10^3 \times (25)^4} = \mathbf{0.039\ W}$$

We know that,

Energy stored by spring = Work done by falling load (P)

$$\therefore \quad \frac{1}{2} \times W \times \delta = P \times (h + \delta) \qquad \left[\because U = \frac{1}{2} \times W \times \delta\right]$$

$$\therefore \quad \frac{1}{2} \times W \times 0.039\ W = 2 \times 10^3 \times (250 + 0.039\ W)$$

$$\therefore \quad 0.0195\ W^2 - 78W - 500 \times 10^3 = 0$$

$$\therefore \quad \mathbf{W} = \frac{-(-78) \pm \sqrt{(-78)^2 - (4 \times 0.0195 \times -500 \times 10^3)}}{2 \times 0.0195} = \mathbf{7444.35\ N}$$

**(i) Maximum shear stress induced :** $\tau = K \times \frac{8WD}{\pi d^3} = 1.184 \times \frac{8 \times 7444.35 \times 200}{\pi \times (25)^3} = \mathbf{287.295\ N/mm^2}$

**(ii) Maximum deflection :** $\delta = 0.039\ W = 0.039 \times 7444.35 = \mathbf{290.33\ mm}$

---

**Problem 5.9 :** *A mechanism used in printing machinery consists of a tension spring assembled with a preload of 30 N. The wire diameter of spring is 2 mm and C = 6. The spring has 18 active coils. The spring wire is hard drawn and oil tempered having following material properties : $\tau$ = 680 MPa, G = 80 kN/mm$^2$.*

*Find (i) Initial shear stress, (ii) Spring rate, (iii) Force to cause the body of spring to its yield strength.*

**Solution : Given data :** $W_1$ = 30 N, d = 2 mm, C = 6, n = 18,

$$\tau = 680 \text{ MPa} = \mathbf{680\ N/mm^2},\ \mathbf{G} = 80 \text{ kN/mm}^2 = \mathbf{80 \times 10^3\ N/mm^2}$$

**Procedure : (i) Initial torsional shear stress induced in the wire ($\tau_1$) :**

We know, $$\mathbf{K} = \frac{4C-1}{4C-4} + \frac{0.615}{C} = \frac{4 \times 6 - 1}{4 \times 6 - 4} + \frac{0.615}{6} = \mathbf{1.2525}$$

We have, $$\tau_1 = K \times \frac{8W_1C}{\pi d^2} = 1.2525 \times \frac{8 \times 30 \times 6}{\pi \times (2)^2} = \mathbf{143.5\ N/mm^2}$$

**(ii) Spring rate (Spring stiffness) :**

We have, $$\delta = \frac{8WD^3n}{Gd^4}$$

$$\therefore \quad \textbf{Spring rate} = \frac{W}{\delta} = \frac{Gd^4}{8D^3n} = \frac{Gd}{8C^3n} = \frac{80 \times 10^3 \times 2}{8 \times (6)^3 \times 18} = \mathbf{5.144\ N/mm}$$

**(iii) Force required (W) :**

We have, $$\tau = K \times \frac{8WC}{\pi d^2}$$

$$\therefore \quad 680 = 1.2525 \times \frac{8 \times W \times 6}{\pi \times (2)^2}$$

$$\therefore \quad W = \mathbf{142.134\ N}$$

---

**Problem 5.10 :** *A railway wagon having 1500 kg mass and moving at 1 m/s velocity dashes against a damper consisting of two helical springs of spring index 6.*

*Springs, which get compressed by 150 mm while resisting a dash, are made of spring steel having allowable shear stress of 360 N/mm$^2$ and modulus of rigidity 8.4 $\times$ 10$^4$ N/mm$^2$. Design the helical coil spring with circular cross-section of spring wire.*

**Solution : Given data :** m = 1500 kg, V = 1 m/s, C = 6, $\delta$ = 150 mm, $\tau$ = 360 N/mm$^2$, G = 8.4 $\times$ 10$^4$ N/mm$^2$

**Procedure : Design of helical spring :**

**(1) Diameter of spring wire :** Let 'W' be the load acting on each spring, which causes a deflection of 150 mm. Since there are two springs, the energy stored in the two springs is given by,

$$\mathbf{U} = 2\left(\frac{1}{2} \times W \times \delta\right) = 2 \times \frac{1}{2} \times W \times 150 = \mathbf{150\ W} \qquad \ldots (1)$$

And, kinetic energy of railway wagon $= \frac{1}{2} \times m \times V^2$

$$\therefore \quad \textbf{K.E.} = \frac{1}{2} \times 1500 \times (1)^2 = 750 \text{ N.m} = \mathbf{750 \times 10^3 \text{ N.mm}} \quad \ldots (2)$$

Equating equations (1) and (2), we get,

$$150 \text{ W} = 750 \times 10^3$$

$$\therefore \quad \mathbf{W} = 5 \times 10^3 \text{ N} = \mathbf{5000 \text{ N}}$$

We know that, torque acting on spring coil is given by,

$$\mathbf{T} = W \times \frac{D}{2} = 5000 \times \frac{D}{2} = \mathbf{2500 \times D} \quad \ldots (3)$$

Also, the torque or twisting moment is given by,

$$\mathbf{T} = \frac{\pi}{16} \cdot \tau \cdot d^3 = \frac{\pi}{\mathbf{16}} \times \mathbf{360 \times d^3} \quad \ldots (4)$$

Equating equations (3) and (4), we get,

$$2500 \text{ D} = \frac{\pi}{16} \times 360 \times d^3$$

$$\therefore \quad 2500 \times 16 \times \frac{D}{d} = \pi \times 360 \times d^2$$

$$\therefore \quad 2500 \times 16 \times C = \pi \times 360 \times d^2 \qquad \left[\because C = \frac{D}{d}\right]$$

$$\therefore \quad 2500 \times 16 \times 6 = \pi \times 360 \times d^2$$

$$\therefore \quad \mathbf{d} = \mathbf{14.56 \text{ mm}}$$

**(2) Number of turns :**

$$\delta = \frac{8W D^3 n}{G \cdot d^4} = \frac{8WC^3 n}{G \cdot d} \qquad \left[\because C = \frac{D}{d}\right]$$

$$\therefore \quad n = \frac{\delta \times G \times d}{8WC^3} = \frac{150 \times 8.4 \times 10^4 \times 14.56}{8 \times 5000 \times (6)^3}$$

$$\therefore \quad \mathbf{n} = 21.23 \text{ turns} \cong \mathbf{22 \text{ turns (say)}}$$

∴ Total number of active turns = n = 22.

Assuming the spring ends to be squared and grounded, **total number of turns** can be calculated by,

$$\mathbf{n'} = n + 2 = 22 + 2 = \mathbf{24}$$

**(3) Solid length :** $\mathbf{L_s} = n' \times d = 24 \times 14.56 = \mathbf{349.44 \text{ mm}}$

**(4) Free length :**

$$\mathbf{L_f} = L_s + \delta_{max} + 0.15 \times \delta_{max} = 349.44 + 150 + (0.15 \times 150) = \mathbf{521.94 \text{ mm}}$$

**(5) Pitch of spring coil :**

$$\mathbf{p} = \frac{\text{Free length}}{n' - 1} = \frac{521.94}{24 - 1} = \mathbf{22.69 \text{ mm}}$$

---

**Problem 5.11 :** *Design a helical compression spring for maximum load of 800 N with a deflection of 25 mm. Spring index is 5 and Wahl's correction factor is 1.3. Maximum permissible shear stress for spring wire is 400 MPa and modulus of rigidity is 84 kN/mm².*

**Solution : Given data :** W = 800 N, C = 5, $\delta$ = 25 mm, $\tau$ = 400 N/mm²,

$$\mathbf{G} = 84 \text{ kN/mm}^2 = \mathbf{84 \times 10^3 \text{ N/mm}^2}, K = 1.3$$

**Procedure : Design of helical spring :**

**(1) Mean diameter of spring coil :** Maximum shear stress induced is given by,

$$\tau = K \times \frac{8WD}{\pi d^3} = K \times \frac{8WC}{\pi d^2} \qquad \left(\because C = \frac{D}{d}\right)$$

$$\therefore \quad d^2 = K \times \frac{8WC}{\pi \times \tau} = 1.3 \times \frac{8 \times 800 \times 5}{\pi \times 400} = 33.0909$$

$\therefore$ $\mathbf{d} = \mathbf{5.75\ mm}$ and $\mathbf{D} = C \times d = 5 \times 5.75 = \mathbf{28.75\ mm}$

Diameter of spring wire = d = **5.75 mm** and Mean diameter of spring coil = D = **28.75 mm**

**(2) Number of turns :** It is calculated from the deflection formula.

$$\delta = \frac{8WD^3 n}{G \times d^4}$$

$$\therefore \quad \mathbf{n} = \frac{\delta \times G \times d^4}{8WD^3} = \frac{25 \times 84 \times 10^3 \times (5.75)^4}{8 \times 800 \times (28.75)^3} = 15.09 \cong \mathbf{16\ (say)}$$

Total number of active turns = n = **16**

Assuming squared and grounded ends, total number of turns is given by,

$$\mathbf{n'} = n + 2 = 16 + 2 = \mathbf{18}$$

**(3) Solid length :** $\mathbf{L_s} = n' \times d = 18 \times 5.75 = \mathbf{103.5\ mm}$

**(4) Free length :** $L_f = n' \times d + \delta_{max} + 0.15 \times \delta_{max}$

$$\therefore \quad \mathbf{L_f} = (18 \times 5.75) + 25 + (0.15 \times 25) = \mathbf{132.25\ mm}$$

**(5) Pitch of the coil :** $\mathbf{p} = \frac{\text{Free length}}{n' - 1} = \frac{132.25}{18 - 1} = \mathbf{7.7794\ mm}$

---

## Numerical Type No. 3 : "Design of Helical Spring Operating between a Load Range"

**Problem 5.12 :** *A helical valve spring is to be designed for an operating load range of approximately 135 N. The deflection of the spring for the load range is 7.5 mm. Assume spring index of 10. Permissible shear stress for the material of the spring = 480 MPa and its modulus of rigidity = 80 kN/mm². Design the spring.*

*Take Wahl's factor* $= \frac{4C - 1}{4C - 4} + \frac{0.615}{C}$, *'C' being the spring index.*

**Solution : Given data :** C = 10, τ = 480 MPa, **G** = 80 kN/mm² = $\mathbf{80 \times 10^3\ N/mm^2}$. Load range is 135 N, for which, deflection of spring is 7.5 mm. Therefore, we can write, W = 135 N, for which, corresponding δ = 7.5 mm.

**Procedure :**

$$\text{Wahl's correction factor, } \mathbf{K} = \frac{4C - 1}{4C - 4} + \frac{0.615}{C} = \frac{(4 \times 10) - 1}{(4 \times 10) - 4} + \frac{0.615}{10} = \mathbf{1.1448}$$

**Design of helical spring :**

**(1) Mean diameter of spring coil :** Maximum shear stress induced is given by,

$$\tau = K \times \frac{8WC}{\pi d^2}$$

$$\therefore \quad d^2 = \frac{K \times 8WC}{\pi \times \tau} = \frac{1.1448 \times 8 \times 135 \times 10}{\pi \times 480} = 10.37$$

$$\therefore \quad \mathbf{d} = \mathbf{3.22\ mm}$$

We know, $\mathbf{D} = C \times d = 10 \times 3.22 = \mathbf{32.2\ mm}$

Thus, diameter of spring wire, d = **3.22 mm,** and Mean diameter of helical spring coil, **D = 32.2 mm**

**(2) Number of turns :** It is calculated from deflection formula.

$$\delta = \frac{8WC^3 n}{G \cdot d}$$

$$\therefore \quad \mathbf{n} = \frac{\delta \times G \times d}{8WC^3} = \frac{7.5 \times 80 \times 10^3 \times 3.22}{8 \times 135 \times (10)^3} = \mathbf{1.788 \cong 4\ (say)}$$

For safe design, minimum number of coils or turns of spring should be taken as 4.

Assuming squared and grounded ends, total number of turns is given by,

$$\mathbf{n'} = n + 2 = 4 + 2 = \mathbf{6}$$

**(3) Solid length :** $\mathbf{L_s} = n' \times d = 6 \times 3.22 = \mathbf{19.32\ mm}$

**(4) Free length :** $\mathbf{L_f} = L_s + \delta_{max} + 0.15 \times \delta_{max} = 19.32 + 7.5 + (0.15 \times 7.5) = \mathbf{27.945\ mm}$

**(5) Pitch of coil :** $\mathbf{p} = \dfrac{\text{Free length}}{n' - 1} = \dfrac{27.945}{6 - 1} = \mathbf{5.589\ mm}$

---

**Problem 5.13 :** *Design a close coiled helical compression spring for service load ranging from 2250 N to 2750 N. The axial deflection of spring for load range is 6 mm. Assume a spring index of 5. Permissible shear stress intensity is 420 N/mm$^2$ and modulus of rigidity is G = 84 kN/mm$^2$. Take design stress 25% excess of permissible stress for severe condition and intermittent operation.*

**Solution : Given data :** $W_1$ = 2250 N, $W_2$ = 2750 N, C = 5,

Permissible shear stress = 420 N/mm$^2$, G = 84 × 10$^3$ N/mm$^2$

Deflection for the load range ($W_2 - W_1$) = (2750 – 2250) = (500 N) is, $\delta$ = 6 mm.

**Procedure :** Wahl's stress concentration factor is,

$$\mathbf{K} = \frac{4C-1}{4C-4} + \frac{0.615}{C} = \frac{4 \times 5 - 1}{4 \times 5 - 4} + \frac{0.615}{5} = \mathbf{1.3105}$$

For severe conditions and intermittent operations, take

Design stress = 1.25 × Permissible stress

$\therefore \quad \tau = 1.25 \times 420 = \mathbf{525\ N/mm^2}$

**(1) Diameter of spring wire :** Maximum shear stress induced due to maximum load ($W_2$) is given by,

$$\tau = K \times \frac{8W_2D}{\pi d^3} = K \times \frac{8W_2C}{\pi d^2} \qquad \left(\because C = \frac{D}{d}\right)$$

$$\therefore \quad d^2 = K \times \frac{8W_2C}{\pi \times \tau} = 1.3105 \times \frac{8 \times 2750 \times 5}{\pi \times 525}$$

Thus, $\mathbf{d} = \mathbf{9.35\ mm}$ and $\mathbf{D} = C \times d = 5 \times 9.35 = \mathbf{46.75\ mm}$

$\therefore$ Diameter of spring wire, d = **9.35 mm**

And, Mean diameter of spring coil, D = **46.75 mm**

**(2) Number of turns :** It is calculated from the deflection formula.

$$\delta = \frac{8WD^3 n}{G \times d^4}$$

$$\therefore \quad \mathbf{n} = \frac{\delta \times G \times d^4}{8WD^3} = \frac{6 \times 84 \times 10^3 \times (9.35)^4}{8 \times 500 \times (46.75)^3} = \mathbf{9.42 \cong 10\ (say)}$$

$\therefore$ Total number of active turns = **n = 10**

Assuming squared and grounded ends, total number of turns is given by,

$$\mathbf{n'} = n + 2 = 10 + 2 = \mathbf{12}$$

**(3) Maximum deflection :** For load range of 500 N, deflection is 6 mm. Therefore, maximum deflection for maximum load of 2750 N can be calculated as follows :

For 500 N → $\delta$ = 6 mm

For 2750 N → $\delta_{max} = \dfrac{2750 \times 6}{500} = \mathbf{33\ mm}$

**(4) Solid length :** $\mathbf{L_s} = n' \times d = 12 \times 9.35 = \mathbf{112.2\ mm}$

**(5) Free length :** $\mathbf{L_f} = L_s + \delta_{max} + 0.15\ \delta_{max} = 112.2 + 33 + (0.15 \times 33) = \mathbf{150.15\ mm}$

**(6) Pitch of coil :** $\mathbf{p} = \dfrac{\text{Free length}}{n' - 1} = \dfrac{150.15}{12 - 1} = \mathbf{13.65\ mm}$

---

**Problem 5.14 :** *Design a spring for balance to measure 0 to 1000 N over a scale of length 80 mm. Spring is to be enclosed in a casing of 25 mm diameter. Approximate number of turns is 30. Modulus of rigidity is 84 kN/mm$^2$. Also, calculate the maximum shear stress, if spring ends are grounded.*

**Solution : Given data :** W = 1000 N, $\delta$ = 80 mm, n = 30, **G** = 84 kN/mm$^2$ = $\mathbf{84 \times 10^3\ N/mm^2}$.

**Procedure :**

**(1) Diameter of spring wire :** As the spring is to be enclosed in a casing of 25 mm diameter, the outer diameter of spring should be less than 25 mm. i.e. $D_o < 25$ mm.

To use trial and error method, **let us assume that, d = 4.1 mm**.

We know that, deflection produced is calculated by,

$$\delta = \frac{8WD^3n}{G \times d^4}$$

$$\therefore \quad D^3 = \frac{\delta \times G \times d^4}{8Wn} = \frac{80 \times 84 \times 10^3 \times (4.1)^4}{8 \times 1000 \times 30} = 7912.13$$

$$\therefore \quad \mathbf{D} = 19.9265 \text{ mm} \cong \mathbf{19.93 \text{ mm}}$$

Thus, D = 20 mm and $D_o = D + d = 19.93 + 4.1 = 24.03$ mm, which is less than 25 mm.

$\therefore$ Diameter of spring wire = d = **4.1 mm is correct assumption,** with

Mean diameter of spring coil = D = **19.93 mm**

$$\therefore \quad \text{Spring index} = \mathbf{C} = \frac{D}{d} = \frac{19.93}{4.1} = \mathbf{4.86}$$

**(2) Total number of turns :**

As the ends are grounded, total number of turns = **n'** = n = **30**. ($\because$ Given that, n = 30)

**(3) Solid length :** $L_s = n' \times d = 30 \times 4.1 = \mathbf{123 \text{ mm}}$

**(4) Free length :** $L_f = n' \times d + \delta_{max} + 0.15 \times \delta_{max} = (30 \times 4.1) + 80 + (0.15 \times 80) = \mathbf{215 \text{ mm}}$

**(5) Pitch of coil :** $\mathbf{p} = \dfrac{\text{Free length}}{n' - 1} = \dfrac{215}{30 - 1} = \mathbf{7.413 \text{ mm}}$

**(6) Maximum shear stress :** Wahl's stress concentration factor is,

$$\mathbf{K} = \frac{4C - 1}{4C - 4} + \frac{0.615}{C} = \frac{4 \times 4.86 - 1}{4 \times 4.86 - 4} + \frac{0.615}{4.86} = \mathbf{1.3208}$$

Maximum shear stress induced is given by,

$$\tau_{max} = K \times \frac{8WD}{\pi d^3} = 1.3208 \times \frac{8 \times 1000 \times 19.93}{\pi \times (4.1)^3} = \mathbf{972.59 \text{ N/mm}^2}$$

---

## Nunerical Type No. 4: "Design of Spring for Safety Valves"

**Problem 5.15 :** *A safety valve of 60 mm diameter is to blow off at a pressure of 1.2 N/mm². It is held on its seat by a close coil helical spring. The maximum lift of valve is 10 mm. Design a suitable compression spring of spring index 5 with an initial compression of 35 mm. The shear stress is limited to 500 MPa. Take G = 80 kN/mm².*

**Solution : Given data :** $D_v = 60$ mm, $p = 1.2$ N/mm², $\tau = 500$ N/mm², $G = 80 \times 10^3$ N/mm², C = 5, Initial compression = 35 mm, Maximum lift of valve = 10 mm.

**Procedure : Design of helical spring :**

**(1) Diameter of spring wire :** Load, at which, the valve blows off can be calculated as,

$$\mathbf{W_1} = \text{Pressure on valve} \times \text{Area of valve} = p \times \left[\frac{\pi}{4} \times (D_v)^2\right] = 1.2 \times \left[\frac{\pi}{4} \times (60)^2\right] = \mathbf{3392.92 \text{ N}}$$

Maximum compression of spring = Initial compression + Maximum lift of valve

$$\therefore \quad \boldsymbol{\delta_{max}} = 35 + 10 = \mathbf{45 \text{ mm}}$$

As the load $W_1$ = 3392.92 N keeps the valve on valve seat with initial compression of 35 mm, therefore, maximum load on spring for maximum deflection, when valve is open, is calculated as below.

Initial compression (35 mm) $\rightarrow W_1$ = 3392.92 N

Maximum compression (45 mm) $\rightarrow$ **Maximum load ($W_2$)** $= \dfrac{3392.92 \times 45}{35} = \mathbf{4362.32 \text{ N}}$

Wahl's stress concentration factor is,

$$\mathbf{K} = \frac{4C - 1}{4C - 4} + \frac{0.615}{C} = \frac{4 \times 5 - 1}{4 \times 5 - 4} + \frac{0.615}{5} = \mathbf{1.3105}$$

Maximum shear stress is given by,

$$\tau_{max} = K \times \frac{8\,W_2 D}{\pi d^3} = K \times \frac{8 W_2\, C}{\pi d^2} \qquad \ldots \left(\because C = \frac{D}{d}\right)$$

$$\therefore \quad d^2 = K \times \frac{8W_2\, C}{\pi \times \tau} = 1.3105 \times \frac{8 \times 4362.32 \times 5}{\pi \times 500} = 145.57$$

$$\therefore \quad d = \mathbf{12.06\ mm}$$

$$\therefore \quad D = C \times d = 5 \times 12.06 = \mathbf{60.3\ mm}$$

Diameter of spring wire = **d = 12.06 mm**, and

Mean diameter of spring coil = **D = 60.3 mm**

**(2) Number of turns :** It is calculated from the deflection formula.

$$\delta_{max} = \frac{8W_2\, D^3\, n}{G \times d^4}$$

$$\therefore \quad n = \frac{\delta_{max} \times G \times d^4}{8W_2\, D^3} = \frac{45 \times 80 \times 10^3 \times (12.06)^4}{8 \times 4362.32 \times (60.3)^3} = \mathbf{9.95 \cong 10\ (say)}$$

∴ Total number of active turns, **n = 10**

Assuming squared and grounded ends, total number of turns is given by,

$$\mathbf{n'} = n + 2 = 10 + 2 = \mathbf{12}$$

**(3) Solid length :** $L_s = n' \times d = 12 \times 12.06 = \mathbf{144.72\ mm}$

**(4) Free length :** $L_f = n' \times d + \delta_{max} + 0.15 \times \delta_{max}$

$$\therefore \quad L_f = (12 \times 12.06) + 45 + (0.15 \times 45) = \mathbf{196.47\ mm}$$

**(5) Pitch of coil :** $p = \frac{\text{Free length}}{n' - 1} = \frac{196.47}{12 - 1} = \mathbf{17.86\ mm}$

---

**Problem 5.16 :** *Design a helical spring for a spring loaded safety valve for the following conditions :*

*Operating pressure = $p_1 = 1\ N/mm^2$*

*Maximum pressure, when the valve blows off freely = $p_2 = 1.075\ N/mm^2$*

*Maximum lift of valve (when the pressure is $1.075\ N/mm^2$) = $\delta = 6$ mm.*

*Diameter of valve seat = $D_v = 100$ mm.*

*Maximum shear stress = $\tau = 400\ N/mm^2$.*

*Modulus of rigidity = $G = 86 \times 10^3\ N/mm^2$*

*Spring index = C = 5.5.*

**Solution : Given data :** $p_1 = 1\ N/mm^2$, $p_2 = 1.075\ N/mm^2$, $\delta = 6$ mm, $D_v = 100$ mm, $\tau = 400\ N/mm^2$, $G = 86 \times 10^3\ N/mm^2$, C = 5.5.

**Procedure :**

**(1) Diameter of spring wire and diameter of spring coil :**

Initial tensile force acting on the spring is,

$$\mathbf{W_1} = \frac{\pi}{4} D_v^2 \times p_1 = \frac{\pi}{4} \times 100^2 \times 1 = \mathbf{7853.98\ N}$$

Maximum tensile force is,

$$\mathbf{W_2} = \frac{\pi}{4} \times D_v^2 \times p_2 = \frac{\pi}{4} \times 100^2 \times 1.075 = \mathbf{8443.03\ N}$$

$$\therefore \quad \textbf{Load range} = \mathbf{W_2 - W_1} = 8443.03 - 7853.98 = \mathbf{589.05\ N}$$

Wahl's factor, $$\mathbf{K} = \frac{4C - 1}{4C - 4} + \frac{0.615}{C} = \frac{4 \times 5.5 - 1}{4 \times 5.5 - 4} + \frac{0.615}{5.5} = \mathbf{1.2785}$$

We have, $$\tau_{max} = K \times \frac{8\,W_2\, C}{\pi d^2} \qquad (\because W_2 \text{ is maximum load acting on spring})$$

$$\therefore \quad 400 = 1.2785 \times \frac{8 \times 8443.03 \times 5.5}{\pi \times d^2}$$

$$\therefore \quad \mathbf{d} = \mathbf{19.44\ mm}$$

$\therefore$ Mean diameter, $\mathbf{D} = C \times d = 5.5 \times 19.44 = \mathbf{106.92\ mm}$

**(2) Number of coils :**

For load range, $W_{range} = 589.05\ N \rightarrow \delta_{range} = 6$ mm

For maximum load, $W_2 = 8443.03\ N \rightarrow \boldsymbol{\delta_{max}} = \dfrac{8443.03 \times 6}{589.05} = \mathbf{85.99\ mm}$

We have,
$$\delta_{max} = \frac{8\ W_{max} \times C^3 \times n}{G \cdot d}$$

$$\therefore \quad 85.99 = \frac{8 \times 8443.03 \times 5.5^3 \times n}{86 \times 10^3 \times 19.44}$$

$\therefore$ Active number of coils = **n** = **12.79 ≅ 14 (say)**

Total Number of coils, **n'** = n + 2 = 14 + 2 = **16** [Assuming squared and grounded ends]

**(3) Solid length,** $\mathbf{L_s} = n' \times d = 16 \times 19.44 = \mathbf{311.04\ mm}$

**(4) Free length,** $\mathbf{L_f} = n' \times d + \delta_{max} + 0.15\ \delta_{max} = (16 \times 19.44) + 85.99 + (0.15 \times 85.99) = \mathbf{409.92\ mm}$

**(5) Pitch of coil :** $\mathbf{p} = \dfrac{L_f}{n' - 1} = \dfrac{409.92}{16 - 1} = \mathbf{27.328\ mm}$

---

**Problem 5.17 :** *Following particulars refer to a valve spring of a petrol engine.*

*(1) Length of spring, when valve is opened = 40 mm*

*(2) Length of spring, when valve is closed = 50 mm*

*(3) Spring load, when valve is opened = 360 N*

*(4) Spring load, when valve is closed = 220 N*

*(5) Maximum inside diameter of spring = 25 mm*

*Calculate size of wire, number of active turns and pitch of coil for maximum operating stress of 400 N/mm² and G = 84 × 10³ N/mm².*

**Solution : Given data :** $L_2 = 40$ mm, $L_1 = 50$ mm, $W_2 = 360$ N, $W_1 = 220$ N, $\tau = 400$ N/mm², $G = 84 \times 10^3$ N/mm²

**Procedure :**

**(i) Diameter of spring wire :**

It is given that, maximum inside diameter of spring = 25 mm

Let us assume that, $D_i = 25$ mm. Therefore, $\mathbf{D} = D_i + d = \mathbf{25 + d}$

Maximum twisting moment or torque applied on spring is given by,

$$\mathbf{T} = W_2 \times \frac{D}{2} = 360 \times \frac{(25 + d)}{2} = \mathbf{4500 + 180\ d} \quad \ldots (1)$$

Also, the twisting moment or torque is given by,

$$T = \frac{\pi}{16} \times \tau \times d^3$$

Substituting the value of T from equation (1),

$$\therefore \quad 4500 + 180\ d = \frac{\pi}{16} \times 400 \times d^3$$

$$\therefore \quad 4500 + 180\ d = 78.54\ d^3$$

$$\therefore \quad \mathbf{d} = \mathbf{4.05\ mm} \quad \ldots (2)$$

$$\therefore \quad \mathbf{D} = D_i + d = 25 + 4.05 = \mathbf{29.05\ mm}$$

$$\therefore \quad \text{Spring index} = \mathbf{C} = \frac{D}{d} = \frac{29.05}{4.05} = \mathbf{7.18}$$

Wahl's stress concentration factor is,

$$K = \frac{4C-1}{4C-4} + \frac{0.615}{C} = \frac{4 \times 7.18 - 1}{4 \times 7.18 - 4} + \frac{0.615}{7.18} = \mathbf{1.207}$$

Maximum shear stress induced in spring is given by,

$$\tau = K \times \frac{8W_2D}{\pi d^3} = K \times \frac{8W_2 C}{\pi d^2} \quad \ldots \left(\because C = \frac{D}{d}\right)$$

$$\therefore \quad d^2 = K \times \frac{8W_2 C}{\pi \times \tau} = 1.207 \times \frac{8 \times 360 \times 7.18}{\pi \times 400}$$

Thus, $\mathbf{d = 4.46\ mm}$ ... (3)

Comparing equations (2) and (3), we will choose larger value for spring wire diameter to have safe design. On comparing equations (2) and (3), we will take,

$\therefore$ $\mathbf{d = 4.46\ mm}$ and $\mathbf{D} = 25 + 4.46 = \mathbf{29.46\ mm}$

Diameter of spring wire = d = 4.46 mm and mean diameter of spring coil = 29.46 mm

**(ii) Number of turns :**

Deflection for load range ($\mathbf{W_2 - W_1}$ = 360 – 220 = **140 N**) is, $\delta = L_1 - L_2 = \mathbf{10\ mm}$.

We have,
$$\delta = \frac{8WD^3 n}{G \times d^4}$$

$$\therefore \quad n = \frac{\delta \times G \times d^4}{8WD^3} = \frac{10 \times 84 \times 10^3 \times (4.46)^4}{8 \times 140 \times (29.46)^3} = \mathbf{11.60 \cong 12\ (say)}$$

$\therefore$ Total number of active turns = **n = 12**

Assuming squared and grounded ends, total number of turns is given by,

$$\mathbf{n'} = n + 2 = 12 + 2 = \mathbf{14}$$

**(iii) Maximum deflection :** For a load range of 140 N, the deflection is 10 mm. Therefore, maximum deflection for maximum load of 360 N can be calculated as follows.

For 140 N $\rightarrow$ $\delta = 10$ mm

For 360 N $\rightarrow$ $\delta_{max} = \frac{360 \times 10}{140} = \mathbf{25.71\ mm}$

**(iv) Solid length :** $\mathbf{L_s} = n' \times d = 14 \times 4.46 = \mathbf{62.44\ mm}$

**(v) Free length :** $\mathbf{L_f} = n' \times d + \delta_{max} + 0.15 \times \delta_{max} = (14 \times 4.46) + 25.71 + (0.15 \times 25.71) = \mathbf{92\ mm}$

**(vi) Pitch of coil :** $\mathbf{p} = \frac{\text{Free length}}{n' - 1} = \frac{92}{14 - 1} = \mathbf{7.08\ mm}$

---

**Problem 5.18 :** *At the bottom of a mine shaft, a group of 10 identical close coiled helical springs are set in parallel to absorb the shock caused by the falling of a cage in case of failure. The loaded cage weighs 75 kN, while counter weight weighs 15 kN. If the loaded cage falls through a height of 50 metres from rest, find the maximum stress induced in each spring. Each spring is made-up of 50 mm diameter steel rod having spring index of 6 and number of active turns in each spring is 20. Modulus of rigidity, G = 80 kN/mm$^2$. Comment, whether the spring design is safe or not, if permissible shearing stress is 2800 N/mm$^2$.*

**Solution : Given data :** $\mathbf{W_1}$ = 75 kN = $\mathbf{75 \times 10^3\ N}$, $\mathbf{W_2}$ = 15 kN = $\mathbf{15 \times 10^3\ N}$, **h** = 50 m = $\mathbf{50 \times 10^3\ mm}$, d = 50 mm, C = 6, n = 20, G = $80 \times 10^3$ N/mm$^2$, $\tau$ = 2800 N/mm$^2$, $\mathbf{P} = W_1 - W_2 = 75 \times 10^3 - 15 \times 10^3 = \mathbf{60 \times 10^3\ N}$

**Procedure :**

Let 'W' = Equivalent static load applied gradually, that shall produce same effect as by falling load 'P'.

We have,
$$\delta = \frac{8WC^3 n}{G \cdot d} = \frac{8 \times W \times (6)^3 \times 20}{80 \times 10^3 \times 50} = \mathbf{8.64 \times 10^{-3}\ 'W'}$$

We know,

Work done by falling load 'P' = Energy stored by 10 springs

$$\therefore \quad P(h + \delta) = 10 \times \left(\frac{1}{2} \times W \times \delta\right)$$

$$\therefore \quad 60 \times 10^3 (50 \times 10^3 + 8.64 \times 10^{-3} W) = 5 \times W \times 8.64 \times 10^{-3} W \qquad [\because \delta = 8.64 \times 10^{-3} W]$$

$$\therefore \quad 3 \times 10^9 + 518.4 W = 0.0432 W^2$$

$$\therefore \quad 0.0432 W^2 - 518.4 W - 3 \times 10^9 = 0$$

$$\therefore \quad \mathbf{W = 269591.43\ N}$$

Now,

$$K = \frac{4C-1}{4C-4} + \frac{0.615}{C} = \frac{4 \times 6 - 1}{4 \times 6 - 4} + \frac{0.615}{6} = \mathbf{1.2525}$$

Maximum shear stress induced,

$$\tau_{max\ ind} = K \times \frac{8WC}{\pi d^2} = 1.2525 \times \frac{8 \times 269591.43 \times 6}{\pi \times (50)^2} = \mathbf{2063.64\ N/mm^2}$$

Since the maximum shear stress induced ($\tau_{max\ ind}$) is less than permissible shear stress (2800 $N/mm^2$), therefore, Design of all 10 springs is safe.

## Nunerical Type No. 5: "Design of Helical Spring Used for Railways"

**Problem 5.19 :** *A rail wagon of mass 20 tonnes is moving with a velocity of 9 kmph. It is brought to rest by using two buffer springs of 360 mm diameter. The maximum deflection of springs is 300 mm. An allowable shear stress for spring material is 800 MPa. Design the spring for buffers.*

**Solution : Given data : m** = 20 tonnes = **20 × 10³ kg**, **V** = 9 km/hr = $\frac{9 \times 10^3}{60 \times 60}$ = **2.5 m/sec**, $\delta_{max}$ = 300 mm, $\tau$ = 800 $N/mm^2$, D = 360 mm

**Procedure : Design of helical spring :**

**(1) Diameter of spring wire :** Let W be the load acting on each spring, which causes a deflection of 300 mm.

Since there are two springs, the energy stored in the two springs is given by,

$$U = 2 \times \frac{1}{2} \times W \times \delta = 2 \times \frac{1}{2} \times W \times 300 = \mathbf{300\ W} \qquad \text{... (1)}$$

We know, kinetic energy of railway wagon = $\frac{1}{2} \times m \times V^2$

$$\therefore \quad K.E. = \frac{1}{2} \times 20 \times 10^3 \times (2.5)^2 = 62.5 \times 10^3\ N\text{-}m = \mathbf{62.5 \times 10^6\ N\text{-}mm} \qquad \text{... (2)}$$

Equating equations (1) and (2), we have,

$$300\ W = 62.5 \times 10^6$$

$$\therefore \quad \mathbf{W = 208.333 \times 10^3\ N}$$

We know that,

$$T = W \times (D/2) = 208.333 \times 10^3 \times (360/2) = \mathbf{37.5 \times 10^6\ N\text{-}mm} \qquad \text{... (3)}$$

Torque or Twisting moment is also given by,

$$T = \frac{\pi}{16} \times \tau \times d^3$$

$$\therefore \quad 37.5 \times 10^6 = \frac{\pi}{16} \times 800 \times d^3 \qquad \text{[Referring equation (3)]}$$

$$\therefore \quad \mathbf{d = 62.03\ mm}$$

Diameter of spring wire = d = **62.03 mm**

**(2) Number of turns :**

$$\delta = \frac{8WD^3 n}{G \times d^4} \qquad [\text{Assume } G = 84 \times 10^3\ N/mm^2]$$

$$\therefore \quad n = \frac{\delta \times G \times d^4}{8WD^3} = \frac{300 \times 84 \times 10^3 \times (62.03)^4}{8 \times 208.333 \times 10^3 \times (360)^3} = 4.79 \cong \mathbf{5\ (say)}$$

Total number of active turns = n = **5**

Assuming squared and grounded ends, total number of turns is given by,

$$n' = n + 2 = 5 + 2 = \mathbf{7}$$

**(3) Solid length :** $L_s = n' \times d = 7 \times 62.03 = \mathbf{434.21\ mm}$

**(4) Free length :** $L_f = n' \times d + \delta_{max} + 0.15 \times \delta_{max} = (7 \times 62.03) + 300 + (0.15 \times 300) = \mathbf{779.21\ mm}$

**(5) Pitch of coil :** $p = \frac{\text{Free length}}{n' - 1} = \frac{779.21}{7 - 1} = \mathbf{129.87\ mm}$

---

**Problem 5.20 :** *A railway wagon weighing 50 kN and moving with a speed of 8 km/hr has to be stopped by 4 buffer springs, in which, the maximum compression allowed is 220 mm. Find number of turns in each spring having D = 150 mm, d = 25 mm, G = 84 kN/mm².*

**Solution : Given data :** Weight = 50 kN; ∴ Mass, $\mathbf{m} = \frac{W}{g} = \frac{50 \times 10^3\ N}{9.81} = \mathbf{5096.84\ kg}$

$V = 8\ km/hr = \frac{8 \times 10^3}{60 \times 60} = \mathbf{2.22\ m/s}$, $\delta = 220$ mm, D = 150 mm, d = 25 mm, $\mathbf{G} = 84\ kN/mm^2 = \mathbf{84 \times 10^3\ N/mm^2}$

**Procedure :**

Kinetic energy, $\mathbf{K.E.} = \frac{1}{2} mV^2 = \frac{1}{2} \times 5096.84 \times (2.22)^2 = 12559.63\ N\text{-}m = \mathbf{12559.63 \times 10^3\ N\text{-}mm}$

This K.E. of railway wagon is absorbed by 4 buffer springs, in the form of strain energy.

∴ Kinetic energy of railway wagon = 4 × Strain energy stored in each spring

$$\therefore \quad K.E. = 4 \times \left(\frac{1}{2} \times W \times \delta\right)$$

$$\therefore \quad 12559.63 \times 10^3 = 4 \times \left(\frac{1}{2} \times W \times 220\right)$$

$$\therefore \quad W = \frac{12559.63 \times 10^3}{220 \times 2} = \mathbf{28.544 \times 10^3\ N}$$

**Number of active turns :** It is calculated by formula of deflection.

$$\delta = \frac{8WC^3 n}{G \cdot d} \qquad \left[\text{where, } C = \frac{D}{d} = \frac{150}{25} = 6\right]$$

$$\therefore \quad 220 = \frac{8 \times 28.544 \times 10^3 \times (6)^3 \times n}{84 \times 10^3 \times 25}$$

$$\therefore \quad n = 9.366 \cong \mathbf{10\ (say)}$$

**Total number of turns :** Assuming squared and grounded ends, we know that,

Total number of turns, $\mathbf{n'} = n + 2 = 10 + 2 = \mathbf{12}$

---

## 5.4 SPRINGS IN SERIES AND PARALLEL

**Springs in Series :** Consider two springs connected in series as shown in Fig. 5.14.

Let, W = Load carried by the springs,

$\delta_1$ = Deflection of spring 1

$\delta_2$ = Deflection of spring 2

$\delta$ = Total deflection of both springs

S = Combined stiffness of both springs = $\frac{W}{\delta}$; Therefore, $\delta = \frac{W}{S}$

$S_1$ = Stiffness of spring 1 = $\frac{W_1}{\delta_1}$; Therefore, $\delta_1 = \frac{W_1}{S_1}$

and $S_2$ = Stiffness of spring 2 = $\frac{W_2}{\delta_2}$; Therefore, $\delta_2 = \frac{W_2}{S_2}$

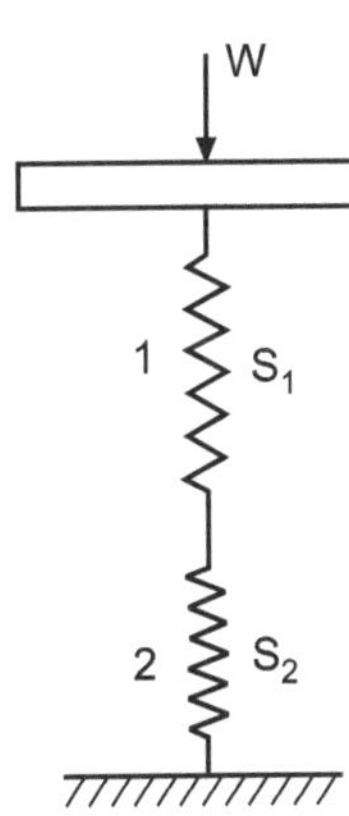

Fig. 5.14 : Springs in series

Here, load acting on individual spring is same and equal to total load 'W'.

$$\therefore \quad W_1 = W_2 = W \qquad \text{... equation (5.16)}$$

A little consideration will show that, *when springs are connected in series, then total deflection of springs is equal to sum of deflections of individual springs.*

∴ Total deflection of springs,

$$\delta = \delta_1 + \delta_2$$

$$\therefore \quad \frac{W}{S} = \frac{W_1}{S_1} + \frac{W_2}{S_2} \qquad \left[\because S = \frac{W}{\delta}; \therefore \delta = \frac{W}{S}\right]$$

$$\therefore \quad \mathbf{\frac{1}{S} = \frac{1}{S_1} + \frac{1}{S_2}} \qquad \text{... equation (5.17)}$$

**Springs in Parallel :** Consider two springs connected in parallel as shown in Fig. 5.15.

Let, $W$ = Load carried by the springs,

$W_1$ = Load shared by spring 1,

$W_2$ = Load shared by spring 2,

$S$ = Combined stiffness of both springs = $\frac{W}{\delta}$; Therefore, $W = \delta \cdot S$

$S_1$ = Stiffness of spring (1) = $\frac{W_1}{\delta_1}$; Therefore, $W_1 = \delta_1 \cdot S_1$

and $S_2$ = Stiffness of spring (2) = $\frac{W_2}{\delta_2}$; Therefore, $W_2 = \delta_2 \cdot S_2$

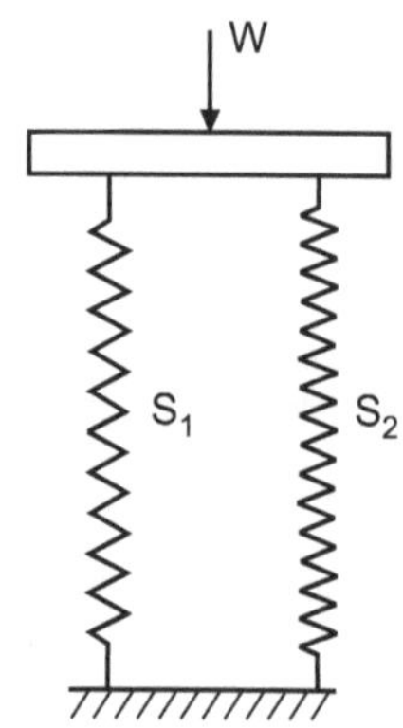

**Fig. 5.15 : Springs in parallel**

A little consideration will show that, *when springs are connected in parallel, then total deflection produced by the springs is same as the deflection of the individual springs.*

$$\therefore \quad \delta = \delta_1 = \delta_2 \qquad \text{... equation (5.18)}$$

Here, total load is shared by both springs.

$$\therefore \quad W = W_1 + W_2$$

$$\therefore \quad \delta \cdot S = \delta_1 \cdot S_1 + \delta_2 \cdot S_2$$

$$\therefore \quad \mathbf{S = S_1 + S_2} \qquad \text{... equation (5.19)} \qquad [\because \delta = \delta_1 = \delta_2 \text{ by equation (5.18)}]$$

**Numerical Type No. 6 : "Design of Helical Springs Arranged in Series or Parallel"**

**Problem 5.21 :** *A helical spring of stiffness 12 N/mm is placed on the top of other spring having stiffness 8 N/mm. Find the force required to give a total deflection of 50 mm.*

**Solution : Given data :** $S_1$ = 12 N/mm, $S_2$ = 8 N/mm, $\delta$ = 50 mm

**Procedure :** Since one spring is placed on the top of second spring, it means that, springs are connected in series.

We have,

$$\frac{1}{S} = \frac{1}{S_1} + \frac{1}{S_2} = \frac{1}{12} + \frac{1}{8}$$

$$\therefore \quad S = \mathbf{4.8\ N/mm}$$

Also,

$$S = \frac{W}{\delta}$$

$$\therefore \quad W = S \times \delta = 4.8 \times 50 = \mathbf{240\ N}$$

---

**Problem 5.22 :** *A closed coil helical spring of 12 active coils has a spring stiffness of K. It is cut into two springs having 5 and 7 turns. Determine the spring stiffness of resultant springs.*

**Solution : Given data :** $n_1$ = 5, $n_2$ = 7, n = 12

**Procedure :** We have, axial deflection,

$$\delta = \frac{8WD^3 n}{G \times d^4} \qquad \text{... (1)}$$

By definition, Stiffness (S) = K = $\frac{W}{\delta}$ [Stiffness is termed as 'K' in problem]

$$\therefore \quad K = \frac{W}{\left(\frac{8\ WD^3 n}{G \cdot d^4}\right)} \qquad \text{... From equation (1)}$$

$$\therefore \quad K = \frac{G \times d^4}{8D^3\,n} = A \times \frac{1}{n} = A \times \frac{1}{12} \qquad \left(\text{Let } A = \frac{G \times d^4}{8D^3}\right) \quad \ldots (2)$$

$$\therefore \quad \mathbf{A = 12\,K} \quad \ldots (3)$$

By referring equation (2), we can write, stiffness of individual spring as,

$$\mathbf{K_1} = A \times \frac{1}{n_1} = \mathbf{A \times \frac{1}{5}} \text{ and } \mathbf{K_2} = A \times \frac{1}{n_2} = \mathbf{A \times \frac{1}{7}}$$

But, from equation (3), we know that, A = 12 K.

$$\therefore \quad \mathbf{K_1} = 12\,K \times \frac{1}{5} = \mathbf{2.4\,K} \text{ and } \mathbf{K_2} = 12\,K \times \frac{1}{7} = \mathbf{1.71\,K}$$

Therefore, $K_1 = 2.4$ K and $K_2 = 1.71$ K are spring stiffness values of two cut down resultant springs.

## 5.5 LEAF SPRING

### 5.5.1 Introduction

- Leaf springs or flat springs are made out of flat plates.
- The ends of leaf spring may be guided along a definite path. Therefore, its deflection makes it possible to act as a structural number, in addition to its function as an energy absorbing device.
- Leaf spring may carry lateral loads, brake torque, driving torque etc. in addition to shocks.

### 5.5.2 Construction

- A leaf spring has semi-elliptical form. It consists of number of flat plates or leaves.
- The longest leaf is called as ***master leaf.*** Two extreme ends of master leaf are shaped like eyes, through which, bolts are passed to secure the spring to its supports.
- Master leaf has to withstand vertical bending loads as well as the loads due to sideways of vehicle (lateral loads). Therefore, one more full length leaf is provided below the master leaf. It is called as **second master leaf having full length**.
- Usually the eyes, through which, spring is attached to shackle, are provided with bushings of anti-friction material.
- Remaining plates or leaves are called as ***graduated leaves***, which are arranged in the order of gradually decreasing length and then clamped to the master leaf with the help of central bolt.

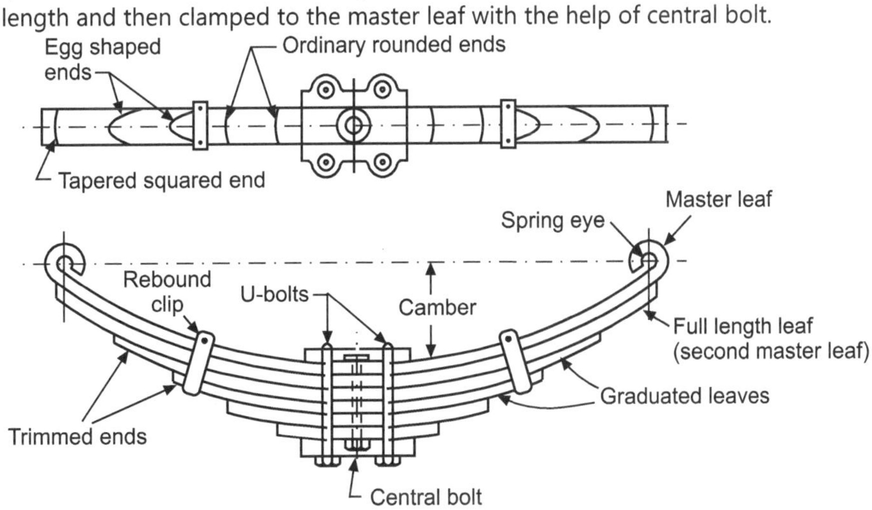


**Fig. 5.16 : Semi-elliptical leaf spring**

### 5.5.3 Functions of Central Bolt, U-clamp, Rebound Clips and Camber in a Leaf Spring

1. **Central bolt :** Leaf spring is made up of number of leaves. These leaves are held together by means of a bolt, positioned at the middle or centre of total span of leaf spring. This bolt is called as **central bolt**.
2. **U-clamp :** Leaf spring is clamped to the axle by means of *U-clamp*.

3. **Rebound clips :** They are located at intermediate positions along the length of spring, so that, graduated leaves can also share the stresses induced in the full-length leaves, when the spring rebounds.
4. **Camber :** Leaves are initially given a curvature, so that, they will tend to straighten under the action of load. This is called as *camber*.

## 5.5.4 Material with Specifications Used for Leaf Spring

- Leaf springs used in automobiles are made up of **'Oil hardened and Tempered alloy steels'** such as,

  1. 50 Cr 1,
  2. 50 Cr 1 V 23, and
  3. 55 Si 2 Mn 90.

## 5.5.5 Applications of Leaf Spring

1. Leaf spring acts as energy absorbing device.
2. Leaf spring also acts as a structural member.
3. Leaf springs are used in automobile suspensions due to their capability of sustaining lateral loads, brake torque and driving torque in addition to shocks.
4. In automobiles, leaf spring is used to damp out vibrations, shocks etc. caused due to road irregularities.

## 5.5.6 Expressions for Stresses and Deflections in Leaf Springs

Let,

$2W$ = Central load

$2L$ = Effective length or Span of spring

$b$ = Width of leaves

$t$ = Thickness of leaves

$n$ = Total number of leaves

$l$ = Length of central band

$n_f$ = Number of full length leaves

$n_g$ = Number of graduated leaves

Then,

(1) Stress in leaf spring :

$$\sigma_b = \frac{6\,WL}{n\,b\,t^2}$$

where,

Effective length of spring $= 2L = 2L_1 - l$ (when Central band is used)

$= 2L = 2L_1 - \frac{2}{3} \cdot l$ (when U-bolt is used)

(2) Deflection in leaf spring :

$$\delta = \frac{6\,WL^3}{n\,E\,b\,t^3}$$

(3) Stress in full length leaves :

$$\sigma_F = \frac{18\,WL}{bt^2\,(2n_g + 3n_f)}$$

(4) Stress in graduated leaves :

$$\sigma_G = \frac{12\,WL}{bt^2\,(2n_g + 3n_f)}$$

(5) Deflection in full length and graduated leaves :

$$\delta = \frac{12WL^3}{E\,b\,t^3\,(2n_g + 3n_f)}$$

## 5.5.7 Nipping in Leaf Springs

- We know that,

Stress in full length leaves: $\sigma_F = \dfrac{18\ WL}{bt^2\ (2n_g + 3n_f)}$

Stress in graduated leaves: $\sigma_G = \dfrac{12\ WL}{bt^2\ (2n_g + 3n_f)}$

We can write, $\sigma_F = \dfrac{18\ WL}{bt^2\ (2n_g + 3n_f)} = 1.5 \times \dfrac{12\ WL}{bt^2\ (2n_g + 3n_f)}$

$\therefore$ $\sigma_F = 1.5 \times \sigma_G$

- It means that, stress induced in full length leaves is 50% greater than the stress induced in graduated leaves. But, in order to utilize the material to the best possible extent/advantage, all the leaves should be equally stressed. This may be achieved by pre-stressing the leaves.

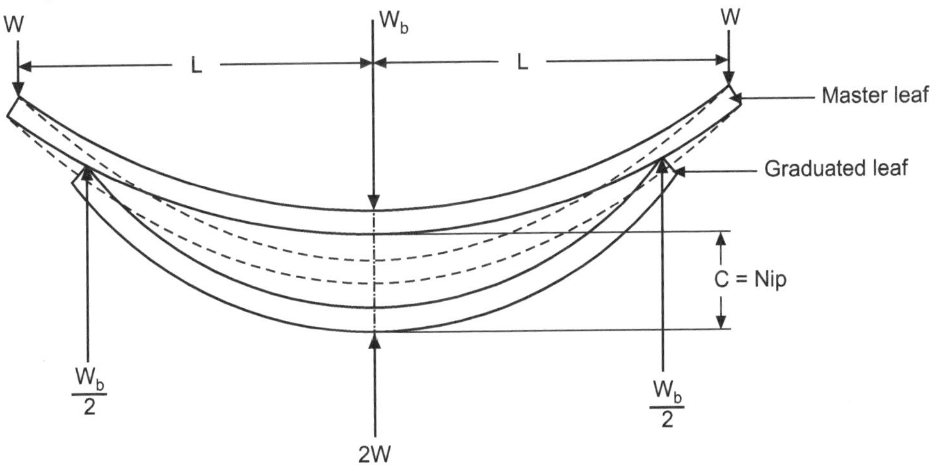


**Fig. 5.17 : Nipping of leaf springs**

- Pre-stressing of spring can be done by giving greater radius of curvature to the full length (Master) leaves than the graduated leaves before assembly as shown in Fig. 5.17.
- Due to this difference in radius of curvature, a gap 'C' exists in between full length leaf and graduated leaf before assembly. This initial gap provided before assembly of leaf spring is called as **'Nip' (C)**.
- When the central bolt holding leaves together is tightened, the master leaf will bend back as shown by dotted lines. Due to this bending, master leaf will have an initial stress in a direction opposite to that of normal load, whereas, the graduated leaves will have an initial stress in the same direction as that of normal load.
- When the load is gradually applied to the spring, the extra full length leaves are first relieved of this initial stress and then stressed in opposite direction. Consequently, the full length leaves will be stressed less than that without nip.
- This process of pre-stressing the spring by giving different radii of curvature before assembly is known as ***nipping.***
- The initial gap between the leaves (i.e. nip) is adjusted in such a way that, stresses in all leaves will be equal under the maximum load condition.
- Then, we can see that, all leaves including full length leaves are equally stressed, when subjected to maximum load.
- Normally, **nip** is adjusted to give stress in full length leaves slightly less than the graduated leaves.
- This is desirable in automobile, because full length leaves are expected to take transverse loads in addition to bending load.

**Numerical Type No. 7 : "Design of Numericals on Leaf Springs"**

**Problem 5.23 :** *A semi-elliptical carriage spring of 1200 mm length withstands a load of 60 kN with maximum permissible deflection of 90 mm. Assume breadth to thickness ratio as 8. Design the spring, if $\sigma_b$ = 540 MPa and E = $2 \times 10^5$ N/mm².*

**Solution : Given data :** Central load on each spring = 2W = 60 kN,

$\therefore \quad W = 30 \text{ kN} = 30 \times 10^3 \text{ N}, \ 2L = 1200 \text{ mm}; \ L = \frac{1200}{2} = 600 \text{ mm},$

$\delta = 90 \text{ mm}, \ \sigma_b = 540 \text{ N/mm}^2, E = 2 \times 10^5 \text{ N/mm}^2, \ \frac{b}{t} = 8 ; \ \because b = 8t$

**Procedure :**

We know that, stress in leaf spring is given by,

$$\sigma_b = \frac{6WL}{nbt^2}$$

$$\therefore \quad n\,b\,t^2 = \frac{6WL}{\sigma_b}$$

$$\therefore \quad n \times b \times t^2 = \frac{6 \times 30 \times 10^3 \times 600}{540} = \mathbf{200 \times 10^3} \quad \ldots (1)$$

Also,

$$\delta = \frac{6WL^3}{nEbt^3}$$

$$\therefore \quad n\,b\,t^3 = \frac{6WL^3}{\delta \times E}$$

$$\therefore \quad n \times b \times t^3 = \frac{6 \times 30 \times 10^3 \times (600)^3}{90 \times 2 \times 10^5} = \mathbf{2.16 \times 10^6} \quad \ldots (2)$$

Dividing equation (2) by equation (1), we get,

$$t = \mathbf{10.8 \text{ mm}}$$

$$\therefore \quad b = 8t = 8 \times 10.8 = \mathbf{86.4 \text{ mm}}$$

To obtain required number of leaves, put the values of b and t in equation (1),

$$n \times b \times t^2 = 200 \times 10^3$$

$$\therefore \quad n \times 86.4 \times (10.8)^2 = 200 \times 10^3$$

$$\therefore \quad n = 19.84 \cong \mathbf{20 \ (say)}$$

---

**Problem 5.24 :** *Design a leaf spring for the following specifications :*

(a) *Total load = 16 kN*

(b) *Number of springs supporting the load = 4*

(c) *Maximum number of leaves = 10*

(d) *Span of the spring = 1000 mm*

(e) *Permissible deflection = 80 mm*

(f) *Young's modulus = $2 \times 10^5$ N/mm²*

(g) *Allowable stress in spring material = 600 N/mm²*

**Solution : Given data :** Central load carried by 4 springs = 16 kN

$\therefore$ Central load carried by each spring = $\frac{16}{4}$ kN = 4 kN

$$\therefore \quad 2W = 4 \text{ kN}$$

$$\therefore \quad W = 2 \text{ kN} = 2 \times 10^3 \text{ N}$$

$2L = 1000 \text{ mm}, \ \delta = 80 \text{ mm}, \ \sigma_b = 600 \text{ N/mm}^2, \ n = 10, \ E = 2 \times 10^5 \text{ N/mm}^2$

**Procedure :** We have, $2L = 1000 \text{ mm}; \ \therefore L = 500 \text{ mm}$

We know that, stress in leaf spring is given by,

$$\sigma_b = \frac{6 \times W \times L}{n \times b \times t^2}$$

Thus,

$$n \times b \times t^2 = \frac{6 \times W \times L}{\sigma_b} = \frac{6 \times 2 \times 10^3 \times 500}{600} = \mathbf{10 \times 10^3} \quad \ldots (1)$$

Also, deflection in leaf spring is given by,

$$\delta = \frac{6 \times W \times L^3}{n \times E \times b \times t^3}$$

Thus, $$n \times b \times t^3 = \frac{6 \times W \times L^3}{\delta \times E} = \frac{6 \times 2 \times 10^3 \times (500)^3}{80 \times 2 \times 10^5} = \mathbf{93.75 \times 10^3} \quad \text{... (2)}$$

Dividing equation (2) by equation (1), we get,

$$t = \mathbf{9.375\ mm}$$

From equation (1),

$$n \times b \times t^2 = 10 \times 10^3$$

$$\therefore \quad 10 \times b \times (9.375)^2 = 10 \times 10^3$$

$$\therefore \quad b = 11.37\ \text{mm} \cong \mathbf{12\ mm\ (say)}$$

---

**Problem 5.25 :** *A semi-elliptical laminated spring 900 mm long and 55 mm wide is held together at the centre by a band 50 mm wide. If the thickness of each leaf is 5 mm, find the number of leaves required to carry a load of 4500 N. Assume a maximum working stress of 490 MPa.*

*If the two of these leaves are full length leaves of spring, find the deflection of the spring. Young's modulus for the spring is 210 kN/mm².*

**Solution : Given data :** Central load = 2W = 4500 N, $\therefore$ $W = \frac{4500}{2} = \mathbf{2250\ N}$,

$2L_1 = 900$ mm, $l = 50$ mm, $b = 55$ mm, $t = 5$ mm, $\sigma_b = 490\ N/mm^2$, $E = 210\ kN/mm^2 = 210 \times 10^3\ N/mm^2$

**Procedure :**

**(i) Number of leaves :** When central band is used, the effective length is,

$$2L = 2L_1 - l = 900 - 50 = 850\ \text{mm}$$

$$\therefore \quad L = \mathbf{425\ mm}$$

We know that, stress in leaf spring is given by,

$$\sigma_b = \frac{6 \times W \times L}{n \times b \times t^2}$$

$\therefore$ Number of leaves,

$$n = \frac{6 \times W \times L}{\sigma_b \times b \times t^2} = \frac{6 \times 2250 \times 425}{490 \times 55 \times (5)^2} = 8.515 \cong \mathbf{10\ turns\ (say)}$$

**(ii) Deflection in leaf spring :**

$$\delta = \frac{6 \times W \times L^3}{n \times E \times b \times t^3} = \frac{6 \times 2250 \times (425)^3}{10 \times 210 \times 10^3 \times 55 \times 5^3} = \mathbf{71.78\ mm}$$

---

**Problem 5.26 :** *A locomotive semi-elliptical laminated spring has overall length of 1 m and sustains a load of 70 kN at its centre. The spring has 3 full-length leaves and 15 graduated leaves with a central band of 100 mm width. All leaves are to be stressed to 400 MPa, when fully loaded. The ratio of the total spring depth to that of width is 2.*

*Take E = 200 kN/mm². Determine :*

*(1) Thickness and width of leaves.*

*(2) Initial gap that should be provided between full length and graduated leaves (Nip).*

*(3) The load exerted on band, after spring is assembled.*

**Solution : Given data :** Central load = 2W = 70 kN; $\therefore$ W = 35 kN = $\mathbf{35 \times 10^3\ N}$, $2L_1 = 1000$ mm, $l = 100$ mm,

$n_f = 3$, $n_g = 15$, $\therefore$ $\mathbf{n = n_f + n_g = 3 + 15 = 18}$, $\sigma_b = 400\ N/mm^2$, $E = 200\ kN/mm^2 = 200 \times 10^3\ N/mm^2$

**Procedure :** When central band is used, the effective length is given by,

$$2L = 2L_1 - l = 1000 - 100 = 900\ \text{mm}$$

$\therefore \quad L = \mathbf{450\ mm}$

It is given that, $\dfrac{\text{Total spring depth}}{\text{Width}} = 2$

$\therefore \quad \text{Total spring depth} = 2 \times \text{Width} = 2 \times b \quad \ldots (1)$

But, by definition,

$\text{Total spring depth} = n \times t = 18 \times t \quad \ldots (2)$

on equating (1) and (2), we get,

$$2 \times b = 18 \times t; \quad \therefore \mathbf{b = 9t}$$

**(i) Thickness and width of leaves :**

We have, $\sigma_b = \dfrac{6 \times W \times L}{n \times b \times t^2}$

$\therefore \quad 400 = \dfrac{6 \times 35 \times 10^3 \times 450}{18 \times 9t \times t^2}$

$\therefore \quad t = 11.34\text{ mm} = \mathbf{12\ mm\ (say)}$ and $b = 9t = 9 \times 12 = \mathbf{108\ mm}$

**(ii) Initial gap provided between full length and graduated leaves (Nip) :**

$$\text{Nip or Initial gap, } C = \frac{2 \times W \times L^3}{n \times E \times b \times t^3} = \frac{2 \times 35 \times 10^3 \times (450)^3}{18 \times 200 \times 10^3 \times 108 \times (12)^3} = \mathbf{9.5\ mm}$$

**(iii) The load exerted on band, after the spring is assembled :**

$$W_b = \frac{2 \times n_f \times n_g \times W}{n \times (2n_g + 3n_f)} = \frac{2 \times 3 \times 15 \times 35 \times 10^3}{18 \times (2 \times 15 + 3 \times 3)} = \mathbf{4487.18\ N}$$

---

**Problem 5.27 :** *A semi-elliptical laminated spring with a span of 600 mm carries a central load of 180 kN. The spring has 2 full-length leaves and 15 graduated leaves having thickness of 12 mm and width 96 mm. U-bolt of width 90 mm is used. Take E = 200 kN/mm². Determine the stresses induced in full-length and graduated leaves. Also find the deflection produced.*

**Solution : Given data :** Central load = $2W = 180$ kN; $\therefore W = 90\text{ kN} = 90 \times 10^3$ N,

$n_f = 2$, $n_g = 15$, $t = 12$ mm, $b = 96$ mm, $2L_1 = 600$ mm, $l = 90$ mm,

$E = 200\text{ kN/mm}^2 = 200 \times 10^3\text{ N/mm}^2$

**Procedure :** When U-bolt is used, the effective length is,

$$2L = 2L_1 - (2/3) \times l = 600 - (2/3) \times 90 = 540\text{ mm}; \quad \therefore L = \mathbf{270\ mm}$$

**(i) Stress induced in full-length leaves :**

$$\sigma_F = \frac{18 \times W \times L}{b \times t^2 \times (2n_g + 3n_f)} = \frac{18 \times 90 \times 10^3 \times 270}{96 \times (12)^2 \times (2 \times 15 + 3 \times 2)} = \mathbf{878.91\ N/mm^2}$$

**(ii) Stress induced in graduated leaves :**

$$\sigma_G = \frac{12 \times W \times L}{b \times t^2 \times (2n_g + 3n_f)} = \frac{12 \times 90 \times 10^3 \times 270}{96 \times (12)^2 \times (2 \times 15 + 3 \times 2)} = \mathbf{585.94\ N/mm^2}$$

**(iii) Deflection in full-length and graduated leaves :**

$$\delta = \frac{12 \times W \times L^3}{E \times b \times t^3 \times (2n_g + 3n_f)} = \frac{12 \times 90 \times 10^3 \times (270)^3}{200 \times 10^3 \times 96 \times (12)^3 \times (2 \times 15 + 3 \times 2)} = \mathbf{17.79\ mm}$$

---

**Problem 5.28 :** *A truck spring has 12 number of leaves, two of which, are full length leaves. The spring supports are 1.05 m apart and the central band is 85 mm wide. The central load is 5.4 kN with permissible stress 280 N/mm². Determine the thickness and width of steel spring leaves. The ratio of total depth to width of the spring is 3. Also, determine the deflection of the spring.*

**Solution : Given data :** Central load = 2 W = 5.4 kN, ∴ W = 2.7 kN = $2.7 \times 10^3$ N, n = 12, $n_f = 2$, $n_g = 10$, $2L_1 = 1050$ mm, $l = 85$ mm, $\sigma_b = 280$ N/mm², E = $2.1 \times 10^5$ N/mm² (Assumed).

**Procedure :** When central band is used, the effective length is,

$$2L = 2L_1 - l = 1050 - 85 = 965 \text{ mm}; \quad \therefore L = \frac{965}{2} = \mathbf{482.5\ mm}$$

It is given that, $\dfrac{\text{Total spring depth}}{\text{Width}} = 3$

∴ Total spring depth = 3 × Width = 3 × b ... (1)

But, by definition, Spring depth = n × t = 12 × t ... (2)

On equating (1) and (2),

$$3 \times b = 12 \times t$$

$$\therefore \quad \mathbf{b = 4\,t}$$

**(1) Thickness and width of leaves :** Assume that, the leaves are not initially stressed. Therefore, maximum bending stress for full length leaves is given by,

$$\sigma_F = \frac{18 \times W \times L}{b \times t^2 \times (2n_g + 3n_f)}$$

$$\therefore \quad 280 = \frac{18 \times 2.7 \times 10^3 \times 482.5}{4t \times t^2 \times (2 \times 10 + 3 \times 2)}$$

$$\therefore \quad 280 = \frac{225.47 \times 10^3}{t^3}$$

$$\therefore \quad t = 9.303 \text{ mm} \cong \mathbf{10\ mm\ (say)}$$

and $\quad b = 4t = 4 \times 10 = \mathbf{40\ mm}$

**(2) Deflection in full length and graduated leaves :**

$$\delta = \frac{12 \times W \times L^3}{E \times b \times t^3 \times (2n_g + 3n_f)} = \frac{12 \times 2.7 \times 10^3 \times (482.5)^3}{2.1 \times 10^5 \times 40 \times (10)^3 \times (2 \times 10 + 3 \times 2)} = \mathbf{16.66\ mm}$$

## Theory Questions for Practice

1. What is function of a spring ? State its applications.
2. Give practical application for each of following springs :
   (i) Helical tension spring. (ii) Leaf spring.
   (iii) Helical torsion spring. (iv) Spiral tension spring.
3. List the material used in springs.
4. Explain the following terms applied to spring :
   (a) Free length, (b) Solid length, (c) Spring index,
   (d) Spring rate, (e) Spring stress factor, (f) Pitch of coil.
5. Explain the stresses induced in helical springs of circular wire.
6. Explain what do you understand by A.M. Wahl's factor ? State its importance in the design of helical springs.
7. Draw a graph of Wahl's stress factor versus spring index for helical compression spring and state the effect of curvature of coil on stress distribution.
8. Explain different end conditions for helical compression springs.

9. Explain springs in series and parallel.
10. Draw a neat sketch of semielliptical leaf spring with construction details. What material is used for leaf spring?

## Numerical Problems for Practice

1. A helical spring is made from a wire of 6 mm diameter and has outside diameter of 75 mm. If the permissible shear stress is 350 MPa and modulus of rigidity is 84 kN/mm², find the axial load, which the spring can carry and deflection per active turn. Take Wahl's factor, $K = \frac{4C - 1}{4C - 4} + \frac{0.615}{C}$ (**Ans.** 382.49 N, 9.233 mm)
2. A helical spring is made from a wire of 6 mm diameter and has outside diameter of 75 mm. If the permissible shear stress is 350 MPa and modulus of rigidity is 84 kN/mm², find the axial load, which the spring can carry and deflection per active turn, neglecting effect and considering effect of curvature. (**Ans.** (i) 412.52 N, 9.958 mm (ii) 382.45 N, 9.232 mm)
3. Design a helical compression spring with squared and grounded ends for operating load range between 80 N and 145 N. The axial deflection of the spring for load range is 6.5 mm, while spring index is 8. The permissible shear stress is 380 N/mm². Take design stress for intermittent operation of spring. (**Ans.** d = 1.94 mm ≅ 2 mm, D = 16 mm, n = 5, n' = 7, $L_s$ = 14 mm, $L_f$ = 30.67 mm, p = 5.11 mm)
4. A helical spring is subjected to a load ranging from 2500 N to 3250 N. Axial compression for above load range is 5 mm. Design the spring, if C = 5, τ = 400 N/mm² and G = 84 kN/mm². (**Ans.** d = 12 mm, D = 60 m, n = 6.52 ≅ 7, n' = 9, $\delta_{max}$ = 21.66 mm, $L_f$ = 132.90 mm, p = 16.61 mm)
5. Design a helical compression valve spring with squared and grounded ends for load range between 80 N to 145 N. The axial deflection of the spring is 6.5 mm. The spring index is 8. Permissible shear stress intensity is 380 N/mm². Take design stress 25% excess of permissible stress for severe condition and intermittent operating of spring. (**Ans.** d = 2.5 mm, D = 20 mm, n = 6, n' = 8, $L_f$ = 36.6 mm, p = 5.21 mm)
6. Design a helical compression spring with squared and grounded ends for operating range 100 N to 135 N. The axial deflection of the spring is 5 mm. Spring index is 8. Permissible shear stress intensity is 330 N/mm². Take design stress 25% excess of permissible stress for severe condition and intermittent operation. (**Ans.** d = 2.81 mm, D = 22.48 mm, n = 9, $\delta_{max}$ = 19.2857 mm, $L_s$ = 30.91 mm, $L_f$ = 53.09 mm, p = 5.309 mm)
7. A spring loaded safety valve for boiler is required to blow off at a pressure of 1.20 N/mm². Diameter of the valve is 65 mm. The maximum lift of the valve is 17.5 mm. Design a suitable compression spring for the safety valve. Assume C = 6 and provide initial compression of 30 mm. The maximum shear stress in the spring material = 450 N/mm² and G = $0.84 \times 10^5$ N/mm². (**Ans.** d = 16.35 mm, D = 98.10 mm, n = 6, n' = 8, $L_f$ = 185.5 mm, p = 26.51 mm)
8. Design a helical spring using following data :
   - Valve seat diameter = 70 mm
   - Maximum pressure, when valve blows off freely = 0.73 N/mm²
   - Valve lift, when pressure rises from 0.7 to 0.73 N/mm² = 3.5 mm
   - Maximum allowable stress = 500 N/mm²
   - Spring index = 6.
   - Modulus of rigidity = $8.3 \times 10^4$ N/mm².

   (**Ans.** d = 10.4 mm, D = 62.4 mm, $\delta_{max}$ = 88.14 mm, n = 15.8 ≅ 16, n' = 18, $L_f$ = 288.57 mm)

❑❑❑

Chapter 6

# DESIGN OF FASTENERS - RIVETED JOINTS

## 6.1 TYPES OF FASTENINGS (OR JOINTS)

- The fastenings (i.e. joints) may be classified into the following two groups:

1. **Permanent fastenings:**
   - o The permanent fastenings are those fastenings, which cannot be disassembled without destroying the connecting components.
   - o The examples of permanent fastenings in increasing order of strength are,
     (i) Soldered joints.
     (ii) Brazed joints.
     (iii) Welded joints.
     (iv) Riveted joints.

2. **Temporary or detachable fastenings:**
   - o The temporary or detachable fastenings are those fastenings, which can be disassembled without destroying the connecting components.
   - o The examples of temporary fastenings are,
     (i) Screwed joints.
     (ii) Keys used to join either a pulley or gear on a shaft.
     (iii) Cotter joints.
     (iv) Pins.
     (v) Splined joints.

## 6.2 INTRODUCTION TO RIVETED JOINTS

- A rivet is a short cylindrical bar with a head integral to it.
- The cylindrical portion of the rivet is called shank or body and lower portion of shank is known as tail, as shown in Fig. 6.1.
- The rivets are used to make permanent fastening between the plates such as in structural work, ship building, bridges, tanks and boiler shells.
- The riveted joints are widely used for joining light metals.

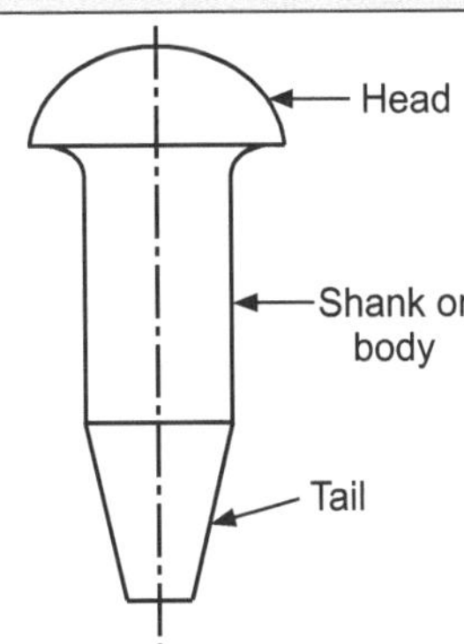


**Fig. 6.1 : Parts of a rivet**

## 6.3 METHODS OF RIVETING

- The function of rivets in a joint is to make a connection that has strength and tightness. The strength is necessary to prevent failure of the joint.
- The tightness is necessary in order to contribute to strength and to prevent leakage as in a boiler or in a ship hull.
- When two plates are to be fastened together by a rivet as shown in Fig. 6.2 (a), the holes in the plates are punched and reamed or drilled.

- Punching is the cheapest method and is used for relatively thin plates and in structural work.
- Since punching injures the material around the hole, therefore drilling is used in most pressure-vessel work.
- In structural and pressure vessel riveting, the diameter of the rivet hole is usually 1.5 mm larger than the nominal diameter of the rivet.

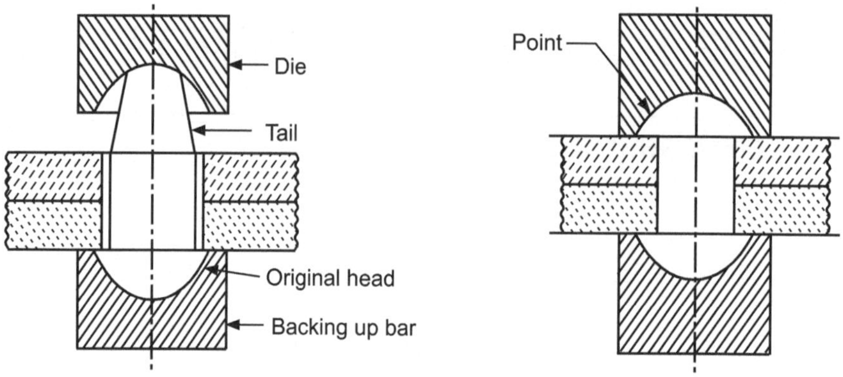


**(a) Initial position** **(b) Final position**

**Fig. 6.2 : Methods of riveting**

- The plates are drilled together and then separated to remove any burrs or chips so as to have **a tight fluid joint** between the plates.
- A cold rivet or a red hot rivet is introduced into the plates and **the point (i.e. second head)** is then formed, When a cold rivet is used, the process is known as **cold riveting** and when a hot rivet is used, the process is known as **hot riveting.**
- The cold riveting process is used for structural joints, whereas, hot riveting is used to make leak proof joints.
- The riveting may be done by hand or by a riveting machine.
- In hand riveting, the original rivet head is backed up by a hammer or heavy bar and then the die or set is placed against the end to be headed and the blows are applied by a hammer. Refer Fig. 6.2 (a). This causes the shank to expand, thus filling the hole and the tail is converted into **a point** as shown in Fig. 6.2 (b). As the rivet cools, it tends to contract. The lateral contraction will be slight, but there will be a longitudinal tension introduced in the rivet, which holds the plates firmly together.
- In machine riveting, the die is a part of the hammer, which is operated by air pressure, hydraulic pressure or steam pressure.

**Notes:**

1. For steel rivets up to 12 mm diameter, the cold riveting process may be used while for larger diameter rivets, hot riveting process is used.
2. In case of long rivets, only the tail is heated and not the whole shank.

## 6.4 MATERIAL OF RIVETS

- The material of the rivets must be tough and ductile.
- They are usually made of steel (low carbon steel or nickel steel), brass, aluminium or copper, but when strength and a fluid tight joint is the main consideration, then the steel rivets are used.

## 6.5 ESSENTIAL QUALITIES OF A RIVET

- According to Indian standard, IS : 2998 - 1982 (Reaffirmed 1992), the material of a rivet must have a tensile strength not less than 40 $N/mm^2$ and elongation not less than 26 %.
- The material must be of such quality that, when in cold condition, the shank shall be bent on itself through 180° without cracking and after being heated to 650°C and quenched, it must pass the same test.
- The rivet, when hot, must flatten without cracking to a diameter 2.5 times the diameter of shank.

## 6.6 MANUFACTURING PROCESS OF RIVETS

- According to Indian standard specifications, the rivets may be made either by cold heading or by hot forging.
- If rivets are made by the cold heading process, they shall subsequently be adequately heat treated, so that, the stresses set up in the cold heating process are eliminated.
- If they are made by hot forging process, care shall be taken to see that, the finished rivets cool gradually.

## 6.7 TYPES OF RIVET HEADS

- According to Indian standard specifications, the rivet heads are classified into the following three types:

### 6.7.1 Rivet Heads for General Purposes (Below 12 mm Diameter)

- **Refer** Fig. 6.3.

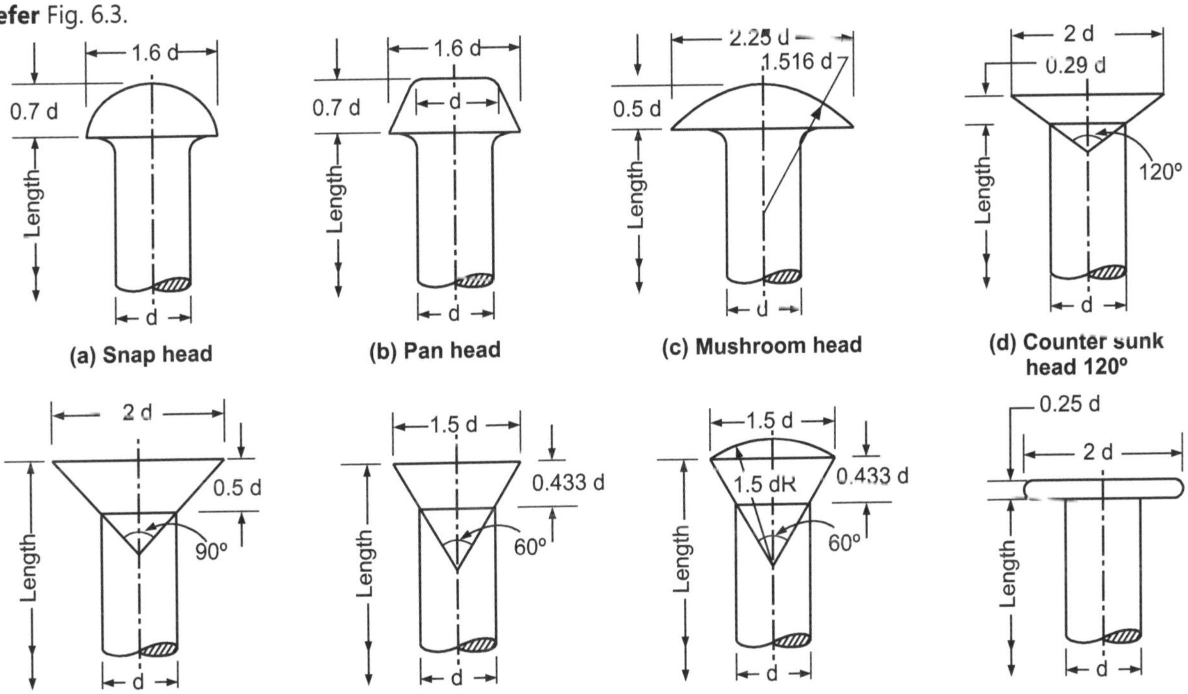


(a) Snap head (b) Pan head (c) Mushroom head (d) Counter sunk head 120°

(e) Flat counter sunk head 90° (f) Flat counter sunk head 60° (g) Round counter sunk head 60° (h) Flat head

**Fig. 6.3 : Rivet heads for general purposes (below 12 mm diameter)**

### 6.7.2 Rivet Heads for General Purposes (From 12 mm to 48 mm Diameter)

- **Refer** Fig. 6.4.

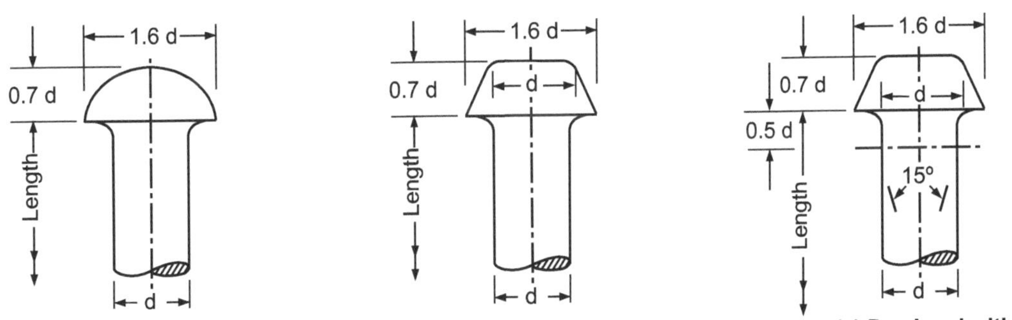


(a) Snap head (b) Pan head (c) Pan head with tapered neck

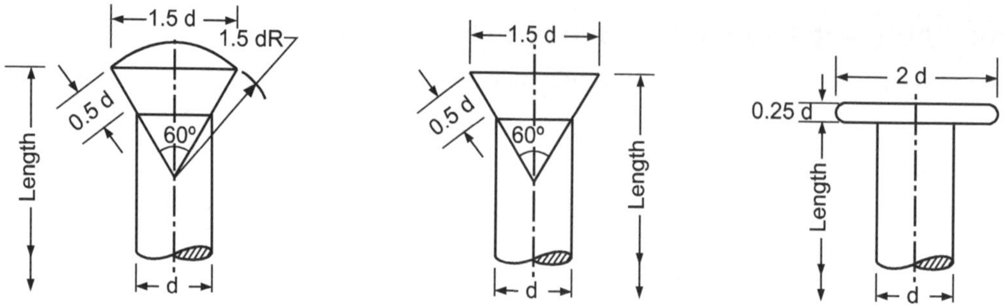


(d) Round counter sunk head 60° (e) Flat counter sunk head 60° (f) Flat head

Fig. 6.4 : Rivet heads for general purposes (from 12 mm to 48 mm diameter)

## 6.7.3 Rivet Heads for Boiler Work (From 12 mm to 48 mm Diameter)

- **Refer** Fig. 6.5.

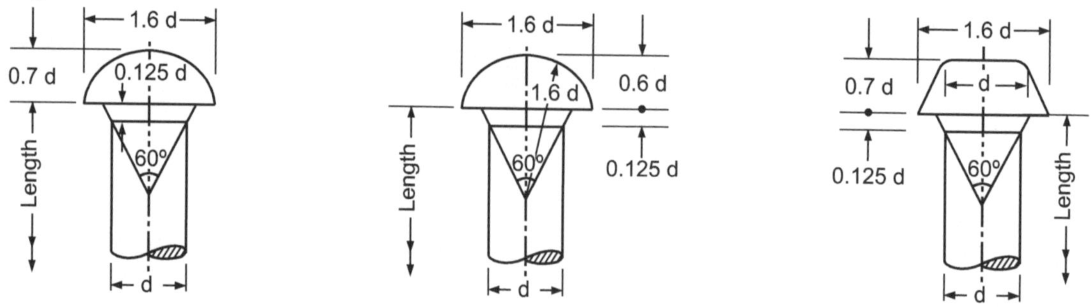


(a) Snap head (b) Ellipsoid head (c) Pan head (Type I)

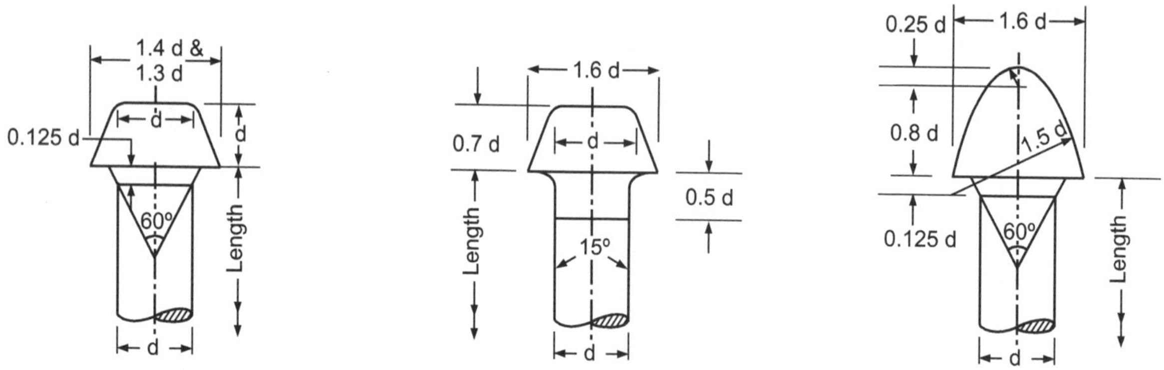


1.4 d for rivets under 24 mm
1.3 d for rivets 24 mm and over
**(d) Pan head (Type II)** (e) Pan head with tapered neck (f) Conical head

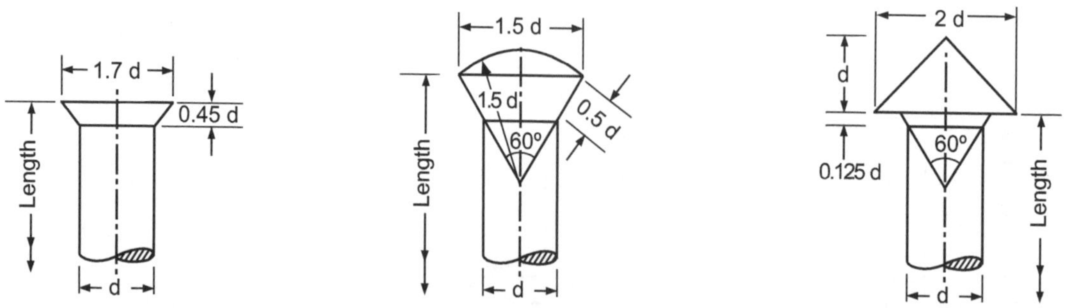


(g) Counter-sunk head (h) Round counter sunk head (i) Steeple head

Fig. 6.5 : Rivet heads for boiler work

- The snap heads are usually employed for structural work and machine riveting.
- The counter sunk heads are mainly used for ship building where flush surfaces are necessary.
- The conical heads (also known as conoidal head) are mainly used in case of hand hammering.
- The pan heads have maximum strength. But, these are difficult to shape.

## 6.8 TYPES OF RIVETED JOINTS

- Riveted joints can be classified as,

**1. On the basis of construction:**

- Following are the two types of riveted joints, depending upon the way, in which, the plates are connected.
  (i) Lap Joint.
  (ii) Butt Joint.

**2. On the basis of number of rows of the rivets:**

- Following are the types of riveted joints depending upon the number of rows of the rivets.
  (i) Single riveted joint.
  (ii) Double riveted joint.

## 6.9 LAP JOINT

- When one plate overlaps the other plate and the two plates are then riveted together, then the type of riveted joint is called as **lap joint.**

## 6.10 BUTT JOINT

- When the main plates are kept in alignment butting (i.e. touching) each other and a cover plate (i.e. strap) is placed either on one side or on both sides of the main plates and then the cover plate is riveted together with the main plates, then the type of riveted joint is called as **butt joint.**
- **Butt joints are further classified** in to following two types:
  1. **Single strap butt joint:** In a single strap butt joint, the edges of the main plates butt (i.e. touch) against each other and only one cover plate is placed on one side of the main plates and then riveted together.
  2. **Double strap butt joint:** In a double strap butt joint, the edges of the main plates butt (i.e. touch) against each other and two cover plates are placed on both sides of the main plates and then riveted together.

## 6.11 SINGLE RIVETED JOINT

- In single riveted joint, there is a single row of rivets in a lap joint as shown in Fig. 6.6 (a) and there is a single row of rivets on each side in a butt joint as shown in Fig. 6.8.

## 6.12 DOUBLE RIVETED JOINT

- In double riveted joint, there are two rows of rivets in a lap joint as shown in Fig. 6.6 (b) and (c) and there are two rows of rivets on each side in a butt joint as shown in Fig. 6.9.

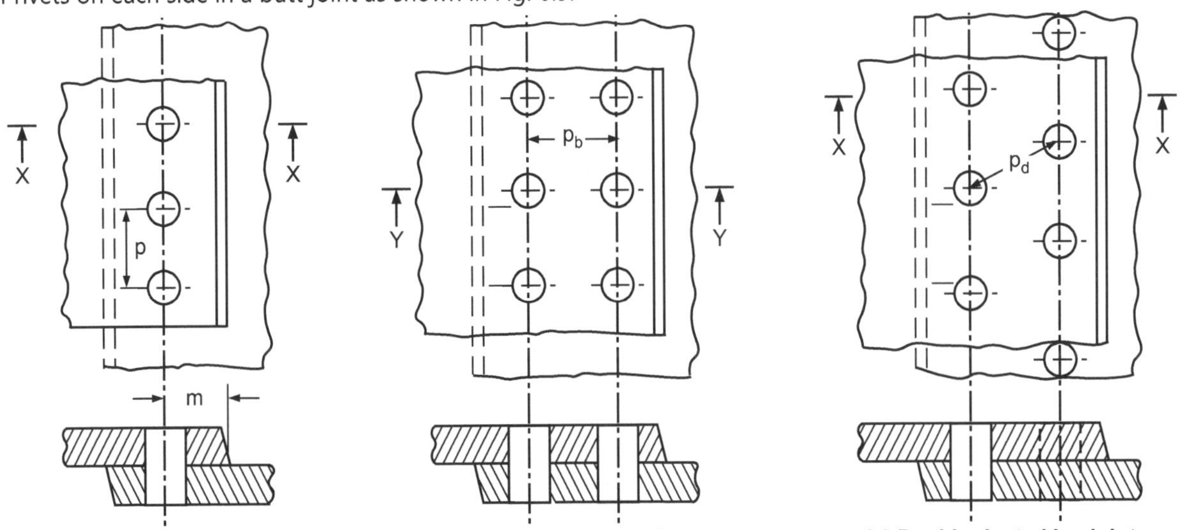


**(a) Single riveted lap joint** **(b) Double riveted lap joint (Chain riveting)** **(c) Double riveted lap joint Zig-zag riveting)**

**Fig. 6.6 : Single and double riveted lap joints**

- Similarly, the joints may be triple riveted or quadruple riveted.

**Notes:**

1. When the rivets in the various rows are opposite to each other, as shown in Fig. 6.6 (b), then the joint is said to be **chain riveted**. On the other hand, if the rivets in the adjacent rows are staggered in such a way that, every rivet is in the middle of the two rivets of the opposite row as shown in Fig. 6.6 (c), then the joint is said to be **zig-zag riveted.**
2. Since the plates overlap in lap joints, therefore the force W, W acting on the plates [See Fig. 6.14 (a)] are not in the same straight line, but they are at a distance equal to the thickness of the plate. These forces will form a couple, which may bend the joint. Hence, the lap joints may be used only, where small loads are to be transmitted. On the other hand, the forces in a butt joint act in the same straight line. Therefore, there will be no couple. Hence, the butt joints are used, where heavy loads are to be transmitted. Refer Fig. 6.14 (b).

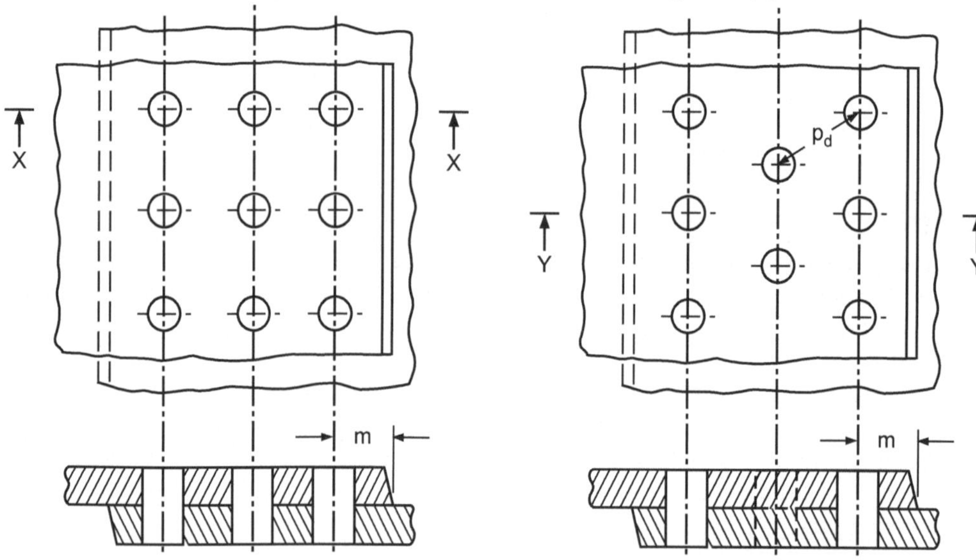


**(a) Chain riveting** **(b) Zig-zag riveting**

**Fig. 6.7 : Triple riveted lap joint**

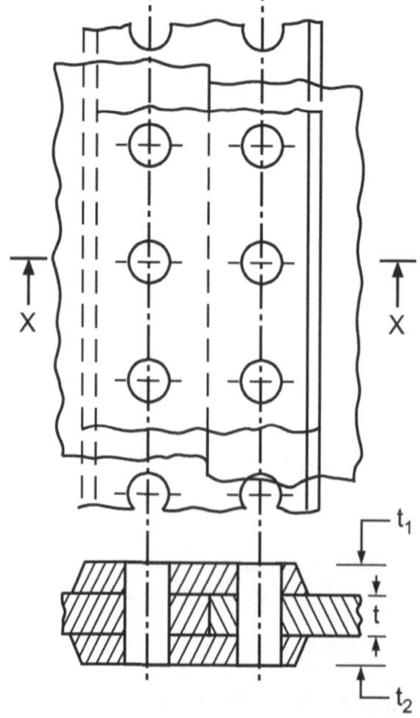


**Fig. 6.8 : Single riveted double strap butt joint**

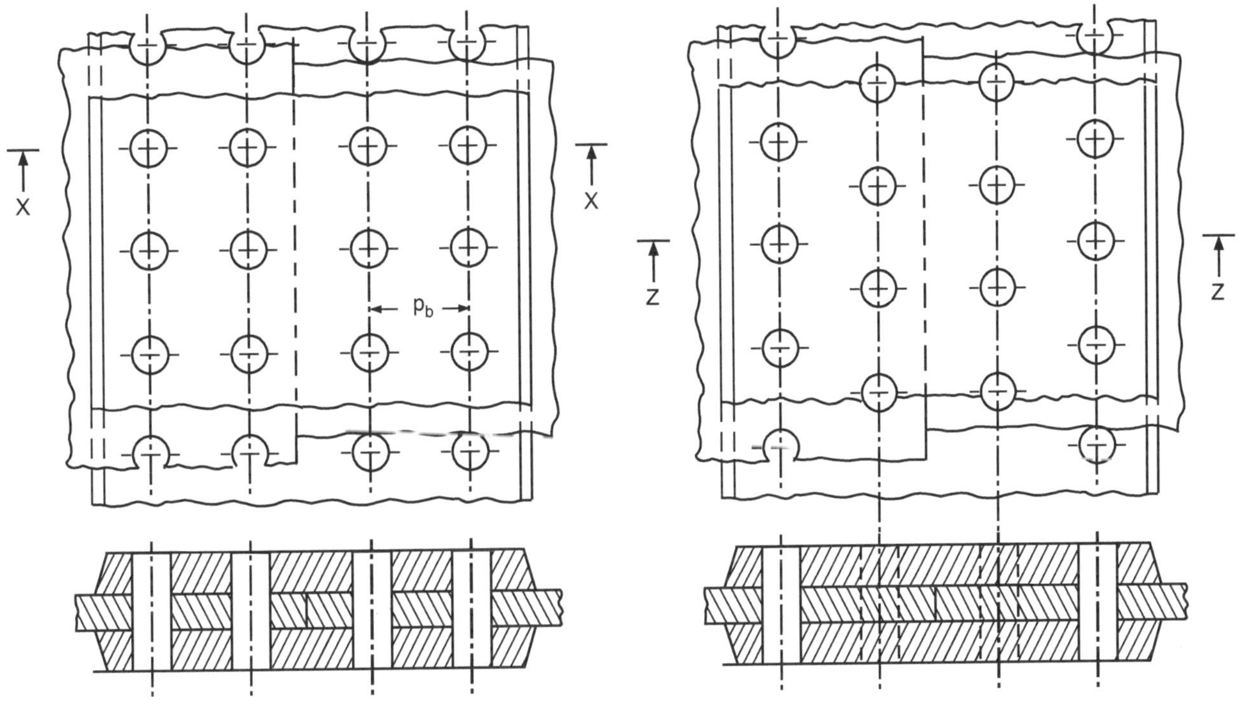


Fig. 6.9 : Double riveted double strap butt joint

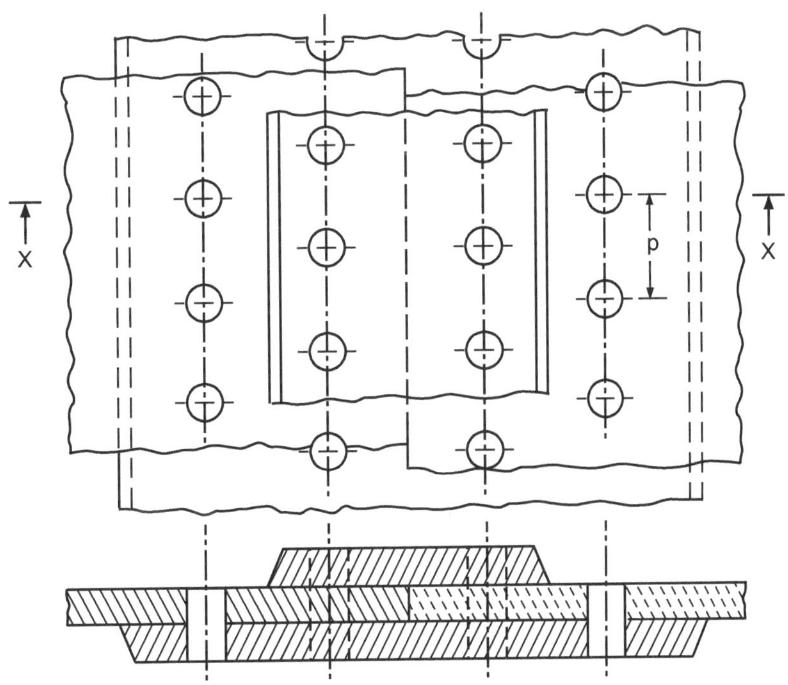


Fig. 6.10 : Double riveted double strap (unequal) butt joint with zig-zag riveting

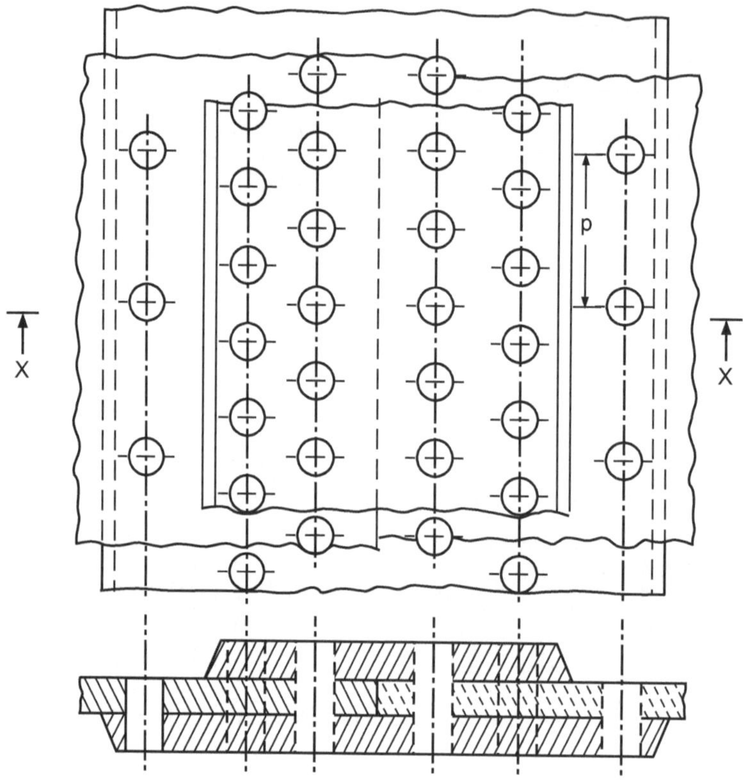


Fig. 6.11 : Triple riveted double strap (unequal) butt joint

## 6.13 IMPORTANT TERMS USED IN RIVETED JOINTS

1. **Pitch:** It is the distance from the centre of one rivet to the centre of the next rivet measured parallel to the seam. Refer Fig. 6.6. It is usually denoted by p.
2. **Back pitch:** It is the perpendicular distance between the centre lines of the successive rows. Refer Fig. 6.6. It is usually denoted by '$p_b$'.
3. **Diagonal pitch:** It is the distance between the centres of the rivets in adjacent rows of zig-zag riveted joint. Refer Fig. 6.6. It is usually denoted by '$p_d$'.
4. **Margin or Marginal pitch:** It is the distance between the centre of rivet hole to the nearest edge of the plate. Refer Fig. 6.6. It is usually denoted by m.
5. **Seam :** It is the line, where two things join. **For example:** Seam is the line of sewing joining two pieces of leather.

## 6.14 FAILURES OF A RIVETED JOINT

A riveted joint may fail in the following ways:

1. **Tearing of the plate at an edge:**

- A joint may fail due to tearing of the plate at an edge as shown in Fig. 6.12.
- This can be avoided by keeping the margin, m = 1.5d, where d is the diameter of the rivet hole.

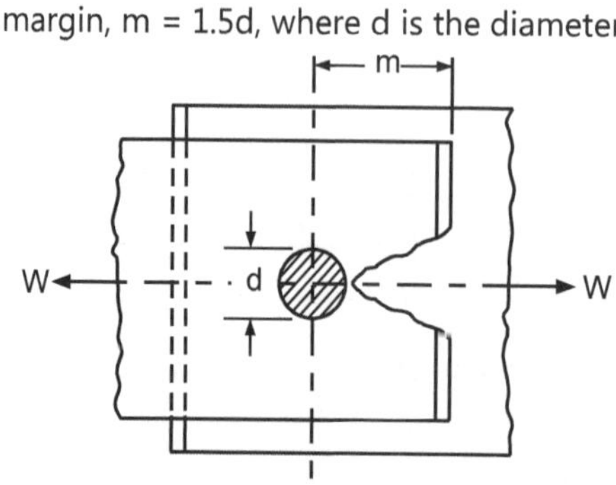


Fig. 6.12 : Tearing or tensile failure of the plate at an edge

**2. Tensile failure or Tearing of the plate across a row of rivets:**

- Due to the tensile stresses in the main plates, the main plate or cover plates may tear off across a row of rivets as shown in Fig. 6.13.
- In such cases, we consider only one pitch length of the plate, since every rivet is responsible for that much length of the plate only.
- The resistance offered by the plate against tearing or Tensile failure is known as tearing resistance or tensile strength of the plate. It is denoted by, $W_t$.
- When the load applied per pitch length becomes greater than tearing resistance or resistance offered by plate against tensile failure ($W_t$), then plate undergoes tensile failure.

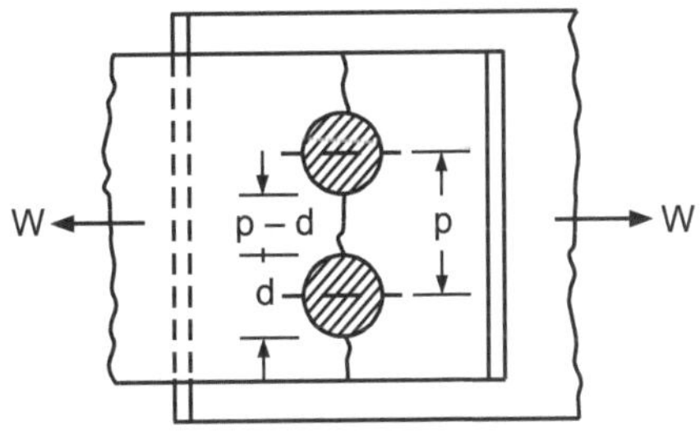


**Fig. 6.13 : Tensile failure or tearing or of the plate across the rows of rivets**

Let $p$ = Pitch of rivets in 'mm'

$d$ = Diameter of rivet or rivet hole in 'mm'

$t$ = Thickness of the plate in 'mm'

$\sigma_t$ = Permissible tensile stresses in plate in 'N/mm²'

Considering failure of plate across a row of rivets in tension,

$$\sigma_t = \frac{\text{Load}}{\text{Area resisting tear or tensile failure}}$$

$$\therefore \quad \sigma_t = \frac{W_t}{(p - d) \times t}$$

**3. Shearing of the rivets:**

- The plates, which are connected by the rivets, exert tensile stress on the rivets, and if the rivets are unable to resist the stress, they are sheared off as shown in Fig. 6.14.

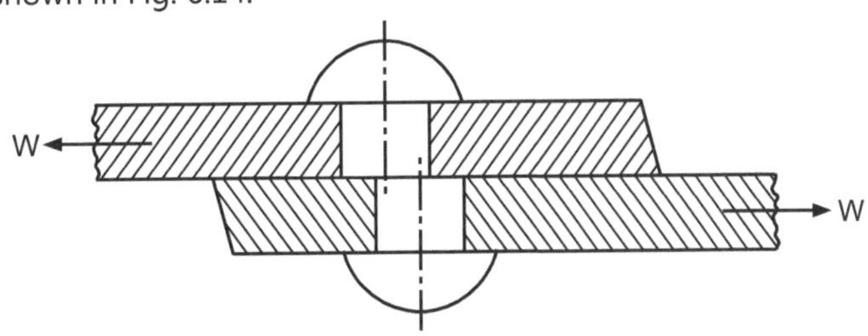


**(a) Shearing off a rivet in a lap joint**

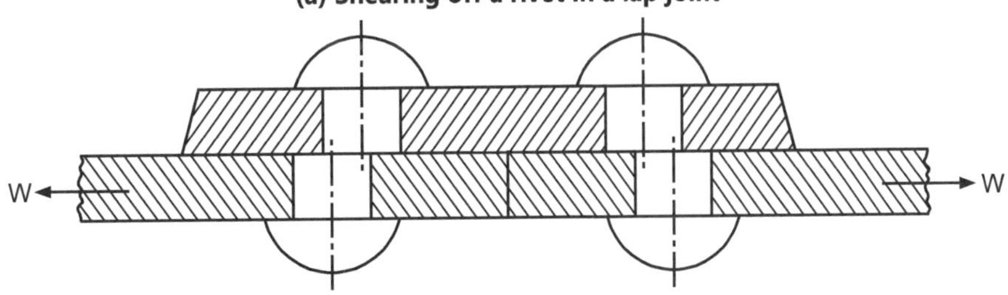


**(b) Shearing off a rivet in a single cover butt joint**

**Fig. 6.14 : Shearing of rivets**

- It may be noted that, the rivets are in **single shear** in a **lap joint** and in a **single cover butt joint**, as shown in Fig. 6.14.
- But the rivets are in **double shear** in a **double cover butt joint** as shown in Fig. 6.15. The resistance offered by a rivet to be sheared off is known as **shearing resistance or shearing strength or shearing value of the rivet.**
- When the load applied per pitch length becomes greater than shearing resistance ($W_s$) offered by rivet against shear failure, then plate undergoes shear failure.
- When the shearing takes place at one cross-section of the rivet, then the rivets are said to be in **single shear**. Refer Fig. 6.14.
- Similarly, when the shearing takes place at two cross-sections of the rivet, then the rivets are said to be in **double shear.** Refer Fig. 6.15.

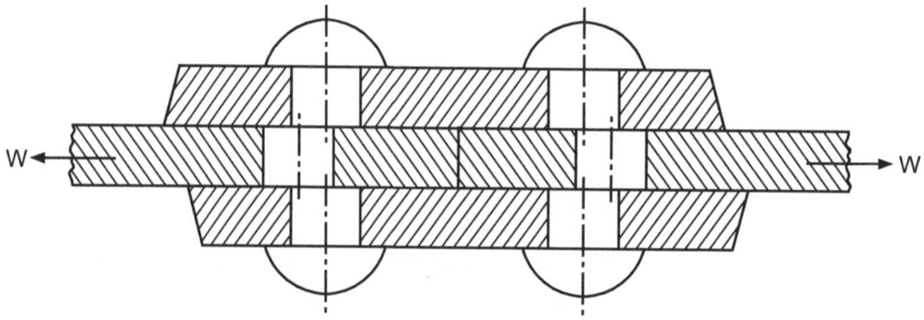


**Fig. 6.15 : Shearing off a rivet in double cover butt joint**

Let, n = Number of rivets per pitch length

$\tau$ = Permissible shear stress for rivet material in 'N/mm$^2$'

Area of rivet resisting shear failure is given by,

(i) $$A = n \times \frac{\pi}{4} \cdot d^2$$ [In single shear]

(ii) $$A = n \times 2 \times \frac{\pi}{4} \cdot d^2$$ [In double shear]

Considering failure of rivet in single shear.

We have, $$\tau = \frac{\text{Load}}{\text{Area of rivet resisting failure in single shear}}$$

$\therefore$ $$\tau = \frac{W_s}{n \times \frac{\pi}{4} \cdot d^2}$$ in 'N/mm$^2$'

Considering failure of rivet in double shear,

We have, $$\tau = \frac{\text{Load}}{\text{Area of rivet resisting failure in double shear}}$$

$\therefore$ $$\tau = \frac{W_s}{n \times 2 \times \frac{\pi}{4} \cdot d^2}$$ in 'N/mm$^2$'

4. **Crushing of the plate or rivets:**

- Sometimes, the rivets do not actually shear off under the tensile stress, but are crushed as shown in Fig. 6.16. Due to this, the rivet hole becomes of an oval shape and hence, the joint becomes loose. The failure of rivets in such a manner is also known as **bearing failure.** The area, which resists this action is the projected area of the hole or rivet on diametral plane.
- The resistance offered by a rivet to be crushed is known as crushing resistance or **crushing strength or bearing value of the rivet**.
- When the load applied per pitch length becomes greater than crushing resistance ($W_{ck}$) offered by rivet against crushing failure, then plate undergoes shear failure.

- **Note:** The number of rivets under shear shall be equal to the number of rivets under crushing.

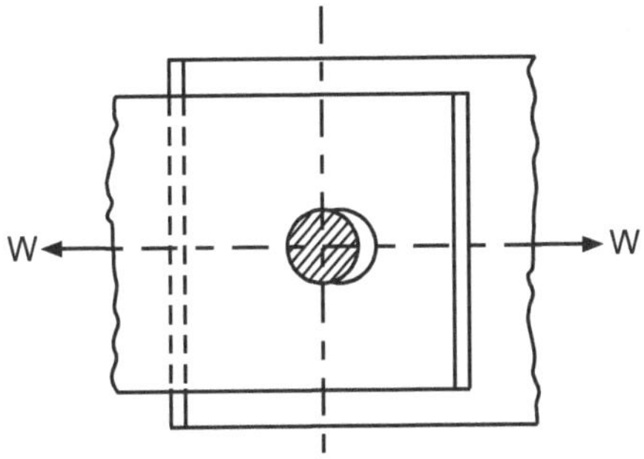


**Fig. 6.16 : Crushing of a rivet**

Let, $\sigma_{ck}$ = Permissible crushing stress for rivet or plate material in 'N/mm$^2$'

Area of rivet resisting crushing failure is given by,

$$A = n \times d \cdot t$$

Considering failure of rivets or plate under crushing,

We have, $$\sigma_{ck} = \frac{W_{ck}}{n \times d \cdot t} \quad \text{in 'N/mm}^2\text{'}$$

**Note: Value of 'n' is taken as,**

$n = 1$ ... For Single Riveted Joint

$n = 2$ ... For Double Riveted Joint

$n = 3$ ... For Triple Riveted Joint

## 6.15 STRENGTH OF A RIVETED JOINT

- **The strength of joint** may be defined as, *"the maximum force, which it can transmit, without causing it to fail."*
- As studied in Article 6.14, we know that, $W_t$, $W_s$ and $W_{ck}$ are the pulls required to tear off the plate (tensile failure), shearing off the rivet (shear failure) and crushing off the rivet (crushing failure) respectively.
- A little consideration will show that, if we go on increasing the pull on a riveted joint, it will fail, when the least of these three pulls is reached, because a higher value of the other pulls will never reach since the joint has failed, either by tearing off the plate, shearing off the rivet or crushing off the rivet.

## 6.16 STRENGTH OF UNRIVETED JOINT

- Strength of un-riveted joint per pitch length is given by,

$$\mathbf{W_p = p \times d \times t} \quad \text{in 'N'}$$

- Strength of un-riveted joint per pitch length can also be referred as **"Strength of solid plate per pitch length".**

## 6.17 EFFICIENCY OF A RIVETED JOINT

- **Efficiency of a riveted joint** is defined as," *the ratio of the strength of riveted joint to the strength of the un-riveted or solid plate."*
- We know that,
  1. Strength of the riveted joint per pitch length = Least of $W_t$, $W_s$ and $W_{ck}$ in 'N'
  2. Strength of un-riveted joint per pitch length = $p \times d \times t$ in 'N'
- Therefore, mathematically, **efficiency of riveted joint is given by,**

$$\eta = \frac{\text{Least value of } W_t, W_s \text{ and } W_{ck}}{\text{Strength of unriveted plate}}$$

$$\therefore \quad \eta = \frac{\text{Least value of } W_t, W_s \text{ and } W_{ck}}{W_p}$$

$$\therefore \quad \eta = \frac{\text{Least value of } W_t, W_s \text{ and } W_{ck}}{p \times t \times \sigma_t}$$

## Numerical Problems on Riveted Joints

**Problem 6.1 :** *For a single riveted lap joint, diameter and pitch of rivet are given as 25 mm and 75 mm respectively. Take thickness of plate as 15 mm. If the permissible stresses are 100 N/mm² in tension, 80 N/mm² in shear and 90 N/mm² in crushing, calculate the minimum value of load per pitch length, which will rupture the joint. Also, calculate the efficiency of joint.*

**Solution : Given data :** d = 2.5 mm, p = 75 mm, t = 15 mm, $\sigma_t$ = 100 N/mm², $\tau$ = 80 N/mm², $\sigma_{ck}$ = 90 N/mm².

**Procedure :**

**1. Minimum load per pitch, which will rupture the joint :** We know that, a riveted joint may rupture or fail under tension, shear or crushing. Therefore, load required per pitch under all 3 conditions must be calculated to find the minimum load per pitch required to rupture the failure.

**(i) Considering tensile failure of plate across a row of rivets :**

We have, $$\sigma_t = \frac{W_t}{(p-d)\times t}$$

$$\therefore \quad \mathbf{W_t} = 100\times(75-25)\times 15 = \mathbf{75000\ N} \quad \ldots (1)$$

**(ii) Considering shear failure of rivets :**

For single riveted lap joint, $$\tau = \frac{W_s}{n\times\frac{\pi}{4}\cdot d^2} \quad (\because \text{ Single shear})$$

$$\therefore \quad \mathbf{W_s} = 1\times 80\times\frac{\pi}{4}\times(25)^2 = \mathbf{39270\ N} \quad (\because \text{ single riveted joint, n=1}) \ldots (2)$$

**(iii) Considering crushing failure of rivets :** For single riveted lap joint.

We have, $$\sigma_{ck} = \frac{W_{ck}}{n\times d\times t}$$

$$\therefore \quad \mathbf{W_{ck}} = 1\times 90\times 25\times 15 = \mathbf{33750\ N} \quad (\because n = 1 \text{ for single riveted joint}) \ldots (3)$$

On comparing equations (1), (2) and (3), we find that, value of $W_{ck}$ (i.e. 33750 N) is minimum.

Therefore, the value of minimum load per pitch, which will rupture the joint, is **33750 N**.

In other words, the riveted joint will failure in crushing.

$$\therefore \quad \mathbf{W_{min}} = W_{ck} = \mathbf{33750\ N}$$

**2. Efficiency of joint :**

We have, $$\eta = \frac{\text{Least value of } W_t, W_s \text{ and } W_{ck}}{\text{Strength of unriveted joint}}$$

$$\therefore \quad \eta = \frac{33750}{p\times t\times\sigma_t}$$

$$\therefore \quad \eta = \frac{33750}{75\times 15\times 100} = \mathbf{0.3 = 30\%}$$

---

**Problem 6.2 :** *Find the efficiency of following riveted joints.*

*(i) Single riveted lap joint of 6 mm plates with 20 mm diameter rivets having a pitch of 50 mm.*

*(ii) Double riveted lap joint of 6 mm plates with 20 mm diameter rivets having a pitch of 65 mm.*

*Assume,*

*Permissible tensile stress in plate = 120 MPa*

*Permissible shearing stress in rivets = 90 MPa*

*Permissible crushing stress in rivets = 180 MPa*

**Solution : Given data :** t = 6 mm, d = 20 mm, $\sigma_t$ = 120 MPa = 120 N/mm², $\tau$ = 90 MPa = 90 N/mm²,
$\sigma_{ck}$ = 180 MPa = 180 N/mm²

**Case (I) : For single riveted lap joint having pitch of 50 mm :**

We have, p = 50 mm

**(i) Considering tensile failure of plate across a row of rivets :**

We have, $\sigma_t = \dfrac{W_t}{(p-d)\times t}$

$\therefore$ $\mathbf{W_t} = 120 \times (50 - 20) \times 6 = \mathbf{21600\ N}$ ... (1)

**(ii) Considering shear failure of rivets :**

We have, $\tau = \dfrac{W_s}{n \times \dfrac{\pi}{4} \times d^2}$ ($\because$ Single shear)

For single riveted lap joint, $\mathbf{n = 1}$

Therefore, above equation gives,

$\mathbf{W_s} = 90 \times 1 \times \dfrac{\pi}{4} \times (20)^2 = \mathbf{28274.33\ N}$ ... (2)

**(iii) Considering crushing failure of rivets :**

We have, $\sigma_{ck} = \dfrac{W_{ck}}{n \times d \times t}$

$\therefore$ $\mathbf{W_{ck}} = 180 \times 1 \times 20 \times 6 = \mathbf{21600\ N}$ ... (3)

On comparing equations (1), (2) and (3), we find that, value of $W_t$ and $W_{ck}$ is equal, i.e. $W_t = W_{ck} = 21600$ N. Also, this value of 21600 N is the minimum value required to rupture the joint. In other words, least value of $W_t$, $W_s$ and $W_{ck}$ is 21600 N.

**Efficiency of riveted joint is given by,**

$$\eta = \frac{\text{Least value of } W_t, W_s \text{ and } W_{ck}}{\text{Strength of unriveted joint}}$$

$\therefore$ $\eta = \dfrac{W_t \text{ or } W_{ck}}{p \times t \times \sigma_t}$

$\therefore$ $\eta = \dfrac{21600}{50 \times 6 \times 120} = \mathbf{0.6 = 60\%}$

**Case (II) : For double riveted tap joint having pitch of 65 mm :**

We have, $p = 65$ mm

**(i) Considering tensile failure of plate across a row of rivets :**

We have, $\sigma_t = \dfrac{W_t}{(p-d)\times t}$

$\therefore$ $\mathbf{W_t} = 120 \times (65 - 20) \times 6 = \mathbf{32400\ N}$ ... (4)

**(ii) Considering shear failure of rivets :**

Since the joint is double riveted, therefore, area of 2 rivets resisting single shear failure is taken.

For double riveted joint, n = 2

We have, $\tau = \dfrac{W_s}{n \times \dfrac{\pi}{4} \times d^2}$

$\therefore$ $\mathbf{W_s} = 90 \times 2 \times \dfrac{\pi}{4} \times (20)^2 = \mathbf{56548.67\ N}$ ... (5)

**(iii) Considering crushing failure of rivets :**

Since joint is double riveted, therefore, area of 2 rivets resisting crushing failure is taken.

We have, $\sigma_{ck} = \dfrac{W_{ck}}{n \times d \times t}$

$\therefore$ $\mathbf{W_{ck}} = 180 \times 2 \times 20 \times 6 = \mathbf{43200\ N}$ ... (6)

On comparing equations (4), (5) and (6), we find that, value of $W_t$ (i.e. 32400 N) is minimum or least value of load per pitch length required to rupture the riveted joint.

∴ Least of $W_t$, $W_s$ and $W_{ck}$ = **32400 N.**

We know that, **Strength of unriveted joint** = $p \times t \times \sigma_\tau = \mathbf{65 \times 6 \times 120 = 46800\ N}$

$$\textbf{Efficiency of riveted joint} = \frac{\text{Least of } W_t, W_s \text{ and } W_{ck}}{\text{Strength of riveted joint}}$$

$$\therefore \quad \eta = \frac{32400}{46800} = \mathbf{0.6923 = 69.23\%}$$

---

**Problem 6.3 :** *A double riveted double cover butt joint in plates 20 mm thick is made with 25 mm diameter rivets at 100 mm pitch. The permissible stresses are:*

*Permissible tensile stress in plate = 120 MPa*

*Permissible shearing stress in rivets = 100 MPa*

*Permissible crushing stress in rivets = 150 MPa*

*Find the efficiency of joint, taking the strength of the rivet in double shear as twice than that of single shear.*

**Solution : Given data :** t = 20 mm, d = 25 mm, p = 100 mm, $\sigma_t$ = 120 MPa = 120 N/mm$^2$, $\tau$ = 100 MPa = 100 N/mm$^2$, $\sigma_{ck}$ = 150 MPa = 150 N/mm$^2$

**Procedure :**

For double riveted double cover butt joint.

We have **n = 2**

**(i) Considering tensile failure of plate across the row of rivets :**

$$\sigma_t = \frac{W_t}{(p - d) \times t}$$

$$\therefore \quad \mathbf{W_t} = 120 \times (100 - 25) \times 20 = \mathbf{180 \times 10^3\ N} \quad \dots (1)$$

**(ii) Considering shear failure or rivets :**

It is given that, the double riveted double cover butt joint undergoes double shear. Therefore, area resisting double shear failure is given by,

$$A = 2 \times \frac{\pi}{4} \cdot d^2$$

We have,

$$\tau = \frac{W_s}{n \times 2 \times \frac{\pi}{4} \cdot d^2}$$

$$\therefore \quad \mathbf{W_s} = 100 \times 2 \times 2 \times \frac{\pi}{4} \times (25)^2 = \mathbf{196.35 \times 10^3\ N} \quad \dots (2)$$

**(iii) Considering crushing failure of rivets :**

We have,

$$\sigma_{ck} = \frac{W_{ck}}{n \times d \times t}$$

$$\therefore \quad \mathbf{W_{ck}} = 150 \times 2 \times 25 \times 20 = \mathbf{150 \times 10^3\ N} \quad \dots (3)$$

On comparing equations (1), (2) and (3) we find that, value of $W_{ck}$ is minimum or least value amongst all loads, i.e. $W_t$, $W_s$ and $W_{ck}$.

Efficiency of riveted joint is given by,

$$\eta = \frac{\text{Least of } W_t, W_s \text{ and } W_{ck}}{\text{Strength of unriveted joint}}$$

$$\therefore \quad \eta = \frac{150 \times 10^3}{p \times t \times \sigma_t}$$

$$\therefore \quad \eta = \frac{150 \times 10^3}{100 \times 20 \times 120} = \mathbf{0.625 = 62.5\%}$$

---

**Problem 6.4 :** *Two plates 10 mm thick are joined by a double riveted lap joint. The pitch of each row of wickets is 90 mm. The rivets are 25 mm in diameter. The permissible stresses are as follows : $\sigma_t$ = 140 MPa, $\tau$ = 110 MPa, $\sigma_{ck}$ = 240 MPa. Find the efficiency of joint.*

**Solution : Given data :** t = 10 mm, p = 90 mm, d = 25 mm, $\sigma_t$ = 140 MPa = 140 N/mm$^2$, $\tau$ = 110 MPa = 110 N/mm$^2$, $\sigma_{ck}$ = 240 MPa = 240 N/mm$^2$

**Procedure :** For double riveted lap joint, **n = 2**.

**(i) Considering tensile failure of plate across the row of rivets :**

We have,

$$\sigma_t = \frac{W_t}{(p - d) \times t}$$

$$\therefore \quad \mathbf{W_t} = 140 \times (90 - 25) \times 10 = \mathbf{91 \times 10^3\ N} \qquad \dots (1)$$

**(ii) Considering shear failure or rivets :**

Since two plates are jointed by means of lap riveted joint, the rivet may fail in single shear. Therefore, area resisting single shear must be taken as $\frac{\pi}{4} \cdot d^2$.

We have,

$$\tau = \frac{W_s}{n \times \frac{\pi}{4} \times d^2}$$

$$\therefore \quad \mathbf{W_s} = 110 \times 2 \times \frac{\pi}{4} \times (25)^2 = \mathbf{108 \times 10^3\ N} \qquad \dots (2)$$

**(iii) Considering crushing failure of rivets :**

We have,

$$\sigma_{ck} = \frac{W_{ck}}{n \times d \times t}$$

$$\therefore \quad \mathbf{W_{ck}} - 240 \times 2 \times 25 \times 10 = \mathbf{120 \times 10^3} \qquad \dots (3)$$

On comparing equations (1), (2) and (3), we find that, value of load '$W_t$' is less than $W_s$ and $W_{ck}$. Therefore, $W_t$ of $91 \times 10^3$ is the least value amongst all load.

$$\textbf{Efficiency of joint} = \frac{\text{Least of } W_t, W_s \text{ and } W_{ck}}{\text{Strength of unriveted joint}}$$

$$\therefore \quad \eta = \frac{91 \times 10^3}{p \times t \times \sigma_t}$$

$$\therefore \quad \eta = \frac{91 \times 10^3}{90 \times 10 \times 140} = \mathbf{0.7222 = 72.22\%}$$

## Theory Questions for Practice

1. Draw a neat and labelled sketch of a rivet.
2. What do you understand by the term "riveted joint"? Explain the necessity of such a joint.
3. Write a short note on permanent fastenings and temporary or detachable fastenings.
4. State four examples of permanent fastenings.
5. State four examples of temporary or detachable fastenings.
6. Write a short note on the material used for rivets.
7. State any four types of riveted joints and draw their neat sketches.
8. Show by neat sketches, the various types of failure, which may fail a riveted joint.
9. State the expression for strength of riveted joint and state the meaning of each term in the expression.
10. Define efficiency of a riveted joint. Give its mathematical expression.

## Numerical Problems for Practice

1. A single riveted lap joint is made in 15 mm thick plates with 20 mm diameter rivets. Determine the strength of the joint, if the pitch of rivets is 60 mm. Take $\sigma_t$ = 120 MPa, $\tau$ = 90 MPa and $\sigma_c$ = 160 MPa. (**Ans.** 28280 N)

2. Two plates 16 mm thick are joined by a double riveted lap joint. The pitch of each row of rivets is 90 mm. The rivets are 25 mm in diameter. The permissible stresses are as follows :

   $\sigma_t$ = 140 MPa, $\tau$ = 110 MPa and $\sigma_c$ = 240 MPa

   Find the efficiency of the joint. (**Ans.** 53.5%)

3. A single riveted double cover butt joint is made in 10 mm thick plates with 20 mm diameter rivets with a pitch of 60 mm. Calculate the efficiency of the joint, if $\sigma_t$ = 100 MPa, $\tau$ = 80 MPa and $\sigma_c$ = 160 MPa. (**Ans.** 53.8%)

4. A double riveted double cover butt joint is made in 12 mm thick plates with 18 mm diameter rivets. Find the efficiency of the joint for a pitch of 80 mm, if $\sigma_t$ = 115 MPa, $\tau$ = 80 MPa and $\sigma_c$ = 160 MPa (**Ans.** 62.6%)

❑❑❑

*Chapter* **7**

# DESIGN OF FASTENERS - WEDLED JOINTS

## 7.1 INTRODUCTION TO WELDED JOINTS

- Welded joint is a permanent joint, which is obtained by fusion of the edges of two parts to be joined together using a filler material.
- Heat required for fusion of material may be obtained by burning of gas or by producing an electric arc. Pressure may or may not be applied during welding process.
- Welding is extensively used in fabrication work, repair work, filling cracks, joining broken parts etc. For more details, refer article 7.1.2.

### 7.1.1 Advantages and Disadvantages of Welded Joints

**Advantages of Welded Joints :**

1. 100% efficiency.
2. More strength and rigidity.
3. Changes and additions to the existing structures can be done easily.
4. It is possible to weld any part of structure to any other point/part.
5. Comparatively less time is required for welding.
6. Welded assembly results in lighter construction.
7. Comparatively cheaper. Cost of welded assembly is lower than cost of bolted joints.
8. Design of welded assemblies can be easily and economically modified to meet the changing product requirements.
9. Welded joints are tight and leak proof as compared to other joints.
10. When two parts are joined by threaded joints (i.e. screwed joints), holes are drilled in the parts to accommodate bolts or screws. These holes reduce the cross-sectional area of parts to be joined, which leads to problem of stress concentration. There is no such problem in case of welded joints.
11. Comparatively high strength. Please note that, strength of weld is more than strength of plates joined together.
12. Machine components of complex shape can be easily welded. For example : Circular steel pipes, nozzles, tapered pipes etc.

**Disadvantages of Welded Joints :**

1. Due to heating required for fusion, temperature stresses are induced in welded parts.
2. Quality and strength of welded joint depends upon the skill of welder. It requires high skilled labour. Therefore, cost of welding increases due to high salaries and wages paid to labours.
3. Difficult to inspect.
4. It cannot be used to join two dis-similar materials (say steel and aluminium), which is possible in riveting (riveted joints).
5. Welded joints are poor in damping out the vibrations.

### 7.1.2 Applications of Welded Joints

1. Welding is extensively used in fabrication as an alternate method of casting or forging.
2. Welding is used as a replacement of bolted joint and riveted joint.
3. Welding is used as a repair medium.

   For example :

   (a) To reunite metal at a crack.

   (b) To build up a small part, which has broken off, such as gear tooth.

   (c) To repair a worn out surface, such as, bearing surface.
4. Preferred in steel structures like bridges, industrial shades, towers, crane and material handling equipments.
5. Solid beams and frame beams under dynamic loads (fluctuating loads).
6. Pressure vessels :

   (a) Unfired pressure vessels like storage tanks, reservoirs, process equipments in chemical, paper and sugar industries.

   (b) Fired pressure vessels like boilers and furnaces.
7. Machine components like flywheels, gears, jigs, fixtures, machine structures, automobile chassis etc. as well as many heavy duty machine frames can be welded suitably.

### 7.1.3 Advantages and Disadvantages of Welded Joints over Riveted Joints

**Advantages of Welded Joints over Riveted Joints :**

1. Welded structures are lighter than riveted structures.
2. Welded joints ensures maximum efficiency as compared to riveted and screwed joints. It may be 100%.
3. Modifications or additions can be done easily in case of welded joints.
4. Welded joint is stronger as compared to riveted joints and screwed joints.
5. Welding can weld or join any part of a structure at any point on that structure. This is not possible in other types of joints.
6. Welding can be accomplished in small duration than riveting.

**Disadvantages of Welded Joints with Respect to Riveted Joints :**

1. Quality and strength of welded joint depends upon skill of welder. In other words, welding needs highly skilled welders. It leads to increased cost of welding process, due to high salaries and wages paid to skilled welders as compared to riveting.
2. Inspection of welding work is difficult as compared to riveted joint.
3. Since the parts to be welded are subjected to uneven heating and cooling during fabrication work, they may get distorted.
4. Due to application of heat required for fusion, parts are subjected to temperature (thermal) stresses. These additional stresses induced in the parts make them weaker to certain extent.
5. In welding, there is no provision for expansion and contraction in the frame. Therefore, there is a possibility of cracks developing in it.

### 7.1.4 Comparison of Welded Joint and Screwed Joint

| Sr. No. | Comparative Point | Welded Joint | Screwed (Bolted) Joint |
|---|---|---|---|
| 1. | **Type of joint** | Rigid and permanent. | Temporary. |
| 2. | **Cost of welded assembly** | Less. | More. |
| 3. | **Strength** | More. | Less. |
| 4. | **Labour requirement** | High skilled labour. | Semi-skilled labour. |
| 5. | **Inspection** | Difficult. | Comparatively much easy. |
| 6. | **Chances of leakage** | Very less due to tight and leak proof joint. | More. |

### 7.1.5 Classification of Welded Joints

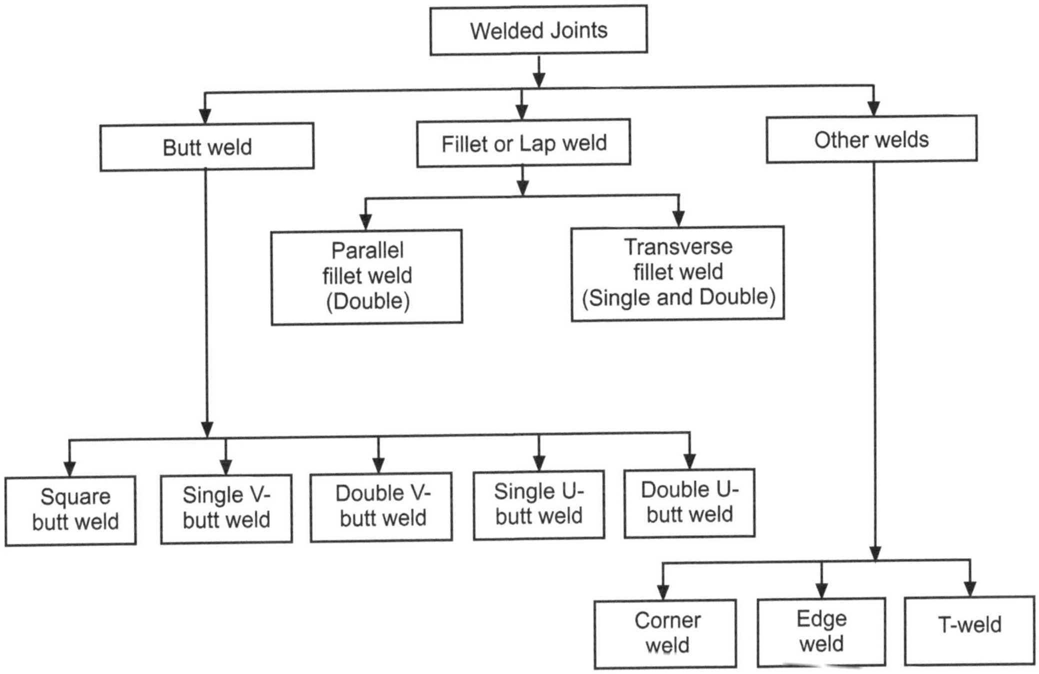


**Chart 1 : Classification/Types of welded joints**

## 7.2 TYPES OF WELDED JOINTS

- Different types of welds (i.e. welded joints), such as, filled weld, butt weld and other welds are briefly explained below.

1. **Lap Joint (Fillet weld) :**
   - Lap joint is a joint between two overlapping components.
   - Lap joint consists of fillet welds.
   - Fillet weld has triangular cross-section joining two surfaces at right angles to each other.
   - Based on the relative positions of load axis with respect to the fillet axis, fillet welds are classified into two types.

   (a) **Parallel fillet weld :** *"If load axis is parallel to axis of fillet, fillet weld is known as parallel fillet weld".*

   (b) **Transverse fillet weld :** *"If load axis is perpendicular to axis of fillet, fillet weld is known as transverse fillet weld".* Transverse fillet weld can be a single transverse fillet weld or a double transverse fillet weld.

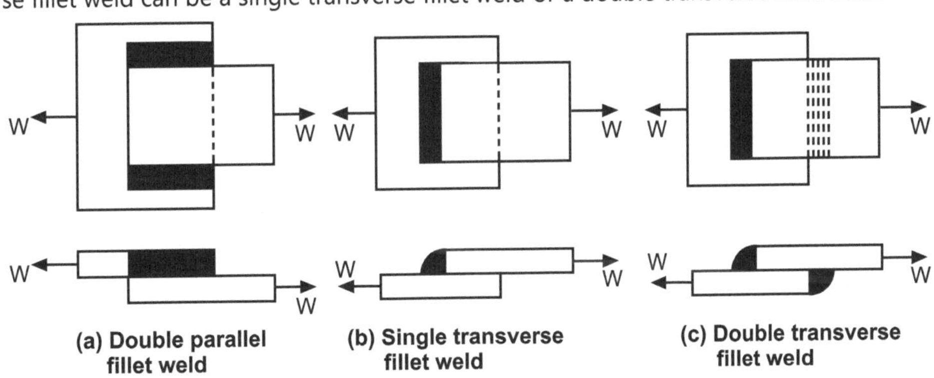


**Fig. 7.1 : Types of fillet weld**

2. **Butt Joint (Butt weld) :**
   - Butt weld is obtained by placing the plates to be joined side by side with their edges nearly touching each other. Refer Fig. 7.2.
   - A small gap is maintained between the edges for filler material.

- **Examples of lap joints :** (i) Square butt (ii) Single V-butt (iii) Double V-butt (iv) Single U-butt (v) Double U-butt.

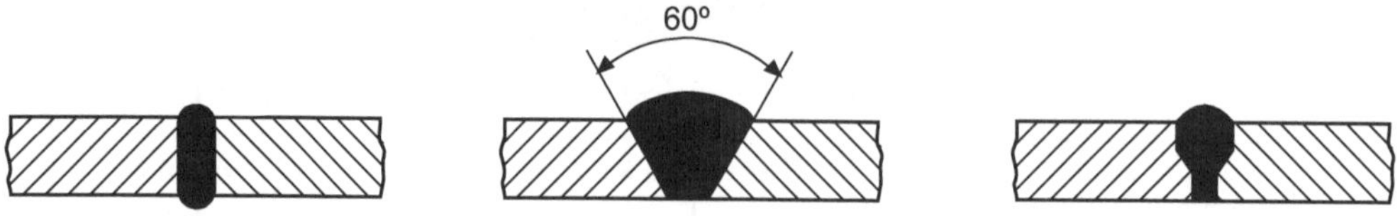


(a) Square butt weld (b) (i) Single V-butt weld (b) (ii) Single U-butt weld

(c) (i) Double V-butt weld (c) (ii) Double U-butt weld

**Fig. 6.2 : Butt welds**

**(a) Square butt weld :** If thickness of plates is less than 5 mm, then edges of plates do not require bevelling and hence, the joint used is known as **Square butt weld**.

**(b) Single V-butt weld or Single U-butt weld :** If thickness of plates lies in between 5 mm and 12.5 mm, then edges of plates are bevelled to 'V' or 'U' groove and hence, joint used is known as **Single V-butt weld** or **Single U-butt weld**.

**(c) Double V-butt weld or Double U-butt weld :** If thickness of plates is more than 12.5 mm, it is necessary to bevel and weld the plates from both sides and hence, joint used is known as **Double V-butt weld** or **Double U-butt weld**.

**3. Other welds :**

- In addition to above mentioned joints, other types of welds used are :

  (i) Corner weld, (ii) Edge weld and (iii) T-weld.

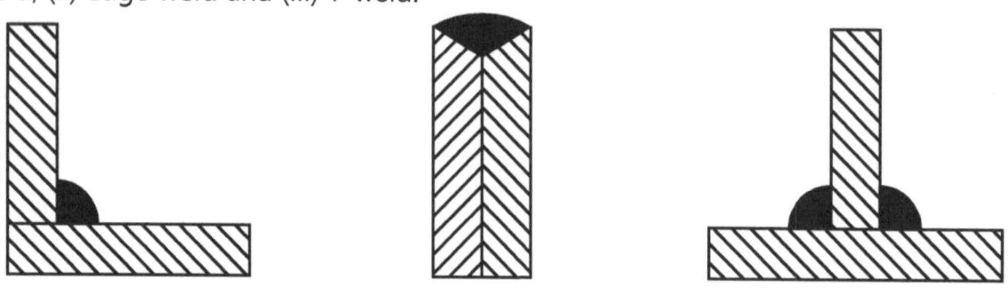

(a) Corner weld (b) Edge weld (c) T-weld

**Fig. 6.3 : Other welds**

## 7.2.1 Standard Location of Welding Symbols

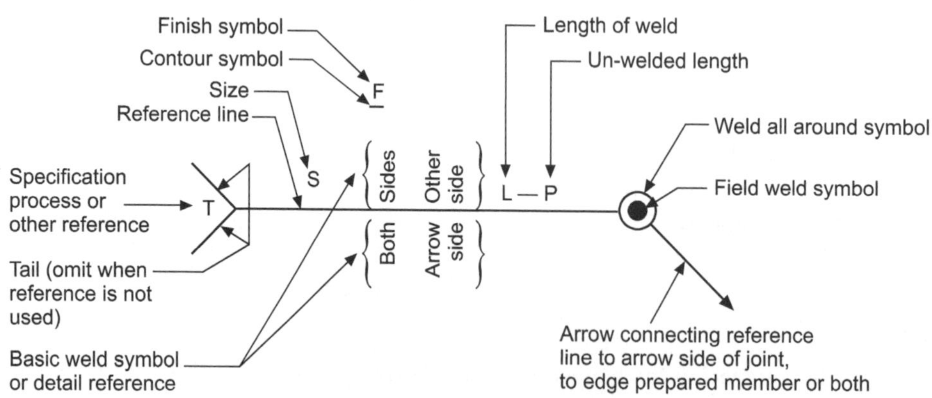


**Fig. 7.4 : Standard location of welding symbols**

## 7.2.2 Various I.S. Symbols Representing Welded Connections

- Welding symbols are used to specify type of weld on working sheet or drawing.

| Sr. No. | Form of weld | Sectional representation | Symbol |
|---|---|---|---|
| 1. | **Fillet** | 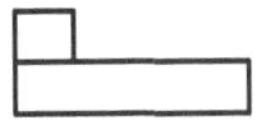 |  |
| 2. | **Square butt** | 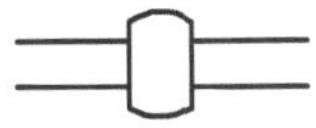 |  |
| 3. | **Single V butt** |  |  |
| 4. | **Double V butt** |  |  |
| 5. | **Single U butt** |  |  |
| 6. | **Double U butt** |  |  |
| 7. | **Single bevel butt** |  |  |
| 8. | **Double bevel butt** |  |  |
| 9. | **Single J butt** |  |  |
| 10. | **Double J butt** |  |  |
| 11. | **Stud** |  |  |
| 12. | **Sealing run** | 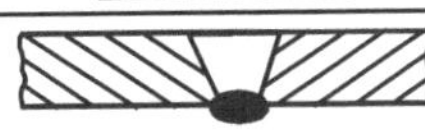 |  |
| 13. | **Spot** | 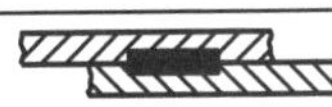 |  |
| 14. | **Seam** | 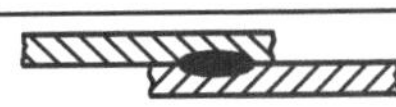 | 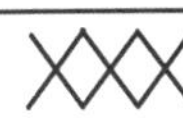 |
| 15. | **Plug** | 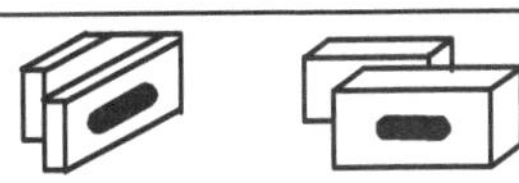 |  |
| 16. | **Projection** |  |  |
| 17. | **Flash** | 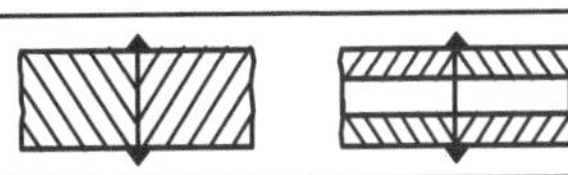 |  |

### 7.2.3 Supplementary I.S. Symbols Representing Welded Connections

| Sr. No. | Particulars | Drawing representation | Symbol |
|---|---|---|---|
| 1. | **Weld all round** |  | ○ |
| 2. | **Field weld** |  | ● |
| 3. | **Flush contour** |  | — |
| 4. | **Convex contour** | 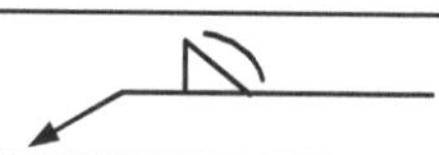 | ⌒ |
| 5. | **Concave contour** | 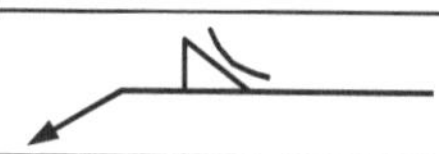 | ( |
| 6. | **Grinding finish** | 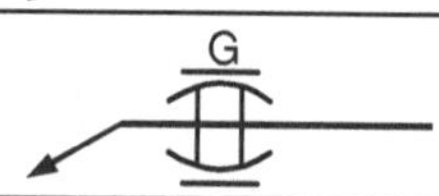  | G |
| 7. | **Machining finish** | 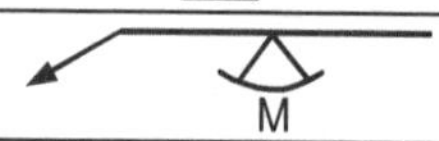  | M |
| 8. | **Chipping finish** | 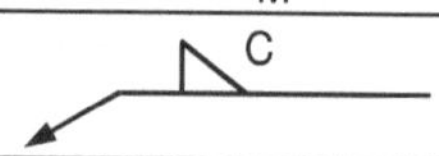  | C |

## 7.3 DESIGN OF SINGLE AND DOUBLE TRANSVERSE FILLET WELD OR LAP JOINTS

- Fillet weld or lap joint is obtained by overlapping the plates and then welding the edges of plates.
- Fig. 7.5 and 7.6 show a single and a double transverse fillet weld respectively.

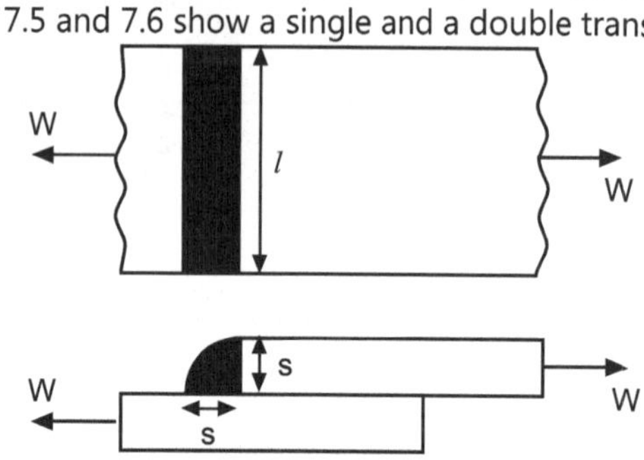


**Fig. 6.5 : Single transverse fillet weld**

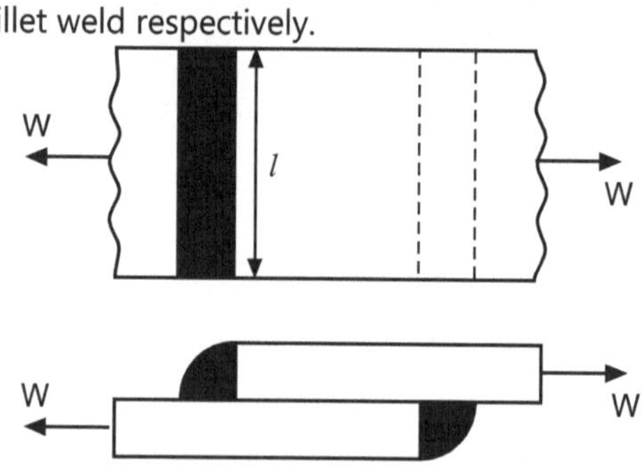


**Fig. 6.6 : Double transverse fillet weld**

- Let us assume that, cross-section of fillet is a right angled triangle ABC with hypotenuse 'AC' making equal angles with other two sides AB and BC. Refer Fig. 7.7.
- We observe that, $l$(AB) = $l$(BC) = Leg or size of weld.
- Also, distance of perpendicular drawn on hypotenuse AC from point B is distance BD. This distance 'BD' is called as **'throat thickness'**.

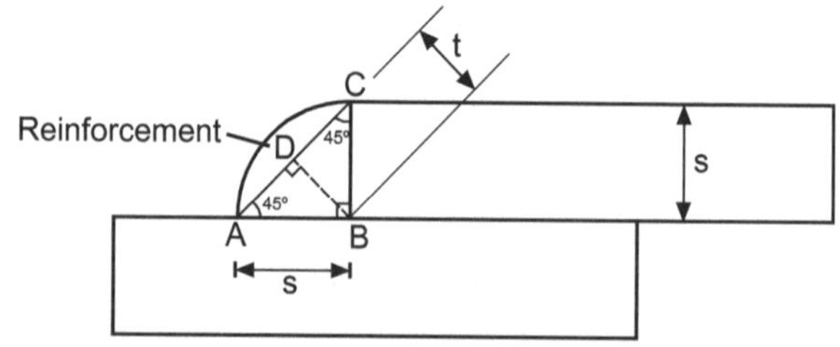


**Fig. 7.7 : Cross-section of fillet weld**

- Let, Size of weld = Thickness of plate = Distance BC = s
  Throat thickness = Distance BD = t
  Length of weld = $l$
- Referring to the Fig. 7.7, from right angled triangle Δ BDC,

We have, $\sin(45°) = \dfrac{BD}{BC}$

$\therefore \quad 0.707 = \frac{t}{s}$

$\therefore \quad \mathbf{t = 0.707\ s}$

$\therefore \quad \begin{bmatrix} \text{Minimum area of} \\ \text{weld or Throat area} \end{bmatrix} = \text{Throat thickness} \times \text{Length of weld}$

$\therefore$ **Minimum area of weld** $= t \times l = \mathbf{0.707\ s \times l}$ … (7.1)

- Transverse fillet welds are designed on the basis of tensile strength.

Therefore, **for single transverse fillet weld,**

$$\sigma_t = \frac{\text{Load}}{\text{Minimum area of weld}} = \frac{W}{0.707 \times s \times l}$$

$\therefore \quad \mathbf{W = 0.707 \times s \times l \times \sigma_t}$ … (7.2)

**For double transverse fillet weld,**

$$\sigma_t = \frac{\text{Load}}{\text{Minimum area of weld}} = \frac{W}{2 \times 0.707 \times s \times l}$$

$\therefore \quad \mathbf{W = 1.414 \times s \times l \times \sigma_t}$ … (7.3)

Above strength equations can be summarized in tabular format as given below.

| Sr. No. | Type of weld | Design equation/Tensile strength |
|---|---|---|
| (a) | For single transverse fillet weld | $W = 0.707 \cdot s \cdot l \cdot \sigma_t$ |
| (b) | For double transverse fillet weld | $W = 2 \times 0.707 \cdot s \cdot l \cdot \sigma_t = 1.414 \cdot s \cdot l \cdot \sigma_t$ |

## 7.4 DESIGN OF SINGLE AND DOUBLE PARALLEL FILLET WELD

- Fig. 7.8 shows a double parallel fillet weld.
- From the previous article, we know that,

  **Minimum area of weld or Throat area** $= 0.707 \times s \cdot l$
- Single or double parallel fillet welds are designed on the basis of shear strength.

  **For single parallel fillet weld,**

$$\tau = \frac{\text{Load}}{\text{Minimum area of weld}} = \frac{W}{0.707 \times s \times l}$$

$\therefore \quad \mathbf{W = 0.707 \times s \times l \times \tau}$ … (7.4)

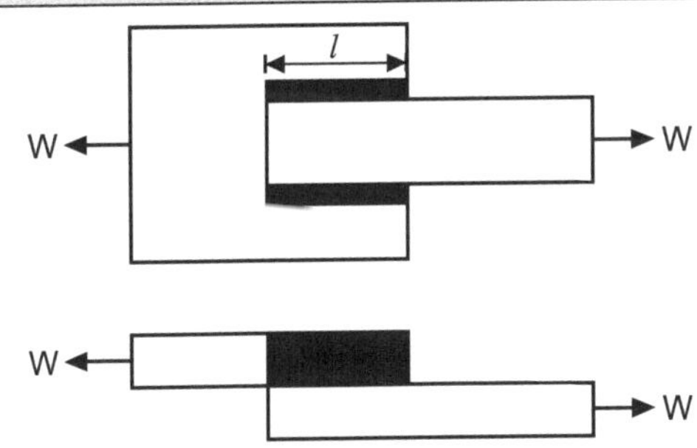

**Fig. 7.8 : Double parallel fillet weld**

**For double parallel fillet weld,** $\tau = \frac{\text{Load}}{\text{Minimum area of weld}} = \frac{W}{2 \times 0.707 \times s \times l}$

$\therefore \quad \mathbf{W = 1.414 \times s \times l \times \tau}$ … (7.5)

Above strength equations can be summarized in tabular format as given below.

| Sr. No. | Type of weld | Design equation/Shear strength |
|---|---|---|
| (a) | Single parallel fillet weld | $W = 0.707 \times s \cdot l \cdot \tau$ |
| (b) | Double parallel fillet weld | $W = 2 \times 0.707 \times s \cdot l \cdot \tau = 1.414 \times s \cdot l \cdot \tau$ |

## 7.5 DESIGN OF COMBINED SINGLE TRANSVERSE AND DOUBLE PARALLEL FILLET WELD

- Combination of single transverse and double parallel fillet weld is also known as **'combined single transverse and double parallel fillet weld'** as shown in Fig. 7.9.
- It is subjected to tensile stress and shear stress.

  Let, $l_1$ = Length of single transverse fillet weld, $l_2$ = Length of each parallel fillet weld,

  s = Size of weld = Thickness of plate, A = Minimum area of weld.

**Tensile strength** $= W_t = \sigma_t \cdot A$

$\therefore \quad W_t = \sigma_t \cdot (0.707 \times s \times l_1)$ [$\because$ A = $0.707 \times s \times l_1$ for single transverse fillet weld]

$\therefore \quad \mathbf{W_t = 0.707 \cdot s \cdot l_1 \cdot \sigma_t}$ … (7.6)

**Shear strength** = $W_s = \tau \cdot A$

$\therefore \quad W_s = \tau \cdot (2 \times 0.707 \times s \cdot l_2)$ [$\because$ $A = 2 \times 0.707 \times s \times l_2$ for double parallel fillet weld]

$\therefore \quad \mathbf{W_s = 1.414 \cdot s \cdot l_2 \cdot \tau}$ ... (7.7)

**Total strength of weld** = $W_t + W_s$

$\therefore \quad \mathbf{W = 0.707 \cdot s \cdot l_1 \cdot \sigma_t + 1.414 \cdot s \cdot l_2 \cdot \tau}$ ... (7.8)

Equation (6.8) represents strength equation of combined single transverse and double parallel fillet weld.

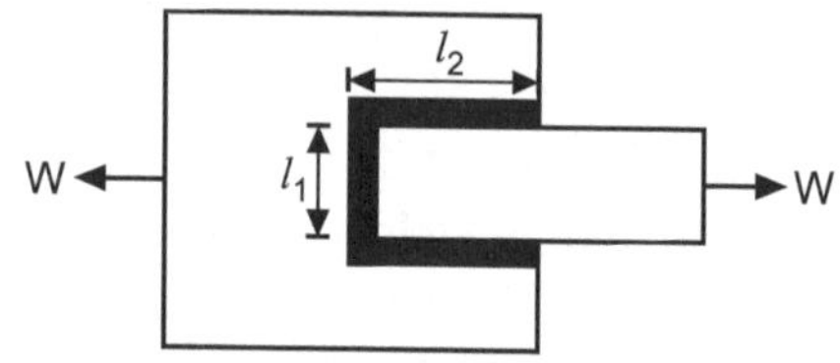

**Fig. 7.9 : Combined single transverse and double parallel fillet weld**

## 7.6 STRESS CONCENTRATION FACTOR FOR WELDED JOINTS

- Stress concentration occurs at the junction of weld and the parent metal.
  Therefore, stress concentration factor should be taken into account in case of fatigue (dynamic or variable) loading.
- Obtain new value of permissible stress for fatigue loading by dividing given permissible stress by respective stress concentration factor and use this new value of permissible stress in design equation.

| Sr. No. | Type of joint | Stress concentration factor ($K_f$) |
|---|---|---|
| 1. | Transverse fillet weld | 1.5 |
| 2. | Parallel fillet weld | 2.7 |

**Note :** In order to avoid starting and stopping of weld run, 12.5 mm should be added to length of each weld.

### 7.6.1 Assumptions Made in Designing Welded Joints

1. The load is distributed uniformly along the entire length of the weld.
2. The stress is spread uniformly over the effective section.
3. Suitable stress concentration factors and factors of safety are employed to account for the unknown factors like,
   (a) Change in physical properties due to high rate of cooling.
   (b) Thermal stresses in welds.
   (c) Homogeneity of weld metal.
4. Proper type of welded joint is used.

**Numerical Type No. 1 :**
**"Design of Double Transverse Fillet Weld (Static Loading)"**

**Problem 7.1 :** *Two steel plates 120 mm wide and 12.5 mm thick are to be connected together by double transverse fillet weld. Maximum tensile stress for plates and welding material is not to exceed 70 N/mm². Find the length of weld required for maximum static loading.*

**Solution : Given data :** Thickness of plate = Size of weld = s = 12.5 mm, $\sigma_t = 70$ N/mm², width = 120 mm

**Procedure :**

Maximum load, which the plate can carry is,

$W = \text{Area} \times \text{Maximum stress}$

$\therefore \quad W = (\text{Width} \times \text{Thickness}) \times \sigma_t$

$\therefore \quad W = (120 \times 12.5) \times 70$

$\therefore \quad \mathbf{W = 105 \times 10^3\ N}$ ... (1)

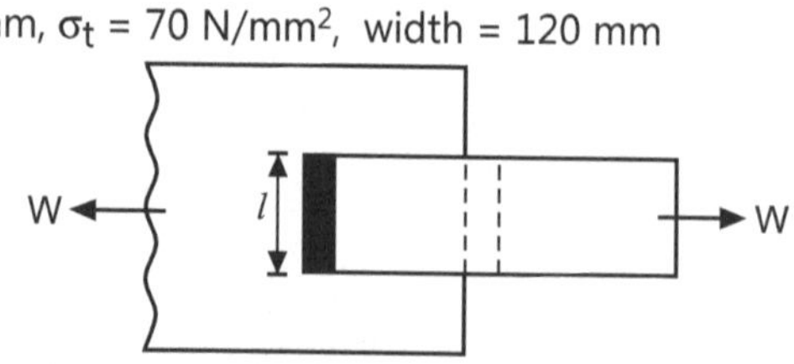

**Fig. 6.10 : Double transverse fillet weld**

For double transverse fillet weld, strength equation is,

$\mathbf{W} = 2 \times 0.707 \times s \times l \times \sigma_t = 1.414 \times 12.5 \times l \times 70 = \mathbf{1237.25 \times l}$ ... (2)

On equating equations (1) and (2), we get,

$105 \times 10^3 = 1237.25 \times l$

$\therefore \quad l = \mathbf{84.86\ mm}$

For starting and stopping of weld run, 12.5 mm is added to the length of weld.

$\therefore$ **Length of weld required,**

$\mathbf{l} = 84.86 + 12.5 = \mathbf{97.36\ mm}$

## Numerical Type No. 2 : "Design of Double Transverse Fillet Weld (Static and Dynamic Loading)"

**Problem 7.2:** *Two steel plates 10 cm wide and 1.25 cm thick are to be connected together by double transverse fillet weld. Take stress concentration factor = 1.5 and tensile stress as 7000 N/cm². Find the length of weld under static as well as dynamic loading.*

**Solution : Given data :** Thickness of plate = Size of weld = **s** = 1.25 cm = **12.5 mm**,

**Width** = 10 cm = **100 mm**, $\sigma_t$ = 7000 N/cm² = **70 N/mm²**, $K_f$ = 1.5

**Procedure :** Maximum load, which the plate can carry is,

$$W = \text{Area} \times \text{Maximum stress} = (\text{Width} \times \text{Thickness}) \times 70$$

$$\therefore \quad \mathbf{W} = 100 \times 12.5 \times 70 = \mathbf{87.5 \times 10^3\ N} \qquad \text{... (1)}$$

**Case (I) : Under static loading :**

For double transverse fillet weld, $\mathbf{W} = 2 \times 0.707 \times s \times l \times \sigma_t = 1.414 \times 12.5 \times l \times 70 = \mathbf{1237.25 \times l}$ ... (2)

On equating equations (1) and (2),

$$87.5 \times 10^3 = 1237.25 \times l$$

$$\therefore \quad l = \mathbf{70.72\ mm}$$

For starting and stopping of weld run, 12.5 mm is added to the length of weld.

$\therefore$ **Length of weld,** $\boldsymbol{l} = 70.72 + 12.5 = \mathbf{83.22\ mm}$

**Case II : Under dynamic loading :** Allowable tensile stress $= \dfrac{70}{K_f} = \dfrac{70}{1.5} = \mathbf{46.67\ N/mm^2}$

For double transverse fillet weld,

$$\mathbf{W} = 2 \times 0.707 \times s \times l \times \sigma_t = 1.414 \times 12.5 \times l \times 46.67 = \mathbf{824.9 \times l} \qquad \text{... (3)}$$

On equating equations (1) and (3),

$$87.5 \times 10^3 = 824.9 \times l$$

$$\therefore \quad l = \mathbf{106.07\ mm}$$

For starting and stopping of weld run, 12.5 mm is added to the length of weld.

$\therefore$ **Length of weld required,**

$$\boldsymbol{l} = 106.07 + 12.5 = \mathbf{118.57\ mm}$$

## Numerical Type No. 3 : "Design of Double Parallel Fillet Weld (Static Loading)"

**Problem 7.3:** *A plate 100 mm wide and 10 mm thick is to be welded to another plate by means of double parallel fillet weld. The plates are subjected to a static load of 80 kN. Find the length of weld, if permissible shear stress in the weld material does not exceed 55 N/mm².*

**Solution : Given data :** Thickness of plate = Size of weld = s = 10 mm,

**W** = 80 kN = **80 × 10³ N**, $\tau$ = 55 N/mm², Width = 100 mm

**Procedure :** For double parallel fillet weld, shear strength is given by,

$$W = 2 \times 0.707 \times s \times l \times \tau$$

$$\therefore \quad 80 \times 10^3 = 1.414 \times 10 \times l \times 55$$

$$\therefore \quad l = \mathbf{102.86\ mm}$$

For starting and stopping of weld run, 12.5 mm is added to the length of weld.

$\therefore$ **Length of weld required,**

$$\boldsymbol{l} = 102.86 + 12.5 = \mathbf{115.36\ mm}$$

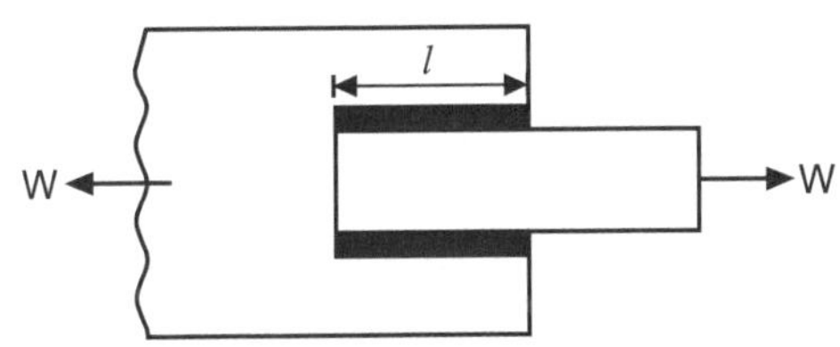


Fig. 7.11 : Double parallel fillet weld

## Numerical Type No. 4 : "Design of Double Parallel Fillet Weld (Static and Dynamic/Fatigue Loading)"

**Problem 7.4 :** *A plate 100 mm wide and 12.5 mm thick is to be welded to another plate by means of double parallel fillet weld. Plates are subjected to a load of 50 kN. Find the length of weld, so that, maximum stress does not exceed 56 N/mm². Consider the joint first under static loading and then under fatigue loading. Take stress concentration factor for parallel fillet welding as 2.7.*

**Solution : Given data :** Thickness of plate = Size of weld = s = 12.5 mm, **W** = 50 kN = $\mathbf{50 \times 10^3}$ **N**, $\tau$ = 56 N/mm², Width = 100 mm, $K_f$ = 2.7

**Procedure :**

**Case I : Under static loading :**

We have, $W = 1.414 \times s \times l \times \tau$ [∵ Double parallel fillet weld]

∴ $50 \times 10^3 = 1.414 \times 12.5 \times l \times 56$

∴ $l$ = **50.51 mm**

Adding 12.5 mm for starting and stopping of weld run, we get,

**Length of weld required,**

$\mathbf{l}$ = 50.51 + 12.5 = **63.01 mm**

**Case II : Under fatigue loading :** Stress concentration factor for parallel fillet weld, $K_f$ = 2.7.

∴ Permissible shear stress, $\tau = \frac{56}{2.7}$ = **20.74 N/mm²**

We have, $W = 1.414 \times s \times l \times \tau$ [∵ Double parallel fillet weld]

∴ $50 \times 10^3 = 1.414 \times 12.5 \times l \times 20.74$

∴ $l$ = **136.39 mm**

Adding 12.5 mm for starting and stopping of weld, we get,

**Length of weld required,**

$\mathbf{l}$ = 136.39 + 12.5 = **148.89 mm**

**Numerical Type No. 5 : "Design of Combined Single Transverse and Double Parallel Fillet Weld (Static Loading)"**

**Problem 7.5 :** *A plate 75 mm wide and 12.5 mm thick is to be joined with another plate by a single transverse and double parallel fillet weld. Maximum tensile and shear stresses are 70 N/mm² and 56 MPa respectively. Find the length of each parallel fillet weld, if joint is subjected to 90 kN.*

**Solution : Given data : W** = 90 kN = $\mathbf{90 \times 10^3}$ **N**, $\sigma_t$ = 70 N/mm², Width = 75 mm

$\tau$ = 56 MPa = **56 N/mm²**, Size of weld = s = 12.5 mm

**Procedure :** Let $l_1$ = length of transverse fillet weld

$l_2$ = Length of each parallel fillet weld

We know that, for calculating length of weld run ($l_2$), **effective length of weld** for single transverse fillet weld ($l_1$) is required. Effective length is obtained by subtracting 12.5 mm from given width of plate.

∴ **Effective length of single transverse fillet weld is,**

$\mathbf{l_1}$ = 75 − 12.5 = **62.5 mm**

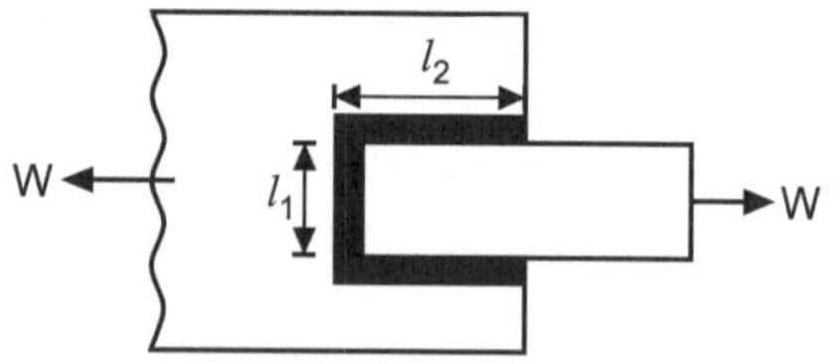

**Fig. 7.12 : Combination of single transverse and double parallel fillet weld**

**Load carried by single transverse fillet weld :**

$W_1 = 0.707 \times s \times l_1 \times \sigma_t$

∴ $\mathbf{W_1} = 0.707 \times 12.5 \times 62.5 \times 70$

= **38664.06 N**

**Load carried by double parallel fillet weld :**

$\mathbf{W_2} = 2 \times 0.707 \times s \times l_2 \times \tau = 1.414 \times 12.5 \times l_2 \times 56 = \mathbf{989.8\, l_2}$

**Total strength of weld (W) :**

We know that, total strength of weld will be equal to sum of load carried by single transverse fillet weld and load carried by double parallel fillet weld.

∴ $W = W_1 + W_2$

∴ $90 \times 10^3 = 38664.06 + 989.8\, l_2$

∴ $l_2$ = **51.86 mm**

For starting and stopping of weld run, 12.5 mm is added to the length of weld,

∴ **Length of each parallel fillet weld required is,**

$\mathbf{l_2}$ = 51.86 + 12.5 = **64.36 mm**

**Problem 7.6 :** *A steel plate 100 mm wide and 10 mm thick is joined with another steel plate by means of a single transverse and double parallel fillet weld, as shown in Fig. 7.13. Strength of welded joint should be equal to strength of plates to be joined. Permissible tensile and shear stresses for weld material and plates are 70 N/mm² and 50 N/mm² respectively. Find the length of weld. Assume tensile force acting on the plates as static.*

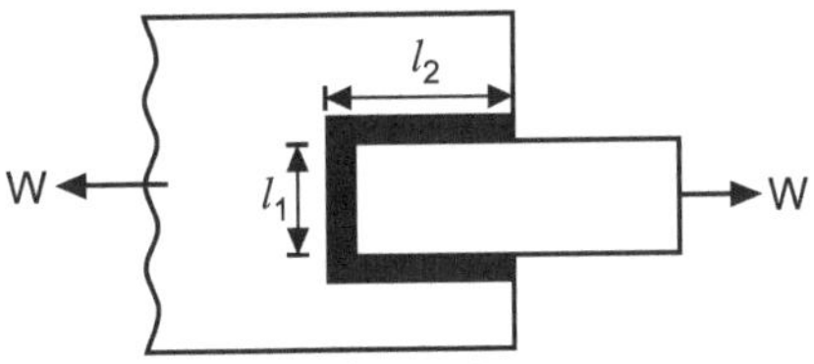


**Fig. 7.13**

**Solution : Given data :** Thickness of plate = s = 10 mm, Width = 100 mm, $\sigma_t = 70\ N/mm^2$, $\tau = 50\ N/mm^2$

**Procedure :** Let, $l_1$ = Length of transverse fillet weld, $l_2$ = Length of double parallel fillet weld

We know that, for calculating length of weld run ($l_2$), effective length of weld for single transverse fillet weld ($l_1$) is required. This effective length for singe transverse fillet weld is obtained by subtracting 12.5 mm from given width of plate.

∴ **Effective length of weld for single transverse fillet weld is,**

$$l_1 = \text{Width} - 12.5 = 100 - 12.5 = \mathbf{87.5\ mm}$$

We know that, maximum load carried by plates can be calculated by,

Strength of plate = Cross-sectional area of plate × Maximum stress

$$\therefore \quad W = (100 \times 10) \times 70 = \mathbf{70 \times 10^3\ N} \qquad [\because \sigma_t\ (70\ N/mm^2) > \tau\ (50\ N/mm^2)]$$

It is given that, strength of plates is equal to strength of weld.

∴ **Total strength of weld, W = $70 \times 10^3$ N** ...(1)

**Load carried by a single transverse fillet weld :**

$$W_1 = 0.707 \times s \times l_1 \times \sigma_t = 0.707 \times 10 \times 87.5 \times 70 = \mathbf{43303.75\ N} \qquad \text{... (2)}$$

**Load carried by a double parallel fillet weld :**

$$W_2 = 1.414 \times s \times l_2 \times \tau = 1.414 \times 10 \times l_2 \times 50 = \mathbf{707 \times l_2} \qquad \text{... (3)}$$

**Total strength of weld :**

We know that, total strength of weld will be equal to sum of loads carried by single transverse fillet weld and load carried by double parallel fillet weld.

$$\therefore \quad W = W_1 + W_2$$

$$\therefore \quad 70 \times 10^3 = 43303.75 + 707 \times l_2 \qquad \text{[From equations (1), (2) and (3)]}$$

$$\therefore \quad l_2 = \mathbf{37.76\ mm}$$

For starting and stopping of weld run, 12.5 mm is added to length of weld.

∴ **Length of each parallel fillet weld required,**

$$l_2 = 37.76 + 12.5 = \mathbf{50.26\ mm}$$

## Numerical Type No. 6 : "Design of Combined Single Transverse and Double Parallel Fillet Weld (Dynamic Loading)"

**Problem 7.7 :** *Find the length of weld run for a plate of size 120 mm wide and 15 mm thick to be welded by means of single transverse and double parallel fillet weld, when subjected to dynamic loading. Take $\sigma_t = 75\ N/mm^2$ and $\tau = 60\ N/mm^2$.*

**Solution : Given data :** Thickness of plate = Size of weld = s = 15 mm, Width = 120 mm,

$$\sigma_t = 75\ N/mm^2 \text{ and } \tau = 60\ N/mm^2$$

**Procedure :** We know that, maximum load, which the plate can carry,

Load = Area × Maximum stress

$$\therefore \quad W = (120 \times 15) \times 75 = \mathbf{135 \times 10^3\ N} \qquad [\because \sigma_t\ (75\ N/mm^2) > \tau\ (60\ N/mm^2)] \ldots (1)$$

Since the joint is subjected to dynamic (variable) loading, let us assume standard values of stress concentration factors as 1.5 and 2.7 for transverse fillet weld and parallel fillet weld respectively, to calculate permissible stresses.

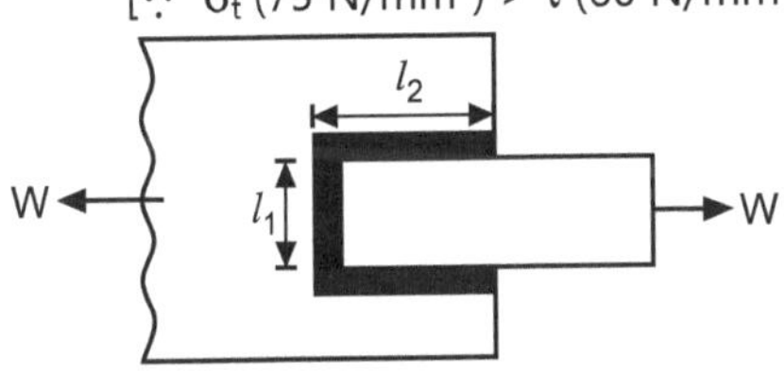


**Fig. 7.14 : Combined single transverse and double parallel fillet weld**

$$\text{Permissible tensile stress, } \sigma_t = \frac{75}{k_f} = \frac{75}{1.5} = \mathbf{50\ N/mm^2}$$

$$\text{Permissible shear stress, } \tau = \frac{60}{K_f} = \frac{60}{2.7} = \mathbf{22.22\ N/mm^2}.$$

Let, $l_1$ = Length of transverse fillet weld

$l_2$ = Length of each parallel fillet weld

We know that, for calculating length of weld run ($l_2$), effective length of weld for single transverse fillet weld ($l_1$) is required. It is obtained by subtracting 12.5 mm from given width of plate.

$\therefore$ **Effective length of weld for single transverse filled weld is,**

$l_1$ = Width – 12.5 = 120 – 12.5 = **107.5 mm.**

**Load carried by single transverse fillet weld :**

$$\mathbf{W_1} = 0.707 \times s \times l_1 \times \sigma_t = 0.707 \times 15 \times 107.5 \times 50 = \mathbf{57001.875\ N} \quad \ldots (2)$$

**Load carried by double parallel fillet weld :**

$$\mathbf{W_2} = 2 \times 0.707 \times s \times l_2 \times \tau = 1.414 \times 15 \times l_2 \times 22.22 = \mathbf{471.28\ l_2} \quad \ldots (3)$$

**Total strength of weld :**

We know that, total strength of weld will be equal to sum of load carried by single transverse fillet weld and load carried by double parallel fillet weld.

$\therefore \quad W = W_1 + W_2$

$\therefore \quad 135 \times 10^3 = 57001.875 + 471.28\ l_2$ [From equations (1), (2) and (3)]

$\therefore \quad l_2 = \mathbf{165.5\ mm}$

For starting and stopping of weld run, 12.5 mm is added to length of weld.

$\therefore$ **Length of each parallel fillet weld required is,**

$$l_2 = 165.5 + 12.5 = \mathbf{178\ mm}$$

## Theory Question for Practice

1. Give merits and demerits of welded joints.
2. Explain strength equations for transverse fillet welds with sketches.
3. Explain strength equations for double parallel fillet weld with neat sketch.
4. State the strength equation of combined double parallel fillet weld and single transverse fillet weld with sketches.

## Numerical Problems For Practice

1. Fig. 7.15 shows a plate 70 mm wide and 12 mm thick joined with another plate by single transverse and double parallel weld. Determine the length of parallel fillet weld of joint subjected to both static and fatigue loading. Given $\sigma_t$ = 80 MN/m$^2$, $\tau$ = 50 MN/m$^2$ and stress concentration factor 1.4 for transverse weld and 2.5 for parallel weld.

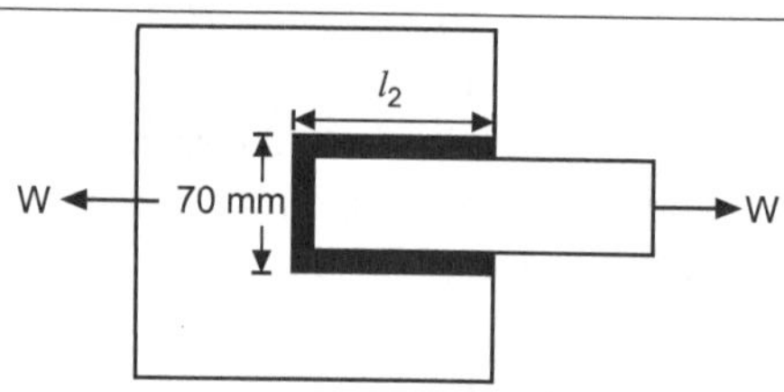


**Fig. 7.15**

(**Note :** $\sigma_t = 80\ \text{MN/m}^2 = \dfrac{80 \times 10^6}{10^3 \times 10^3\ \text{mm}^2} = 80\ \text{N/mm}^2$)

(**Ans.** (i) $l_2$ = 45.3 mm (ii) $l_2$ = 110.38 mm)

2. A plate, 75 mm wide and 10 mm thick, is joined with another steel plate by means of single transverse and double parallel fillet welds, as shown in Fig. 7.16. The joint is subjected to maximum tensile force of 55 kN. The permissible tensile and shear stresses in the weld material are 70 N/mm$^2$ and 50 N/mm$^2$ respectively. Determine the required length of each parallel fillet weld.

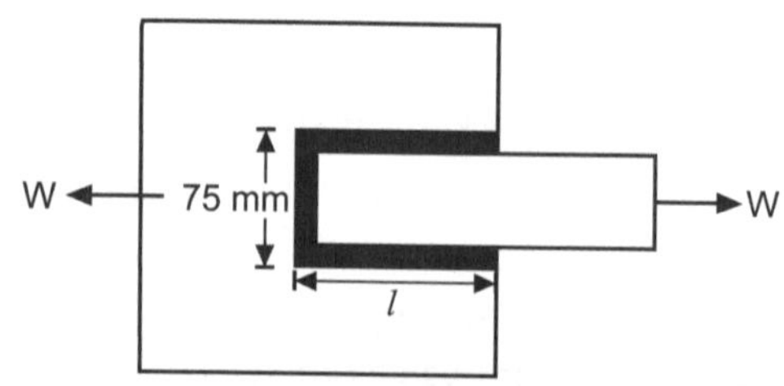


**Fig. 7.16**

(**Ans.** $l_2$ = 46.55 mm)

3. A plate of 100 mm wide and 12.5 mm thick is to be welded to another plate by means of parallel fillet welds. The plates are subjected to load of 50 kN. Find the length of the weld, so that, the maximum stress does not exceed 56 MPa. Consider the joint under fatigue loading. (**Ans.** $l$ = 148.896 mm)
4. A plate 90 mm wide and 10 mm thick is to be joined with a single transverse and double parallel fillet weld. The maximum tensile and shear stress are 70 N/mm$^2$ and 55 N/mm$^2$ respectively. Find the length of each parallel fillet weld, if joint is subjected to 70 kN. (**Ans.** 53.19 mm)

❑❑❑

*Chapter* **8**

# DESIGN OF FASTENERS - SCREWED JOINTS

## 8.1 INTRODUCTION TO SCREWED JOINTS

- **A screw thread** is formed by cutting a continuous helical groove on a cylindrical surface.
- A screw made by cutting a single helical groove on the cylinder is known as *'single start threaded screw'*. When a second thread is cut in the space between grooves of first thread, then screw is called as *'double start threaded screw'*. Similarly, triple and quadruple threads can be manufactured or produced.
- Helical grooves cut on cylindrical surface may be either right hand or left hand.
- **Screwed joint** is mainly composed of two elements : **Screw (bolt) and nut**.
- Screwed joints are used, where the machine parts are required to be readily connected or disconnected without causing any damage to the machine or screw fastened with nut.
- Screwed joints are also known as **'Threaded joints'** or **'Bolted joints'**.

### 8.1.1 Functions of Scewed (Threaded) Joints with Applications

1. **To connect the components together, which should be readily disassembled :**
   - For this purpose, 'V' threads having high strength and low efficiency are used to avoid loosening, when the machine or any application is running.
   - **Applications :** (i) Fasteners, (ii) Machine bolts, (iii) Stove bolts, (iv) Wood screws etc.
2. **To transmit power or energy :**
   - For this purpose, square threads having high transmission efficiency are used for power transmission. Use of square threads reduces loss of power during transmission as compared to Buttress threads, Acme threads etc.
   - **Applications :** (i) Screw Jack, Toggler jack used in Automotive applications.
     (ii) Lead screw on a Lathe machine transmitting motion.

### 8.1.2 Examples, Where Screwed Joints are Preferred Over Welded Joints

- Following are few examples, where screwed joints are preferred over welded joints.

1. Fasteners.
2. Stove bolts.
3. Machine bolts.
4. Automobile components.
5. Cylinder head of engine.
6. Structural bridges.
7. Pressure vessels.

### 8.1.3 Advantages and Disadvantages of Screwed Joints

**Advantages of Screwed Joints :**

1. Screwed Joints are convenient to assemble and dissemble in case of breakage or maintenance/repair.
2. Screwed Joints are highly reliable in operation.
3. A wide range of screwed joints can be used for different types of operating conditions.
4. Low production cost due to benefit of standardization.
5. Screwed Joints offer high clamping force.

**Disadvantages of Screwed Joints :**

1. The main disadvantage in screwed Joints is stress concentration in the threaded portion. Problem of stress concentration becomes severe under conditions of variable loading.
2. When screwed joints, such as, nut and bolt assembly or any other screw fasteners are used to join the parts together, then the parts should have holes for the provision of inserting bolts or screws. Due to presence of these holes, the screwed joints get weakened.
3. Screwed joints are weaker than welded joints.
4. Locking arrangement is required in case of applications subjected to vibrations or shocks.

## 8.2 VARIOUS FORMS OF THREAD PROFILES

Screw threads have been standardized according to their cross-sectional form. Most commonly used screw threads are,

**(a) 'V' threads:**

- All bolts and similar fasteners have V-shaped threads.
- V-threads are used in automotive and aeronautical industries, because, if V-threads are used to form screwed joints, then chances of loosening due to vibrations are greatly reduced.

**(b) Whitworth threads :**

- Whitworth threads are better, because they are filleted at top and bottom to eliminate sharp corners and roots. Due to fillets, strength of threads is increased.

**(c) Square threads:**

- Square threads are widely used to transmit power, because they offer highest power transmission efficiency than any other thread.

  Also, square threads can transmit power in either direction.
- In square threads, the sides of threads are parallel.
- **Applications :** (i) Screw jack, (ii) Lead screw of lathe, (iii) Toggle jack.

**(d) Acme threads:**

- Acme threads are easier to cut, cheaper to manufacture and stronger than square threads.
- Also, they can transmit power in either direction like square threads.
- But, Acme threads offer less power transmission efficiency as compared to square threads. Therefore, acme threads are not preferred.

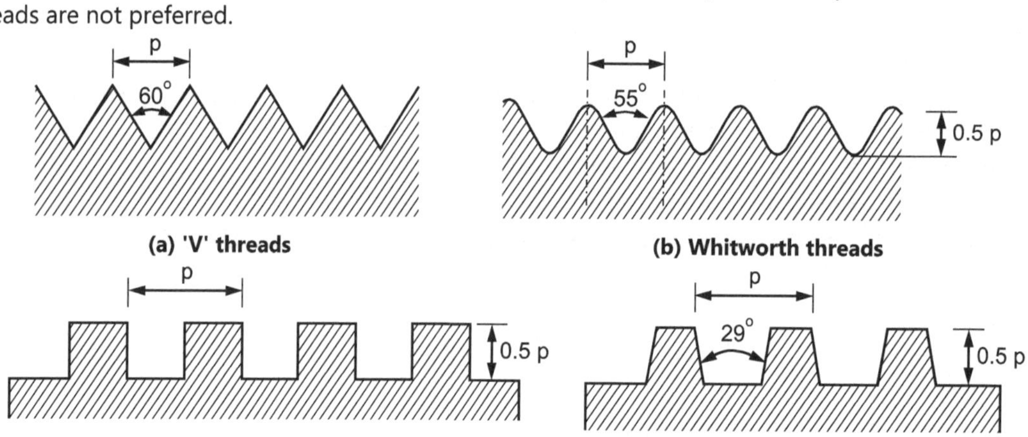


**(a) 'V' threads** **(b) Whitworth threads**

**(c) Square threads** **(d) Acme threads**

**Fig. 8.1: Various forms of screw threads**

### 8.2.1 Advantages and Disadvantages of 'V' Threads

**Advantages of 'V' Threads :**

1. 'V' threads are used in bolts, studs, nuts, tap bolts etc. to tighten the parts together.
2. 'V' threads offer greater frictional resistance to motion (while rotating screw or nut in fasteners) than square threads. Therefore, 'V' threads are preferred over square threads for fastening purpose.
3. 'V' threads are stronger than square threads.
4. 'V' threads are cheaper, because, they are easy to cut using a die or machine.

**Disadvantages of 'V' Threads :**

1. 'V' threads are not suitable for power transmission.
2. 'V' threads have one component of force, which acts perpendicular to the screw axis. It causes bursting action on the nut and increase in friction, whereas, square threads are ideal for power transmission, because they do not have any component of force perpendicular to axis.

## 8.3 TERMINOLOGY FOR SCREW THREADS

1. **Crest :**
   - **Crest** is defined as, "*the prominent part of thread*".
2. **Root :**
   - **Root** is defined as, "*the bottom formed between two flanks of thread*".
3. **Flank :**
   - **Flank** is defined as, "*a straight edge, which connects crest and root*".
4. **Thread angle or Included angle :**
   - **Thread angle** is defined as, "*the included angle between two flanks of thread*".
5. **Major diameter :**
   - **Major diameter** is defined as, "*the largest diameter of an external or internal screw thread*".
   - Major diameter is also called as **Nominal diameter**.
   - Any screw or bolt is always specified by this nominal diameter.
   - Major diameter or Nominal diameter is denoted by $d_o$.
6. **Minor diameter :**
   - **Minor diameter** is defined as, "*the smallest diameter of an external or internal screw thread*".
   - It is also known as **root diameter or core diameter**.
   - Minor diameter or root diameter or core diameter is denoted by $d_c$.
   - For 'V' threads, $d_c = 0.84\ d_o$.
7. **Pitch circle diameter :**
   - **Pitch circle diameter** is defined as, "*the diameter of an imaginary cylinder, which is co-axial with screw axis and intersects the flanks of threads in such a way, so that, width of thread is equal to width of space between threads*".
   - It is also known as **effective diameter**. It is denoted by $d_p$.
8. **Pitch :**
   - **Pitch** is defined as, "*the axial distance between a point on one thread to corresponding point on next (adjacent) thread*". It is measured parallel to thread axis. It is denoted by p.
   - Mathematically, $\text{Pitch} = \dfrac{1}{\text{No. of threads per unit length of screw}}$

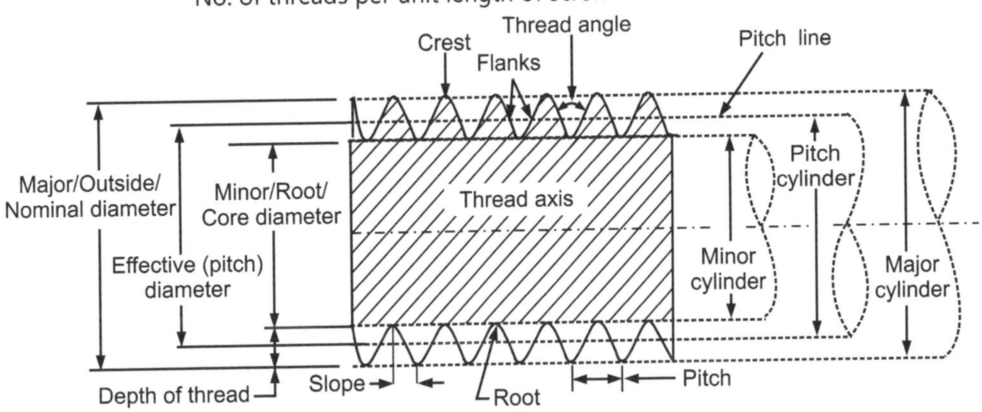


**Fig. 8.2 : Terminology of external thread**

9. **Lead :**
   - **Lead** is defined as, "*the linear distance travelled by screw in upward or downward direction, when it completes one revolution or rotation*".
   - In short, lead is the axial advancement of screw in vertical direction (upward or downward), when it completes one revolution.
   - Lead = Pitch (in single start threads) = 2 × Pitch (in double start threads) and so on.

**10. Hand of threads :**

**(a) Right hand threads :** If the helix on outer surface ascends from left to right, the thread is called as right hand thread.

**OR**

When the nut assembled on screw is given clockwise rotation leading to **tightening** of joint, then thread is called as **right hand thread.** On rotating in anticlockwise direction, it **loosens** the joint. The standard screwed joint employs right hand threads. Refer Fig. 8.3 (a).

**(b) Left hand threads :** If the helix on outer surface ascends from right to left, the thread is called as left hand thread.

**OR**

When the nut assembled on screw is given clockwise rotation leading to **loosening** of joint, then thread is called as **left hand thread.** On rotating in anticlockwise direction, it **tightens** the joint. Refer Fig. 8.3 (b).

**(a) Right hand threads** **(b) Left hand threads**

**Fig. 8.3 : Hands of thread**

## 8.3.1 Types of Screw Fastening

**1. Through bolts :**

- A through bolt is also called as a bolt.
- It is a cylindrical bar with a head at one end and the threads at another end.
- The cylindrical part is called as **shank.**
- The bolt is passed through a drilled hole in the two parts to be fastened together and the nut is screwed and subsequently tightened on to the threaded end to clamp the two parts securely to each other.
- The nut and head are made hexagonal or sometimes square.

**2. Tap bolt :**

- A tap bolt or screw differs from a bolt.
- It is screwed into a tapped hole of one of the parts to be fastened, without nut.
- Tap bolt is used, when the arrangement does not provide any space to accommodate the nut.
- Frequent insertion or removal of the tap bolt is likely to damage the threads. Therefore, tap bolts are used, only when parts are not to be disconnected very often.

**3. Stud :**

- Stud is a round bar threaded at both ends.
- One end of stud is screwed into tapped hole of parts to be fastened, while the other end receives nut on it.
- **Applications:** For connecting the covers, like vessel containing high pressure steam, cylinder head of an engine etc.

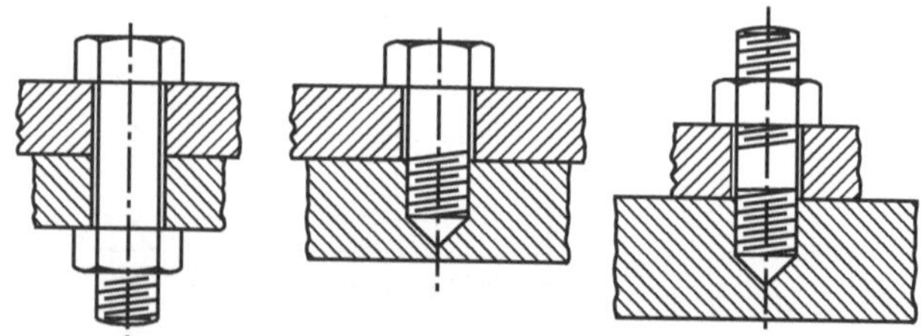

**(a) Through bolt** **(b) Tap bolt** **(c) Stud**

**Fig. 8.4 : Use of bolt, tap bolt and stud**

**4. Cap screws:**

- Cap screw is similar to tap bolt. Therefore, it is screwed into a tapped hole of one of the parts to be fastened, without nut.
- It is of small size as compared to tap bolt.

- A variety of shapes of heads are available as shown in Fig. 8.5.

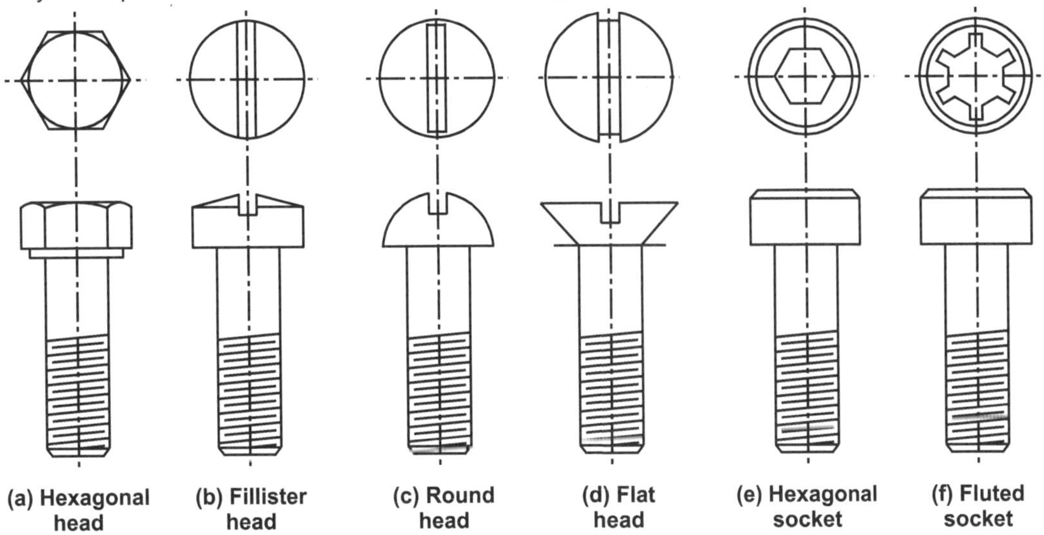


**Fig. 8.5 : Cap screws**

5. **Set screw:**
   - A set screw is screwed through a tapped hole in one part, until the end of the screw is pressed against the other part. The friction between the screw end and the part prevents the relative motion.
   - The set screws, shown in Fig. 8.6, are very similar to cap screws, but are threaded practically throughout the length and are still smaller in size.
   - **Use:** Set screws are used to prevent the relative motion between the two parts.
   - **Application:** Set screws can be used instead of key for transmission of low power. In other words, they can be used instead of key in light power transmission.

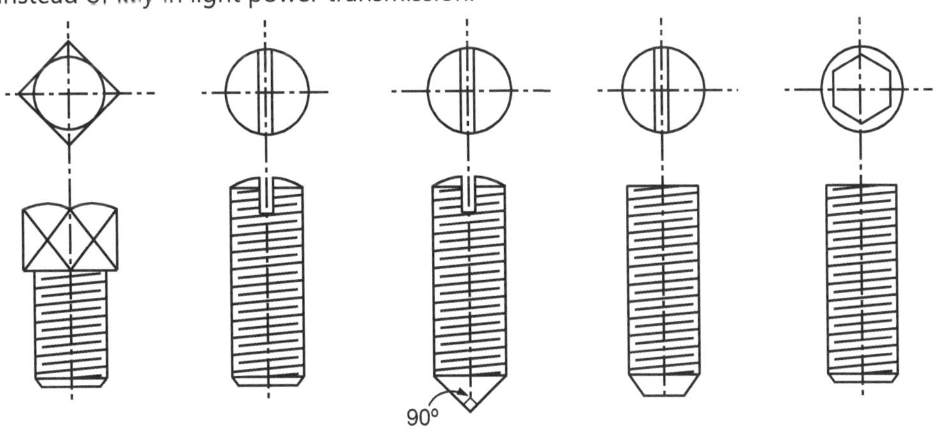


**Fig. 8.6 : Set screws**

## 8.3.2 Use of Cap Screws and Set Screws

**Cap screws :**

- They are used to fasten covers, cases etc.
- Nut is not required during fastening.

**Set screws :**

- They are used to prevent relative motion between two parts.
- They resist the relative motion between two parts by means of friction between the point of screw and one part.

### 8.3.3 Preloading of Bolts

- The material of bolt has elastic as well as plastic range of loading.
- **'Preloading of bolts'** means initial tightening of bolts, which gives strength to the joint.
- Good quality bolts can be preloaded into the plastic range to develop more strength.

**Advantages of Preloading :**

1. It prevents leakages in high-pressure vessels.
2. It decreases fluctuations in loads, which reduces the chances of fatigue failure.

### 8.3.4 Use of Spring Washer and Gaskets in Screwed Joints

**Spring Washers :**

- Washers are thin circular plates having a central hole slightly larger than the bolt diameter.
- Usually, dimensions of washers are taken as,
    1. **Thickness of washer** = $0.15 \times d_o$
    2. **Outside diameter of washer** = $2 \times d_o$

    where, $d_o$ is the nominal diameter or major diameter of bolt or stud.

**Uses of Washers :**

1. If two thin plates are joined by nut and bolt assembly along with washers placed in between them, then washers increase the area, over which, the pressure of nut gets uniformly distributed. Thus, washers are used to prevent distortion of thin plate. On rough surface, washer becomes a flat seat for nut.
2. Washers are also used, when the hole, through which, bolt passes is much larger than size of bolt.
3. Washers provide sufficient bearing area.

**Gaskets :**

- When a screwed joint is to be made leak proof, a metal gasket or asbestos packing is inserted between the surfaces to be screwed.
- The other words, gasket is used to make the screwed joint leakproof.
- Materials used for gaskets should have low stiffness, i.e. minimum deflection should occur for maximum tightening load.

## 8.4 STRESSES IN SCREWED FASTENING UNDER STATIC LOADING

Following stresses are induced in screwed fastening under static loading :

1. Initial stresses due to screwing up forces,
2. Stresses due to external forces, and
3. Stress due to combination of stresses at (1) and (2).

### 8.4.1 (a) Initial Stresses Due to Screwing Up Forces

Following stresses are induced in a bolt, screw or stud, when it is screwed up tightly.

1. Tensile stress due to stretching of bolt.
2. Torsional shear stress caused by frictional resistance offered by threads of screw during its tightening.
3. Shear stress across threads.
4. Compression or crushing stress on threads.
5. Bending stress, if surfaces under head of bolt or nut are not perfectly parallel to bolt axis.

**Note :** Core diameter ($d_C$) may be taken as $0.84\ d_o$, where $d_o$ is the major diameter.

**Brief description of above stresses induced due to screwing up forces :**

**1. Tensile stress due to stretching of bolt :**

- In actual practice, it is not possible to determine above mentioned stresses accurately. Therefore, bolts are designed on the basis of direct tensile stress considering a large factor of safety, in order to account for the un-determined stresses.

- **Initial tension in a bolt**, based on experiments, may be found by the relation,

$$W_i = 2840\, d_o \qquad \text{in 'N'}$$

where, $d_o$ = Nominal diameter of bolt in 'mm'

- The above relation is used for making a joint fluid-tight like cover bolted on top of engine cylinder containing high pressure steam, flanged joints in case of long pipelines, through which, high pressure water/liquid flows and many more applications etc. Sometimes, $W_i$ is also denoted as $W_1$.
- When the joint is not required as tight as fluid-tight joint, then the initial tension in a bolt may be reduced from ($2840\, d_o$) to ($1420\, d_o$). In such cases,

$$W_i = 1420\, d_o \qquad \text{in 'N'}$$

- Bolts of smaller diameter (less than M 16 or M 18) are not permitted in making fluid-tight joints, because they may fail during tightening due to small size and hence, low strength.
- If bolt is not initially stressed, then the maximum safe axial load, which may be applied to it, is given by,

$$W = \text{Permissible stress} \times \text{Cross-sectional area at bottom of thread (i.e. stress area)}$$

- Stress area may be obtained by using the following relation,

$$\text{Stress area} = \frac{\pi}{4}\left(\frac{d_p + d_c}{2}\right)^2 \qquad \text{in 'mm}^2\text{'}$$

where, $d_p$ = Pitch diameter or Mean diameter of screw in 'mm'

and $d_c$ = Core or Minor or Root diameter of screw in 'mm'

**2. Torsional shear stress caused due to frictional resistance of threads of screw during its tightening :**

- Torsional shear stress induced due to frictional resistance of threads of screw during its tightening may be obtained by using basic torsion equation mentioned below.

$$\frac{T}{J} = \frac{\tau}{r}$$

$$\therefore \quad \tau = \frac{T}{J} \times r = \frac{T}{\frac{\pi}{32}(d_c)^4} \times \frac{d_c}{2} = \frac{16\,T}{\pi\,(d_c)^3}$$

where, $\tau$ = Torsional shear stress induced in 'N/mm²'

$J$ = Polar moment of inertia = $\frac{\pi}{32}(d_c)^4$ in 'mm⁴'

$T$ = Torque applied in 'N·mm'

and $d_c$ = Minor or Core diameter of thread in 'mm'

**Note :** In the above equation, least diameter, i.e. $d_c$ is considered for design purpose, which is weakest section/dimension.

**3. Transverse shear stress across threads :**

- Average thread shear stress for screw ($\tau_s$) is,

$$\tau_s = \frac{W}{\pi\, d_c \cdot n \cdot t}$$

where, $d_c$ = Core diameter, t = Thickness of thread section at the root

- Average thread shear stress for nut ($\tau_n$) is,

$$\tau_n = \frac{W}{\pi\, d_o \cdot n \cdot t}$$

where, $d_o$ = Major diameter

**4. Compression or Crushing stress on threads :**

- Compression or Crushing stress between screw threads is obtained by,

$$\sigma_{ck} = \frac{W}{\frac{\pi}{4}[(d_o)^2 - (d_c)^2] \cdot n}$$

where, $d_o$ = Major diameter, $d_c$ = Minor diameter,

and $n$ = Number of threads of screw in engagement with nut.

## 8.4.1 (b) Stresses Due to External Forces

Following stresses are induced in a bolt, when it is subjected to an external load.

1. Tensile stress. 2. Shear stress. 3. Combined tensile and shear stress.

**1. Tensile stress :**

- Bolts, studs and screws usually carry a load in the direction of bolt axis, which induces tensile stress in them.
- If $d_c$ = Root or Core diameter of thread and W = Externally applied load, then, permissible tensile stress for the bolt material ($\sigma_t$) is given by,

$$\sigma_t = \frac{W}{\frac{\pi}{4} \cdot (d_c)^2} \quad \text{or} \quad d_c = \sqrt{\frac{4W}{\pi \cdot \sigma_t}}$$

- Now, the value of nominal diameter of bolt can be taken from design data book or it can be calculated by using the following relation.

$$d_c = 0.84\, d_o \qquad \therefore\ d_o = \frac{d_c}{0.84}$$

**Note :**

(a) External load (W) is taken up by 'n' number of bolts. It means, load W is equally shared by all bolts.

$\therefore$ Load on each bolt = $\frac{W}{n}$

Tensile stress induced in each bolt can be calculated by,

$$\therefore \quad \sigma_{t\,ind.} = \frac{\text{External load on each bolt}}{\text{Weakest cross-sectional area of bolt}} = \frac{\frac{W}{n}}{\frac{\pi}{4} \times d_c^2} = \frac{W}{\frac{\pi}{4}(d_c)^2 \times n}$$

(b) If standard table or design data book is not available, then for coarse threads, we use, $d_c = 0.84\, d_o$, where $d_o$ is nominal diameter of bolt.

**2. Shear stress :**

- Take an example of **flange coupling**, where bolts are used to prevent the relative movement between two parts.
- Shear stress gets induced in these bolts. This shear stress should be avoided as far as possible. It should be noted that, when the bolts are subjected to direct shearing load, then they should be located, in such a way that, the shearing load comes upon the body (i.e. shank) of bolt and not upon the threaded portion.
- Methods to reduce shear stress are,

(i) Bolts may be relieved of shear load by using shear pins.

(ii) When 'n' number of bolts are used to share the shearing load, the finished bolts should be fitted to the reamed holes.

If shear load ($W_s$) is shared equally by 'n' number of bolts, then load shared by each bolt will be $\frac{W_s}{n}$.

Induced shear stress can be calculated by,

$$\tau_{ind.} = \frac{\text{Load}}{\text{Area under shear}} = \frac{\frac{W_s}{n}}{\frac{\pi}{4} d_o^2} = \frac{W_s}{\frac{\pi}{4} \cdot (d_o)^2 \times n}$$

where, $d_o$ = Nominal or Major diameter of bolt.

**3. Combined tensile and shear stress :**

- Take an example of **coupling bolt**, where bolt is subjected to combined tensile and shear loads. *Diameter of shank* of bolt is obtained from *shear load,* whereas, *diameter of threaded part of bolt* is obtained from *tensile load.*
- A diameter slightly larger than required diameter may be assumed and stresses due to combined load should be checked for the following principal stresses.

(a) Maximum principal shear stress, $\tau_{max} = \frac{1}{2}\sqrt{(\sigma_t)^2 + 4\tau^2}$

(b) Maximum principal tensile stress, $\sigma_{t\ max} = \frac{\sigma_t}{2} + \frac{1}{2}\sqrt{(\sigma_t)^2 + 4\tau^2}$

- Calculated values of these induced stresses should not exceed more than permissible stress values for safe design.

## 8.4.1 (c) Stress Due to Combined Forces

- Resultant axial load acting on a bolt depends upon the following factors :
  1. Initial tension due to tightening of the bolt,
  2. External load, and
  3. Relative elastic yielding (extensibility) of bolt and connected members.
- Actual conditions of loading usually lie between two extremes explained below.

(i) When connected members or parts are very yielding as compared to bolt, as shown in Fig. 8.7 (a), then resultant load on the bolt is approximately equal to sum of initial tightening load and external load.

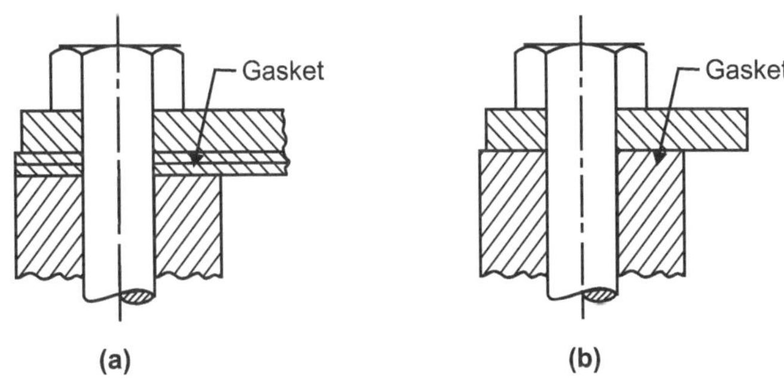


Fig. 8.7

(ii) On the other hand, if bolt is very yielding as compared to connected members or parts, as shown in Fig. 8.7 (b), then resultant load will be either initial tightening load or external load, whichever is greater.

- In order to determine the resultant axial load (W) acting on bolt, following equation may be used :

$$W = W_1 + \left(\frac{a}{1 + a}\right) \times W_2 = W_1 + K \cdot W_2 \qquad \ldots \left(\text{Substituting } \frac{a}{1 + a} = K\right)$$

where, $W_1$ = Initial tightening load or

Initial tension due to tightening of the bolt = $W_i = 2840\ d_o$

$W_2$ = External load on the bolt,

and $a$ = Ratio of Elasticity of connected parts to Elasticity of bolt = $\frac{E_p}{E_b}$

- Value of $K = \frac{a}{1 + a}$ is always less than unity.
- If combination of soft gasket and large sized bolts is taken, then value of 'a' is high. Hence, value of $K = \frac{a}{1 + a}$ will be larger and very close to unity. Due to higher value of K, value of term $(K \cdot W_2)$ will be more as compared to initial tightening load $(W_1)$.
- If combination of hard gasket and small sized bolts is taken, then value of 'a' is low. Hence, value of $K = \frac{a}{1 + a}$ will be smaller. Due to lower value of 'K', value of term $(K \cdot W_2)$ will be very small as compared to initial tightening load $(W_1)$. Therefore, in such a combination, resultant load will be mainly due to the initial tightening load.
- Value of 'a' may be estimated by design engineer to obtain an approximate value of 'K' to find the resultant load.
- Values of $\frac{a}{1 + a}$ (i.e. K) for various types of joints are given in Table 8.1.

**Table 8.1 : Values of 'K' for various types of joints**

| Sr. No. | Type of joint | $K = \frac{a}{1 + a}$ |
|---|---|---|
| 1. | Metal to metal joint with through bolts | 0.00 to 0.10 |
| 2. | Hard copper gasket with long through bolts | 0.25 to 0.50 |
| 3. | Soft copper gasket with long through bolts | 0.50 to 0.75 |
| 4. | Soft packing with through bolts | 0.75 to 1.00 |
| 5. | Soft packing with studs | 1.00 |

- Design engineer has control over the effect of resultant load acting on a bolt by taking sizes of connected parts and bolts in suitable proportions, and thereby, specifying initial tension in the bolt.

## 8.5 BOLTS OF UNIFORM STRENGTH

- Take an example of bolts of cylinder head of an I.C. engine. Here, we find that, bolts are subjected to shock loading. Therefore, resilience of bolt should be considered in order to prevent breakage at threaded portion.
- When ordinary bolt of uniform diameter is subjected to shock loads, then severe stress concentration occurs at the weakest part of bolt i.e. threaded portion. This threaded portion has an area equal to cross-sectional area available at the root of threads $\left(\frac{\pi}{4} d_c^2\right)$, which is smaller than cross-sectional area of shank. Therefore, more stress will be induced in the threaded portion as compared to portion of shank.
- Due to this, a great portion of energy will be absorbed at the region of threaded portion, which may lead to its fracture.

**Methods to Obtain Bolts of Uniform Strength are explained below :**

**Method (1) :**

- If diameter of shank of bolt is made equal or even slightly lesser than the root diameter (i.e. core diameter) of thread ($d_c$) as shown in Fig. 8.8 (b), then stress induced in the shank portion will be higher than stress induced in threaded portion. Therefore, shank will absorb a large portion of energy. In this way, the bolt becomes lighter in weight and shock absorbing capacity of bolt gets increased due to an increased modulus of resilience. Such bolts are called as *bolts of uniform strength*. **Resilience of a bolt** may also be increased *by increasing its length.*

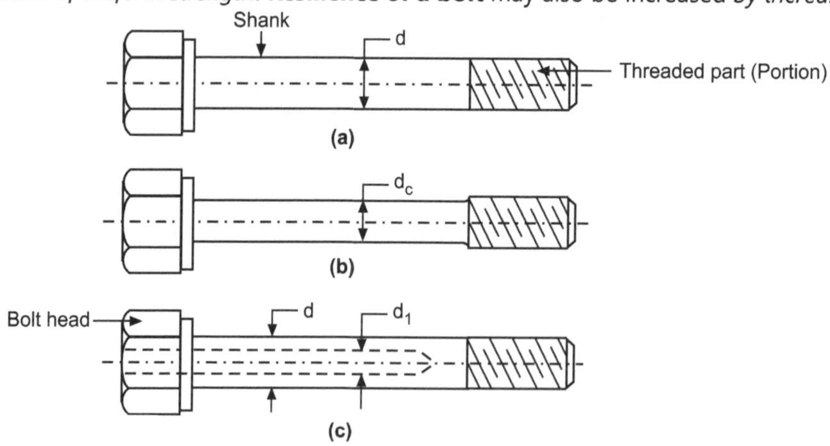


**Fig. 8.8 : Bolts of uniform strength**

**Method (2) :**

- In this method, an axial hole is drilled through the bolt head, to make shank area at least equal to root area of thread. Refer Fig. 8.8 (c).

If $d_1$ = Diameter of hole/Inner diameter of drilled shank,

$d$ = Outer diameter of drilled shank,

$d_o$ = Outer diameter/nominal diameter of thread,

$d_c$ = Root or Core diameter of thread,

Then, cross-sectional area of drilled shank = $\frac{\pi}{4}(d^2 - d_1^2)$

- For bolt of uniform strength, following condition must be satisfied

$$\frac{\pi}{4} \times (d^2 - d_1^2) = \text{Root or core area of threaded portion}$$

$$\therefore \quad \frac{\pi}{4} \times (d^2 - d_1^2) = \frac{\pi}{4} \times d_c^2$$

- In order to find out, diameter '$d_1$' of hole to be drilled for obtaining bolt of uniform strength, following concept is used.

$$\begin{bmatrix} \text{Material removed, while drilling} \\ \text{hole in shank portion} \end{bmatrix} = \begin{bmatrix} \text{Material removed, while cutting} \\ \text{threads in threaded portion} \end{bmatrix}$$

$$\therefore \quad \frac{\pi}{4} d_1^2 = \frac{\pi}{4}\left(d_o^2 - d_c^2\right)$$

$$\therefore \quad d_1^2 = d_o^2 - d_c^2$$

$$\therefore \quad \mathbf{d_1 = \sqrt{d_o^2 - d_c^2}}$$

Since, values of $d_o$ and $d_c$ are known, we can calculate value of $d_1$. Now, an axial hole having diameter $d_1$ should be drilled through bolt head to make bolt of uniform strength.

## 8.6 CYLINDER COVER WITH BOLTS

- Fig. 8.9 shows a cylinder cover fastened to the cylinder (or vessel/container/tank) with number of bolts.
- Here,

| | | | |
|---|---|---|---|
| $D$ | = | Diameter of valve or cylinder | in 'mm' |
| $d_o$ | = | Major or Nominal diameter of bolt | in 'mm' |
| $d_c$ | = | Core or Root diameter of bolt | in 'mm' |
| $D_p$ | = | Pitch circle diameter (P.C.D.) of bolts | in 'mm' |
| $t$ | = | Thickness of cylinder cover | in 'mm' |
| $n$ | = | Number of bolts | |
| $A$ | = | Area of cylinder = $\frac{\pi}{4} D^2$ | in 'mm$^2$' |
| $p$ | = | Pressure of stream inside the cylinder | in 'N/mm$^2$' |
| $p_c$ | = | Circular pitch = $\frac{\pi \cdot D_p}{n}$ | in 'mm' |

- Relation between Nominal diameter ($d_o$) and Core diameter ($d_c$) is given by,

$$d_c = 0.84\, d_o$$

- Bolts or studs for cylinder cover are subjected to tensile load. Therefore, they are designed by considering failure under tension, using the following equation.

$$\text{Permissible tensile stress} = \frac{\text{Load}}{\text{Minimum cross-sectional area of bolt}}$$

$$\therefore \quad \sigma_t = \frac{W}{\frac{\pi}{4} \cdot d_c^2}$$

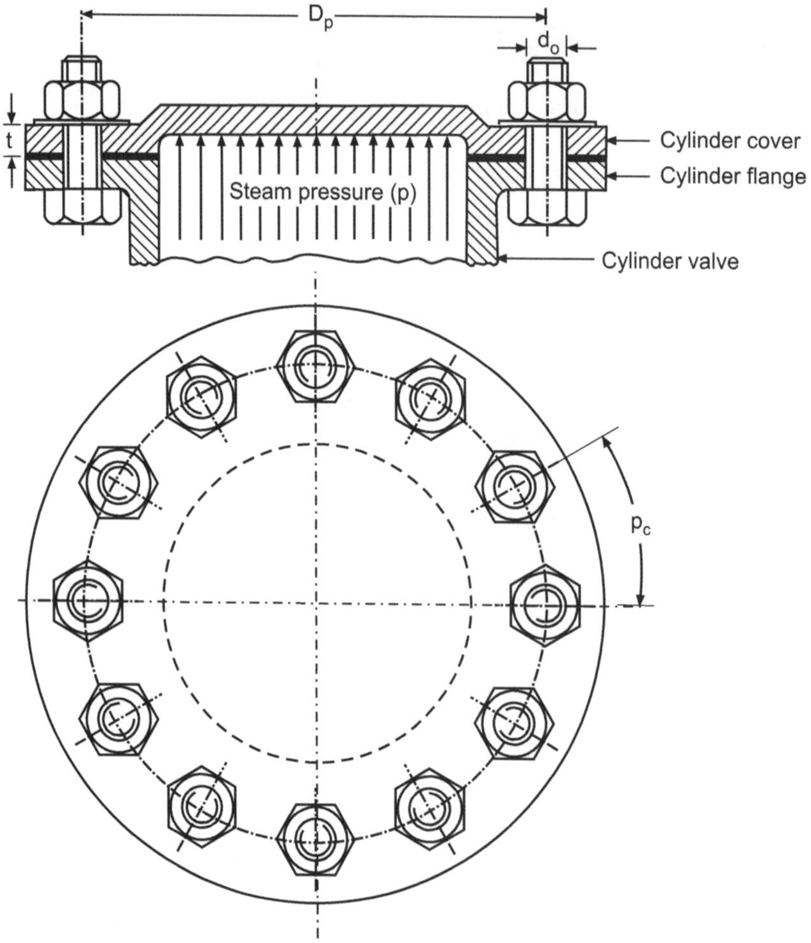


**Fig. 8.9 : Design of bolts for cylinder cover**

> **Important Note :** Bolts or studs are always specified by nominal diameter ($d_o$).
> For example : Bolt having nominal diameter 24 mm is represented as M24 bolt.

## Numerical Type No. 1 : "Design of Bolts, if Bolts are not Initially Stressed"

**Problem 8.1 :** *Determine safe tensile load for bolts of size M20 and M36. Assume that, bolts are not initially stressed. Take safe tensile stress as 200 MPa.*

**Solution : Given data :** $\sigma_t$ = 200 MPa = **200 N/mm²**.

**Procedure :**

**Case I :** It is given that, bolt size is M20.

$\therefore \quad d_o = 20$ mm; $\therefore$ $\mathbf{d_c} = 0.84\, d_o = 0.84 \times 20 = \mathbf{16.8\ mm}$

Considering failure of bolt in tension, we have,

Tensile stress, $$\sigma_t = \frac{W}{\frac{\pi}{4} \cdot d_c^2}$$

$$\therefore \quad 200 = \frac{W}{\frac{\pi}{4} \times (16.8)^2}$$

$$\therefore \quad W = \mathbf{44334.15\ N}$$

**Case II :** It is given that, bolt size is M36.

$\therefore \quad d_o = 36$ mm; $\therefore$ $\mathbf{d_c} = 0.84\, d_o = 0.84 \times 36 = \mathbf{30.24\ mm}$

Considering tensile failure of bolt in tension, we have,

Tensile stress, $$\sigma_t = \frac{W}{\frac{\pi}{4} \cdot d_c^2}$$

$$\therefore \quad 200 = \frac{W}{\frac{\pi}{4} \times (30.24)^2}$$

$$\therefore \quad W = \mathbf{143642.66\ N}$$

---

**Problem 8.2 :** *A cylinder head of steam engine is held in position by M20 bolts. Effective diameter of cylinder is 350 mm and steam pressure is 0.75 N/mm². If bolts are not initially stressed, find the number of bolts required. Take working stress for bolt material as 20 N/mm².*

**Solution : Given data :**

Diameter of cylinder = D = 350 mm

Pressure inside the cylinder = p = 0.75 N/mm²

Nominal diameter of bolts = $d_o$ = 20 mm

Permissible stress = $\sigma_t$ = 20 N/mm²

**Procedure :** We have, $\mathbf{d_c} = 0.84\ d_o = 0.84 \times 20 = \mathbf{16.8\ mm}$

Total load acting on the cylinder cover,

$$W_n = \text{Pressure of steam inside the cylinder} \times \text{Area of cylinder}$$

$$\therefore \quad \mathbf{W_n} = p \times \frac{\pi}{4} \times (D)^2 = 0.75 \times \frac{\pi}{4} \times (350)^2 = \mathbf{72.1584 \times 10^3\ N}$$

This load is taken up by 'n' number of bolts.

∴ Load on each bolt,

$$W = \frac{W_n}{n} = \frac{72.1584 \times 10^3}{n}$$

Considering failure of bolt in tension,

$$\sigma_t = \frac{W}{\frac{\pi}{4} \times (d_c)^2}$$

$$\therefore \quad 20 = \frac{72.1584 \times 10^3}{n \times \frac{\pi}{4} \times (16.8)^2} \qquad \left(\because W = \frac{72.1584 \times 10^3}{n} \text{ and } d_c = 16.8 \text{ mm}\right)$$

$$\therefore \quad n = \frac{72.1584 \times 10^3 \times 4}{\pi \times (16.8)^2 \times 20} = 16.2760 \cong \mathbf{18\ (say)}$$

∴ **Number of bolts required = 18**

---

**Problem 8.3 :** *A steam engine cylinder has effective diameter of 250 mm. It is subjected to maximum steam pressure of 1.5 MPa. A cylinder cover is fixed to the cylinder with the help of 12 bolts. Pitch circle diameter of bolts is 400 mm. Permissible tensile stress of bolt material is 30 MPa. Determine nominal diameter of bolts.*

**Solution : Given data :** D = 250 mm, p = 1.5 MPa or N/mm², n = 12, $\sigma_t$ = 30 MPa or N/mm², $D_p$ = 400 mm

**Procedure :**

Total load acting on cylinder cover is given by,

$$\mathbf{W_n} = p \times \frac{\pi}{4} \times (D)^2 = 1.5 \times \frac{\pi}{4} \times (250)^2 = \mathbf{73631.07\ N}$$

This load is taken up by 12 number of bolts.

$$\therefore \quad \text{Load on each bolt, } \mathbf{W} = \frac{W_n}{n} = \frac{73631.07}{12} = \mathbf{6135.923\ N}$$

Considering failure of bolt in tension,

$$\sigma_t = \frac{W}{\frac{\pi}{4} \times d_c^2}$$

$$\therefore \quad 30 = \frac{6135.923 \times 4}{\pi \times d_c^2}$$

$$\therefore \quad \mathbf{d_c = 16.137\ mm}$$

Bolt or stud is always specified by its Nominal diameter (i.e. Major diameter). Therefore, major diameter ($d_o$) is calculated by using the following relation of $d_c$ and $d_o$.

We know, $\quad d_c = 0.84\ d_o$

$$\therefore \quad \mathbf{d_o} = \frac{d_c}{0.84} = \frac{16.137}{0.84} = 19.21\ \text{mm} \cong \mathbf{20\ mm\ (say)}$$

If $d_o$ = 20 mm, then, $\mathbf{d_c} = 0.84 \times d_o = 0.84 \times 20 = \mathbf{16.8\ mm}$

∴ **We shall use 12 bolts of size M20 having $d_c$ as 16.8 mm.**

---

**Problem 8.4 :** *An engine cylinder is 30 cm in diameter and steam pressure is 1 N/mm² absolute with 0.3 N/mm² back pressure. If cylinder cover is held by 12 studs, find size of stud. Assume that, studs are not initially stressed. Take permissible stress for stud material as 28 MPa.*

**Solution : Given data : D** = 30 cm = **300 mm**, n = 12, $\sigma_t$ = 28 N/mm²

Pressure inside the cylinder, **p** = Absolute pressure – Back pressure = 1 – 0.3 = **0.7 N/mm²**

**Procedure :** Total load acting on the cylinder cover is given by,

$$W_n = \text{Pressure} \times \text{Area of cylinder}$$

$$\therefore \quad \mathbf{W_n} = p \times \frac{\pi}{4} \times (D)^2 = 0.7 \times \frac{\pi}{4} \times (300)^2 = \mathbf{49.48 \times 10^3\ N}$$

This load is taken up by 12 number of studs (i.e. Bolts without head).

$$\therefore \text{ Load on each stud,} \quad \mathbf{W} = \frac{W_n}{n} = \frac{49.48 \times 10^3}{12} = \mathbf{4123.34\ N}$$

Considering the failure of stud in tension,

$$\sigma_t = \frac{W}{\frac{\pi}{4} \times (d_c)^2}$$

$$\therefore \quad d_c^2 = \frac{4 \times W}{\pi \times \sigma_t} = \frac{4 \times 4123.34}{\pi \times 28}$$

$$\therefore \quad d_c = \mathbf{13.69\ mm}$$

But, $\quad d_c = 0.84\ d_o$

$$\therefore \quad \mathbf{d_o} = \frac{13.69}{0.84} = 16.29\ \text{mm} \cong \mathbf{18\ mm\ (say)}$$

Therefore, $\quad \mathbf{d_c} = 0.84\ d_o = 0.84 \times 18 = \mathbf{15.12\ mm}$

∴ **Studs of size M 18 should be used, having core diameter 15.12 mm.**

---

**Problem 8.5 :** *Determine size of bolts used to fix cylinder head of a steam engine. Engine cylinder has a bore of 400 mm and maximum steam pressure inside cylinder is 1.5 N/mm². Cylinder head is held on the cylinder by 16 number of bolts. Permissible tensile stress for the bolt material is 25 N/mm².*

**Solution : Given data :** D = 400 mm, p = 1.5 N/mm², n = 16, $\sigma_t$ = 25 MPa

**Procedure :**

Total load exerted by steam pressure on the cylinder head (cover) is,

$$W_n = \text{Pressure of steam inside cylinder} \times \text{Area of cylinder}$$

$$\therefore \quad \mathbf{W_n} = p \times \frac{\pi}{4} \times (D)^2 = 1.5 \times \frac{\pi}{4} \times (400)^2 = \mathbf{188495.56\ N}$$

This load is taken up by 16 number of bolts.

∴ Load on each bolt, $\mathbf{W} = \frac{W_n}{n} = \frac{188495.56}{16} = \mathbf{11780.97\ N}$

Considering the failure of bolt in tension,

$$\sigma_t = \frac{W}{\frac{\pi}{4} d_c^2}$$

$$\therefore \quad 25 = \frac{11780.97 \times 4}{\pi \times (d_c)^2}$$

$$\therefore \quad d_c = \mathbf{24.499\ mm}$$

Bolt or stud is always specified by its nominal diameter or major diameter ($d_o$). Therefore, we use,

$$d_c = 0.84\ d_o$$

$$\therefore \quad \mathbf{d_o} = \frac{d_c}{0.84} = \frac{24.499}{0.84} = \mathbf{29.16\ mm \cong 30\ mm\ (say)}$$

∴ We will use 16 bolts of size M30 (having core diameter $d_c = 0.84 \times 30 = 25.2$ mm) for tightening cylinder head/cover.

---

**Numerical Type No. 2 : "Design of the Bolts for Cylinder Cover Considering the Stress due to Combined Forces"**

**Problem 8.6 :** *Cylinder head of a steam engine is subjected to steam pressure of 0.7 N/mm². It is held in position by means of 12 bolts. A soft copper gasket is placed to make the joint leak proof. Effective diameter of cylinder is 300 mm. Find size of bolt, so that, stress in the bolt is not to exceed 100 N/mm².*

**Solution : Given data :** $p = 0.7$ N/mm², $n = 12$, $D = 300$ mm, $\sigma_t = 100$ N/mm²

**Procedure :** Initial tightening load acting on each bolt = $\mathbf{W_1 = 2840\ d_o}$ ... (1)

Total load acting on cylinder head (cover) due to steam pressure is calculated as,

$$\mathbf{W_n} = p \times \frac{\pi}{4} \times (D)^2 = 0.7 \times \frac{\pi}{4} \times (300)^2 = \mathbf{49.48 \times 10^3\ N}$$

This load is taken up by 12 bolts.

∴ Load acting on each bolt due to steam pressure is calculated by,

$$\mathbf{W_2} = \frac{W_n}{n} = \frac{49.48 \times 10^3}{12} = \mathbf{4123.34\ N} \quad \text{... (2)}$$

**Resultant load acting on each bolt is calculated by,**

$$W = W_1 + K \cdot W_2$$

where, K = ratio of elastic limits of connected parts = 0.6 for copper gasket (Assumed).

$$\therefore \quad \mathbf{W} = 2840\ d_o + (0.6 \times 4123.34) = \mathbf{2840\ d_o + 2474.004} \quad \text{[From (1) and (2)]}$$

Considering the failure of bolt in tension,

$$\sigma_t = \frac{W}{\frac{\pi}{4} \times (d_c)^2}$$

$$\therefore \quad 100 = \frac{2840\ d_o + 2474.004}{\frac{\pi}{4} \times (0.84\ d_o)^2} \quad [\because d_c = 0.84\ d_o]$$

$$\therefore \quad (0.84\ d_o)^2 \times 100 \times \frac{\pi}{4} = 2840\ d_o + 2474.004$$

$$\therefore \quad 55.41\ d_o^2 - 2840\ d_o - 2474.004 = 0$$

$$\therefore \quad \mathbf{d_o} = 52.11\ \text{mm} \cong \mathbf{54\ mm\ (say)}$$

$$\therefore \quad \mathbf{d_c} = 0.84\ d_o = 0.84 \times 54 = \mathbf{45.36\ mm}$$

**We should use 12 bolts of size M 54, having core diameter 45.36 mm.**

**Problem 8.7 :** *Cylinder head of steam engine is subjected to a steam pressure of 0.7 N/mm$^2$. It is held in position by means of 12 bolts. Bolts are tightened with an initial preload of 1.5 times the steam load. A soft copper gasket is used to make the joint leak proof. Effective diameter of cylinder is 300 mm. Find the size of bolts, so that, stress in bolts is not to exceed 100 MPa.*

**Solution : Given data :** $p = 0.7$ N/mm$^2$, $n = 12$, $D = 300$ mm, $\sigma_t = 100$ MPa

**Procedure :** Steam load acting on cylinder head $= p \times \frac{\pi}{4} \times D^2$

$$\therefore \quad \mathbf{W_n} = 0.7 \times \frac{\pi}{4} \times (300)^2 = \mathbf{49.48 \times 10^3\ N}$$

This load is taken up by 12 bolts.

$$\therefore \text{ Steam load acting on each bolt} = \mathbf{W_2} = \frac{W_n}{n} = \frac{49.48 \times 10^3}{12} = \mathbf{4123.34\ N} \qquad \ldots (1)$$

In the numerical problem, initial tightening load acting on all bolts is given in terms of Initial preload due to tightening. Both terms have same meaning. Therefore, we can write,

$$\text{Initial preload on all 12 bolts due to tightening} = 1.5 \times \text{Steam load}$$

$$= 1.5 \times 49.48 \times 10^3 = \mathbf{74220\ N}$$

$$\therefore \quad \text{Initial preload on each bolt} = \mathbf{W_1} = \frac{74220}{12} = \mathbf{6185\ N} \qquad \ldots (2)$$

**Resultant load acting on each bolt is calculated by,**

$$W = W_1 + K \cdot W_2$$

where, $K = 0.6$ for soft copper gasket **(Assumed)**

$$\therefore \quad \mathbf{W} = 6185 + (0.6 \times 4123.34) = \mathbf{8659\ N} \qquad \text{[From (1) and (2)]}$$

Considering the failure of bolt in tension,

$$\sigma_t = \frac{W}{\frac{\pi}{4} \cdot d_c^2}$$

$$\therefore \quad d_c^2 = \frac{W \times 4}{\pi \times \sigma_t} = \frac{8659 \times 4}{\pi \times 100}$$

$$\therefore \quad d_c = \mathbf{10.49\ mm}$$

$$\therefore \quad \mathbf{d_o} = \frac{d_c}{0.84} = \frac{10.49}{0.84} = 12.49 \text{ mm} \cong \mathbf{14\ mm\ (say)}$$

$$\therefore \quad \mathbf{d_c} = 0.84 \times d_o = 0.84 \times 14 = \mathbf{11.76\ mm}$$

**We should use 12 bolts of size M14, having core diameter 11.76 mm.**

---

**Problem 8.8 :** *Determine the diameter of hole, that must be drilled in M 60 bolt, to make the bolt of uniform strength.*

**Solution : Given data :** $d_o = 60$ mm.

**Procedure :** We have, $\mathbf{d_c} = 0.84\, d_o = 0.84 \times 60 = \mathbf{50.4\ mm}$

For bolt of uniform strength, diameter of hole to be drilled through bolt head is calculated by,

$$\mathbf{d_1} = \sqrt{d_o^2 - d_c^2} = \sqrt{(60)^2 - (50.4)^2} = \mathbf{32.55\ mm \cong 33\ mm\ (say)}$$

---

**Problem 8.9 :** *Two machine parts are fastened together tightly by means of a 22 mm tap bolt. If the load tending to separate these parts is neglected, find the stress induced in the bolt due to initial tightening.*

**Solution : Given data :** $d_o = 22$ mm

**Procedure :**

We have, $\mathbf{d_c} = 0.84\, d_o = 0.84 \times 22 = \mathbf{18.48\ mm}$

Load acting on each bolt due to initial tightening is calculated by,

$$\mathbf{W_1} = 2840\, d_o = 2840 \times 22 = \mathbf{62480\ N}$$

We know that, resultant load acting on each bolt is given by,

$$W = W_1 + K \cdot W_2$$

where, $W_2$ is the load acting on each bolt, which tends to separate parts. Since, this load $W_2$ is asked to be neglected in numerical problem, therefore, $W_2 = 0$. Above equation becomes,

$$\therefore \quad \mathbf{W} = 62480 + (K \times 0) = \mathbf{62480\ N}$$

Considering failure of bolt in tension,

$$\text{Induced tensile stress, } \sigma_t = \frac{W}{\frac{\pi}{4} d_c^2} = \frac{62480 \times 4}{\pi \times (18.48)^2} = \mathbf{232.94\ N/mm^2 \text{ or } MPa}$$

**Numerical Type No. 3 : "Design of the Bolts subjected to Fatigue Loading for Cylinder Cover Considering the Stress due to Combined Forces"**

**Problem 8.10 :** *Cylinder head of steam engine is subjected to a steam pressure of 0.7 N/mm². It is held in position by means of 12 bolts. Bolts are tightened with an initial preload of 1.5 times the steam load. A soft copper gasket is used to make the joint leak proof. Effective diameter of cylinder is 300 mm. Find the size of bolts, so that, stress in bolts is not to exceed 100 MPa.*

**Solution : Given data :** $p = 0.7\ N/mm^2$, $n = 12$, $D = 300$ mm, $\sigma_t = 100$ MPa

**Procedure :** Steam load acting on cylinder head $= p \times \frac{\pi}{4} \times D^2$

$$\therefore \quad \mathbf{W_n} = 0.7 \times \frac{\pi}{4} \times (300)^2 = \mathbf{49.48 \times 10^3\ N}$$

This load is taken up by 12 bolts.

$$\therefore \text{ Steam load acting on each bolt} = \mathbf{W_2} = \frac{W_n}{n} = \frac{49.48 \times 10^3}{12} = \mathbf{4123.34\ N}$$

But, it is given that, bolts are subjected to fatigue loading. Therefore, the steam load acting on each bolt should be divided by a loading factor of 0.85 to get the new value of steam load to compensate for variable load due to fatigue.

$$\therefore \text{ Steam load acting on each bolt} = \mathbf{W_2} = \frac{4123.34}{0.85} = \mathbf{4850.98\ N} \qquad \ldots (1)$$

In the numerical problem, initial tightening load acting on all bolts is given in terms of Initial preload due to tightening. Both terms have same meaning. Therefore, we can write,

$$\text{Initial preload on all 12 bolts due to tightening} = 1.5 \times \text{Steam load}$$
$$= 1.5 \times 49.48 \times 10^3 = \mathbf{74220\ N}$$

$$\therefore \quad \text{Initial preload on each bolt} = \mathbf{W_1} = \frac{74220}{12} = \mathbf{6185\ N} \qquad \ldots (2)$$

**Resultant load acting on each bolt is calculated by,**

$$W = W_1 + K \cdot W_2$$

where, $K = 0.6$ for soft copper gasket **(Assumed)**

$$\therefore \quad \mathbf{W} = 6185 + (0.6 \times 4850.98) = \mathbf{9095.59\ N} \qquad \text{[From (1) and (2)]}$$

Considering the failure of bolt in tension,

$$\sigma_t = \frac{W}{\frac{\pi}{4} \cdot d_c^2}$$

$$\therefore \quad d_c^2 = \frac{W \times 4}{\pi \times \sigma_t} = \frac{9095.59 \times 4}{\pi \times 100} = 115.81$$

$$\therefore \quad d_c = \mathbf{10.76\ mm}$$

$$\therefore \quad \mathbf{d_o} = \frac{d_c}{0.84} = \frac{10.76}{0.84} = 12.81 \text{ mm} \cong \mathbf{14\ mm\ (say)}$$

$$\therefore \quad \mathbf{d_c} = 0.84 \times d_o = 0.84 \times 14 = \mathbf{11.76\ mm}$$

**We should use 12 bolts of size M14, having core diameter 11.76 mm.**

## Theory Questions for Practice

1. Give advantages and disadvantages of screwed joints.
2. Sketch and explain any two thread profiles used for screwed joints.
3. Define the following terms related to screwed joints:

   (i) Major diameter, (ii) Minor diameter, (iii) Pitch and (iv) Lead.
4. What are the bolts of uniform strength? Explain any two methods of obtaining bolts of uniform strength.
5. What are different stresses induced in screwed fastening?
6. Give two examples, where screwed joints are preferred over welded joints.

## Numerical Problems for Practice

1. Determine safe tensile load for a bolt of M 30, assuming a safe tensile stress of 45 MPa. (**Ans.** W = 22444.16 N)
2. Effective diameter of cylinder is 400 mm. Maximum pressure of steam acting on cylinder cover is 1.12 N/mm$^2$. Find the number of M 24 size bolts required to fix the cover. Stress in bolts should not exceed 33 MPa. Assume bolts are not initially stressed. (**Ans.** n = 14)
3. Cylinder head of a steam engine is held in position by M 20 bolts. Effective diameter of cylinder is 300 mm and steam pressure is 0.6 N/mm$^2$. Bolts are not initially stressed. Find the number of bolts required, if working stress for the bolt material is not to exceed 20 MPa. (**Ans.** n = 10)
4. A double acting steam engine having 360 mm cylinder diameter works on maximum steam pressure of 1 N/mm$^2$ absolute with 0.03 N/mm$^2$ back pressure inside cylinder. Determine the diameter of screwed end of piston rod, where permissible safe stress in tension is 60 MPa. (**Ans.** M 56 size)
5. A steam engine cylinder of 300 mm diameter is supplied with steam at 1.5 N/mm$^2$. Cylinder cover is fastened by means of 8 bolts of size M 20. The joint is made leak-proof by means of suitable gaskets. Find stress produced in bolts. (**Ans.** $\sigma_t$ = 202.85 N/mm$^2$)

❑❑❑

Chapter 9

# ERGONOMICS AND AESTHETIC CONSIDERATION IN DESIGN

## 9.1 MAN-MACHINE RELATIONSHIP

- Machine has separate identity in itself in machine design. However, it cannot perform well without man.
- **Egronomists** consider man-machine as a joint system, to perform combination of activities to obtain desired output from the given input.
- From display instruments, operator gets information about the operations of machine.
- If he feels that, a correction is necessary, he will operate the levers or controls manually. After doing corrections, he can observe the change in performance of machine, which will be indicated on display panels. Refer fig. 9.1.
- Contact between man and machine in this closed-loop system arises at two places - **display instruments**; which give information to the operator, and **control**; with which, the operator adjusts the machine.

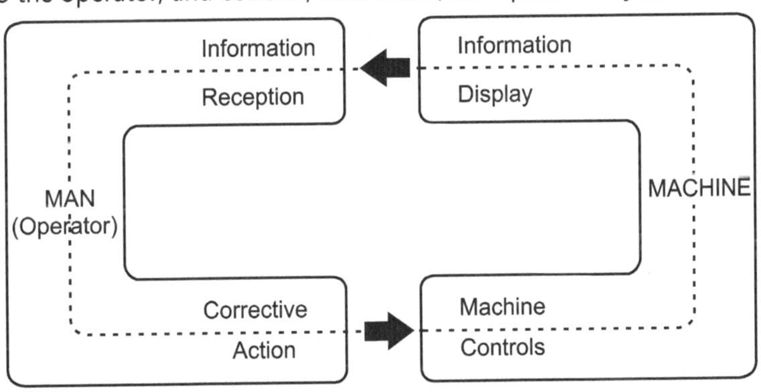


**Fig. 9.1 : Man-machine closed loop system**

## 9.2 INTRODUCTION TO ERGONOMICS

- Ergonomics is concerned with optimizing the overall and detailed design of controls, operations, and maintenance interfaces between machines and men, in both working conditions, normal and emergency.
- Men and machines interact simultaneously with environment, surrounding work place etc. in terms of efficiency, reliability, safety on low cost-effective basis.
- This involves a knowledge of human characteristics such as :
  (a) Anatomy and physiology – structure and function of human body, senses, mental processes.
  (b) Anthropometry – information on body size, field of movement, field of view, speed of response, forces exerted.
  (c) Psychology – knowledge of brain, nervous system and human behaviour.
  (d) Conditions, which improve efficiency and comfort, by minimizing fatigue.
- Therefore, **Ergonomics** is defined as *'the relationship between man and machine and the application of anatomical, physiological and psychological principles to solve the problems arising from man-machine relationship'*.

- The word **'ergonomics'** is derived from two Greek words – **'ergon' = work** and **'nomos' = natural laws**. Ergonomics means the natural laws of work.
- From design considerations, the topics included in ergonomic studies are,
  (a) Anatomical factors in design of driver's seat.
  (b) Layout of instrument dials and display panels for accurate perception by the operators.
  (c) Design of hand levers and hand wheels.
  (d) Energy expenditure in hand and foot operations.
  (e) Lighting, noise and climatic conditions in machine environment (i.e. surrounding work-place conditions).
- **Ergonomic** information applied to design ensures comfort, safety etc. to human beings, due to best possible environments and designs prepared.
- While designing equipments for human use, it is important to realize that, man has limited power. Hence, the designer should design the most suitable equipment considering human health and efficiency.
- **Ergonomics** can also be defined as *"a systematic study of relationship between people and their occupation, equipment and work environment for the purpose of increasing human efficiency and removing those features of design, which tend to cause inefficiency or physical disability"*.

### 9.2.1 Importance of Ergonomics in Design

- Ergonomics reduces physical and mental stresses.
- Ergonomics helps to study the man-machine relationship.
- Working environment can be improved to increase human efficiency.
- Reliability, safety etc. can be improved on low cost-effective basis.

## 9.3 ERGONOMIC CONSIDERATIONS WHILE DESIGNING A MACHINE

- Major ergonomic considerations in the process of designing a machine are,
  (a) Design of displays. (b) Design of controls. (c) Environment and safety.

### 9.3.1 Design of Displays

- Displays are devices, through which, the man (user) can receive the information from the machine.
- A good display device is one, which allows proper combination of speed, accuracy and sensitivity of display.
- Display devices can be mainly classified into two groups.

**1. Qualitative Displays :**

- Qualitative displays indicate only the condition or state without giving any values. Refer Fig. 9.2.
- Qualitative display or signal is used to indicate an 'on' or 'off' response such as valve open or closed, power 'ON' or 'OFF'.
- **For example :** Red, yellow and green lights used as traffic signals, ON-OFF indicators of home appliances etc.

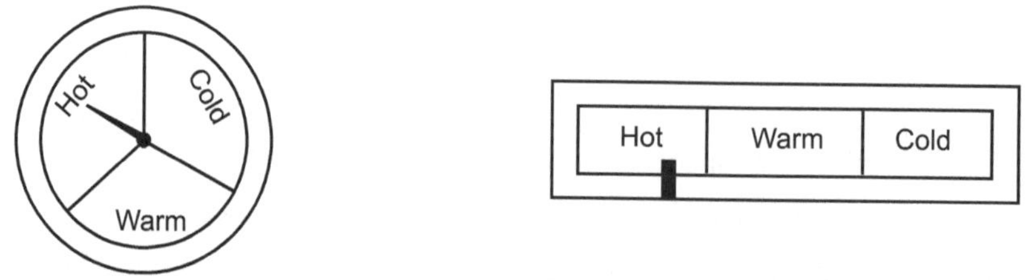


**(a) Circular dial** **(b) Straight legend**

**Fig. 9.2 : Qualitative displays**

**2. Quantitative Displays :**

- Quantitative displays give numerical information about the equipment or machine, as required by the user.
- Information may be displayed as a value or number. It can also be displayed by pointer-scale arrangement etc. Refer Fig. 9.3.
- **For example :** Voltmeters, Ammeters, Speedometers, Watches, Energy meters etc.

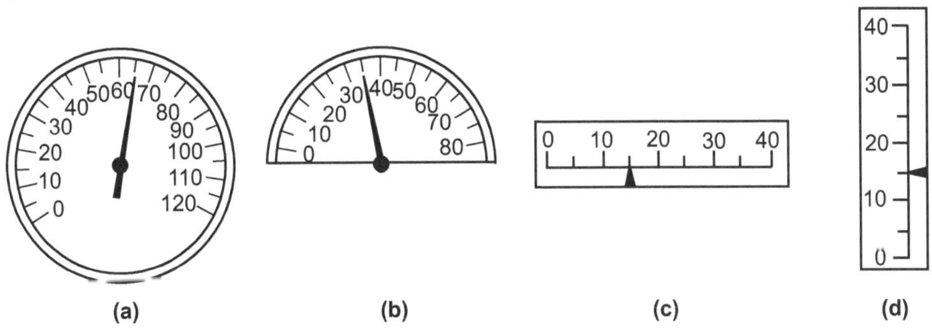


**Fig. 9.3 : Quantitative displays**

### 9.3.1.1 Ergonomic Considerations in Design of Displays

Ergonomic considerations in design of displays are as follows:

- Scale on the dial indicator should be divided in suitable linear progression such as 0-10-20-30.
- Number of subdivisions between the divisions should be minimum.
- Vertical figures should be used for stationary dials, while the radially oriented figures should be used for rotating dials.
- Height of letters or numbers on displays should be $\geq \frac{\text{Reading distance}}{200}$.
- Pointer should have knife-edge with a mirror placed or located behind the graduated scale to minimize parallax error.
- Numbering should increase in clockwise direction on a circular scale, rightwards on a horizontal scale and upwards on a vertical scale.
- Distinction in display groups should be made with the help of colour, shape and size. Important displays, like warnings, fire alarms, urgent attention etc. can be made more effective by use of flashing lights or a light of varying brightness.

## 9.3.2 Design of Equipment for Controls

- Controls are the devices, through which, man conveys instructions to the machine.
- Selection of control device depends upon the following factors:
  (a) Required speed of operation.
  (b) Required accuracy of control.
  (c) Required operating force.
  (d) Required range of control.
  (e) Required direction of control.
  (f) Convenience of user.
- Control device is usually rotated in clockwise direction or moved to the right for 'on' or 'increase'. However, different movements are adopted for special circumstances.
- Various types of controls used in machines are: Crank, Handwheel, Knob, Lever, Push button, Foot pedal etc. Refer Fig. 9.4.

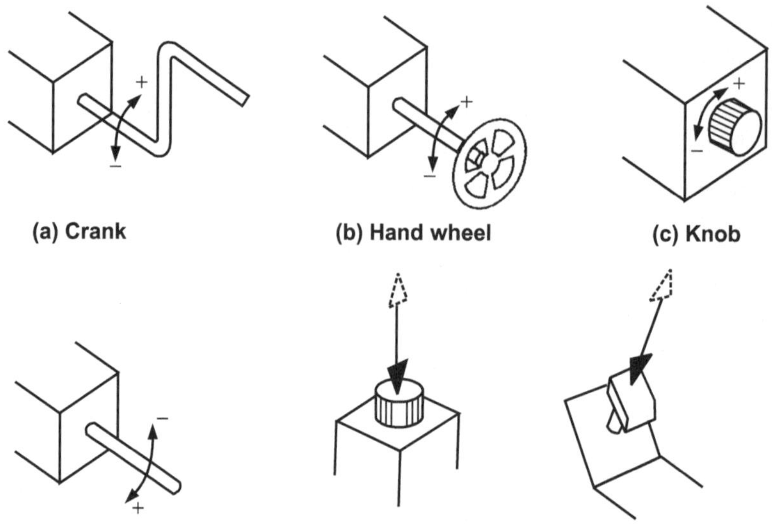

(a) Crank (b) Hand wheel (c) Knob

(d) Horizontal lever (f) Push button (g) Foot pedal

**Fig. 9.4 : Equipment for controls**

### 9.3.2.1 Ergonomic Considerations in Design of Controls

Ergonomic considerations in design of controls are as follows :

- Controls should be easily accessible and properly positioned.
- Control operation should involve minimum motions.
- Shape of control device should be selected in such a way that it satisfies conformity with the anatomy of hands and feet.
- Control should be painted in proper colour to attract the attention.

## 9.3.3 Design of Environment and Safety

- Working environment affect significantly the man-machine relation. It affects the efficiency, health and safety of the operator.
- Varieties of functions are performed by human beings in industrial environment such as controlling, supervising, maintaining the work and equipment etc.
- Most important factors affecting efficiency of human being are heat, humidity, light, air movement, noise, acceleration including vibrations.

**(a) Lighting :**

- Amount of light required to enable a task to be efficiently performed depends upon the degree of precision required with the work, time allowed to see the task, reflective characteristics of the equipment involved and the sharpness of vision of human beings.
- The surrounding area including walls, ceiling, floor and other objects should be bright and more coloured than the workplace, where he should concentrate.
- Lighting should match the needs of task as far as illumination is concerned.
- Glare causes discomfort and reduce visibility of the task. Glare can be minimized or eliminated by careful design of lighting sources and their positions and changes in texture of the working surfaces.
- Workers will become less tired, if the lighting and the colour scheme are arranged, so that, there is a gradual change in brightness.

**(b) Noise :**

- It produces three effects – Annoyance, Loss of hearing power and Reduced efficiency at work.
- High pitched noise, interrupted and sudden noise, unexpected noise produces maximum discomfort than steady noise environment.

- Experimental studies indicate that, the workers do not work slower, when exposed to noise, but it appears that, loud noise causes them to be less accurate and liable to make mistakes during work.
- If the noise level is too high, it can be stopped at source by better maintenance of equipment, pluging ears and providing sound insulating walls.

**(c) Temperature :**

- In order to perform task efficiently, the operator should feel neither too hot nor too cold, but comfortable. Optimum air temperature for most of the manufacturing operation ranges in between 18°C and 23°C.
- When heavy work is to be done, the temperature should be lowered, and when office work is to be done, it should be little higher.
- Radiant temperature comfort level lies between 16°C to 20°C, which is most important safety measure to protect workers from radiant heat, when high temperature processes are carried out.

**(d) Humidity and air circulation :**

- Humidity has little effect at ordinary temperatures, but low humidity may cause discomfort through drying of nose and throat.
- Air humidity and air velocity become important at high temperature, because they influence the amount of sweat, which should be evaporated from body surface to produce cooling effect.

**Ideal Working Conditions :**

| Property | Range or value |
|---|---|
| Temperature | 18°C to 23°C. |
| Relative humidity | 30% to 70%. |
| Illumination level | Better than 200 lux (lighting intensity on floor/$m^2$). |
| Ventilation rate of air | 6.6 $m^3$/hr at sea level per person and light work. |
| Noise level | Not to be more than 10 dB. |

### 9.3.4 An Example Illustrating Role of Erognomic Considerations, While Designing a Lathe Machine

- While designing a lathe machine, a design engineer must consider following ergonomic criteria, so that, a lathe machine operator can operate and control the machine effectively, comfortably and without any fatigue leading to poor work, more scrap or wastages etc.

(1) Controls on lathe should be easily accessible and properly positioned.

(2) Control operation should involve minimum motions.

(3) Height of lathe machine should match with average height of workers to have ease in operation.

(4) Lathe machine should make less noise during operation.

(5) Effort and power required in turning any hand-wheel as per operation should be less.

(6) They should give desired accuracy in operation.

## 9.4 AESTHETICS

- **Aesthetics** is defined as *'a set of principles of appreciating a beauty'*. It deals with external appearance of product.
- Appearance is an outward expression of quality of any product, which makes first communication of product with the consumer.
- Although function, cost, safety and physical aspects of ergonomics may be important in the initial stages of design process, appearance of a product plays a major role in sale-ability of product in market.
- An engineering design with good aesthetics will be pleasing to the eye and gives a visual impression of functioning efficiently.

### 9.4.1 Importance of Aesthetics in Design

- Each product is to be designed to perform a specific function or a set of functions for satisfying needs of customers.
- Parameters considered by a customer, while selecting a product are :
  1. Functional performance,
  2. Durability,
  3. Initial and running cost,
  4. Ability to withstand in adverse conditions,
  5. Service support available,
  6. Comfort to the user, and
  7. Appearance.
- In a present days of buyer's market, a number of products available in the market are having most of the parameters identical. Thus, the *appearance of the product* becomes a major factor in attracting the customer.
- This is particularly true for automobiles, daily needs and home appliances like domestic refrigerators, television sets, music systems etc.
- At any stage in the product life, the aesthetic quality cannot be separated from the product quality.
- Growing importance of aesthetic considerations on product design has given rise to a separate discipline, known as *'industrial design'*.
- Now-a-days, job of an industrial designer is to create new shapes and forms for any product, which are aesthetically appealing to consumers.

## 9.5 CONSIDERATIONS IN AESTHETIC DESIGN

Various aspects of aesthetic design are,

1. Shape (Form)
2. Symmetry and Balance
3. Colour
4. Continuity
5. Variety
6. Proportion
7. Size
8. Contrast
9. Impression and Purpose
10. Style
11. Material and Surface Finish
12. Tolerance
13. Noise

**1. Shape (Form) :**

There are *five basic shapes* (forms) of the products, such as step, taper, shear, streamline and sculpture, as shown in Fig. 9.5. The external shape of any product is based on one or combination of these basic shapes.

**(i) Step form :** It is a stepped structure having vertical accent. It is similar to the shape of multistorey building.

**(ii) Taper form :** It consists of tapered blocks or tapered cylinders.

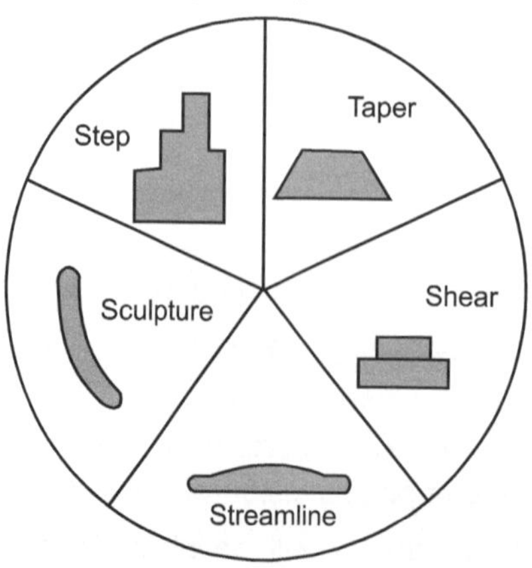


**Fig. 9.5 : Basic types of product shapes (forms)**

**(iii) Shear form :** It has a square outlook.

**(iv) Streamline form :** It has a streamlined shape having a smooth flow as seen in automobile and aeroplane structures.

**(v) Sculpture form :** It consists of ellipsoids, paraboloids and hyperboloids.

**2. Symmetry and Balance :**

- Most of the life forms in nature are approximately symmetrical about at least one axis.
- Human eye is thus conditioned to see the things in symmetrical form [Refer Fig. 9.6 (a)] and tends to reject asymmetrical shapes as ugly [Refer Fig. 9.6 (b)].
- In many artificial or man made products, symmetry about at least one axis improves the aesthetic appeal of the product.
- However, if functional requirement demands for asymmetrical shape, then balance in the product improves the aesthetic feeling. Refer Fig. 9.6 (c).

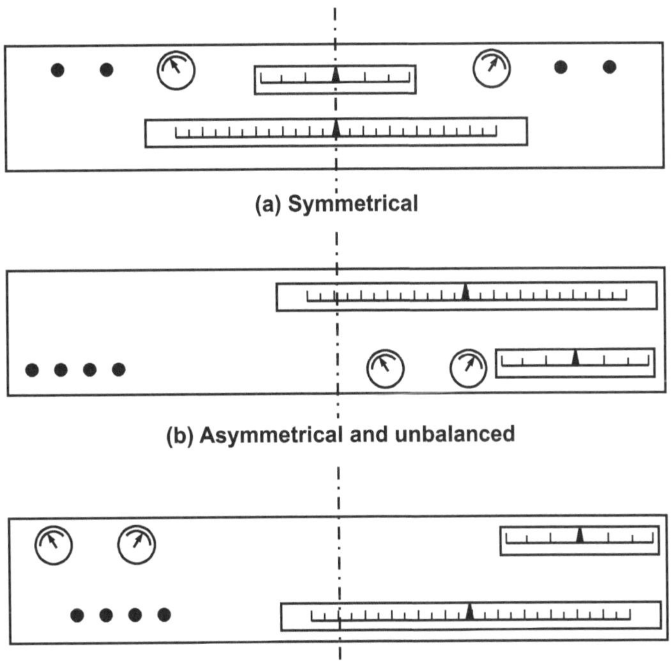

(a) Symmetrical

(b) Asymmetrical and unbalanced

(c) Asymmetrical and balanced

**Fig. 9.6 : Arrangement of a control panel**

**3. Colour (Morgan's Colour Code) :**

- *Colour* is one of the major contributors to the aesthetic appeal of product. Many colours are linked with different moods and conditions.
- Selection of colour should be compatible with conventions. 'Morgan' has suggested the following colour codes with their meanings.

| Colour | Meaning |
|---|---|
| Red | Danger, Hot |
| Orange | Possible Danger |
| Yellow | Caution |
| Green | Safe |
| Blue | Cold |
| Gray | Dull |

4. **Continuity :**
   - A product, which has good *continuity of elements*, is aesthetically appealing.
   - **For example :** At a point of changing cross-sectional area, if fillet radius is provided, then it shows continuity of product, thereby, improving its appearance. Refer Fig. 9.7 (b).

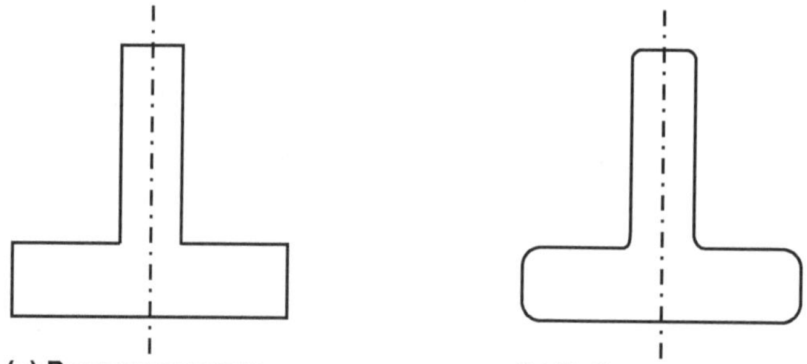

(a) Poor appearance (b) Better appearance

**Fig. 9.7 : Continuity**

5. **Variety :**
   - *Variety* is particularly useful, while marketing wide range of products.
   - **For example :** Daily appliances used by consumers like mobile, laptop, television, automobiles etc. are manufactured in number of varieties by a single manufacturer, even though, all items are functioning in same manner.
6. **Proportion :**
   - *Proportion* is concerned with the relationship in size amongst parts or components of product.
   - The product, which is out of proportion, is not aesthetically pleasing.
   - Spanner shown in Fig. 9.8 (a) satisfies the functional ability and it is also easy to manufacture. But, it is out of proportion and hence poor in appearance.
   - Spanner shown in Fig. 9.8 (b) is in proportion and aesthetically pleasing.

(a) Poor appearance (b) Better appearance

**Fig. 9.8 : Spanner**

7. **Size :**
   - Due to advancements in electronic field, designers can use previously unaccepted housing for integrated items, so freeing them from many of design constraints.
   - Now, design of telephone is an example of integrating the entire telephone circuitry in a single component providing good balance, proportion and ergonomic styling.
8. **Contrast :**
   - *Contrast* is a distinction between the adjacent elements of any product, which have clearly different characteristics and functions.
   - Contrast improves the appearance of product.
9. **Impression and Purpose :**
   - Product should give the *impression* of satisfactory performance.
   - Taper shape gives the impression of strength and stability as shown in Fig. 9.9 and 9.10 respectively.

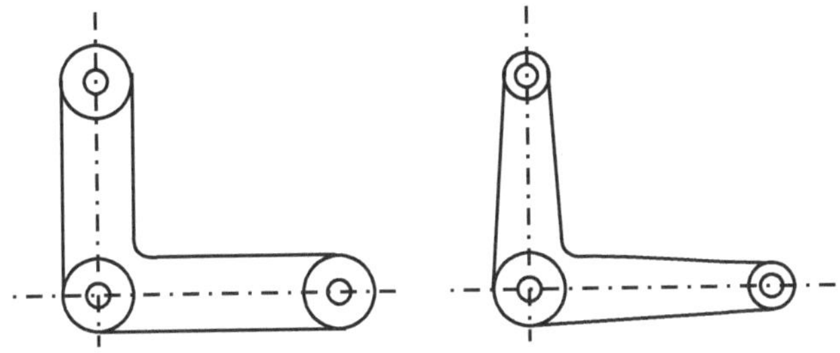

(a) Impression of weakness (b) Impression of strength

Fig. 9.9

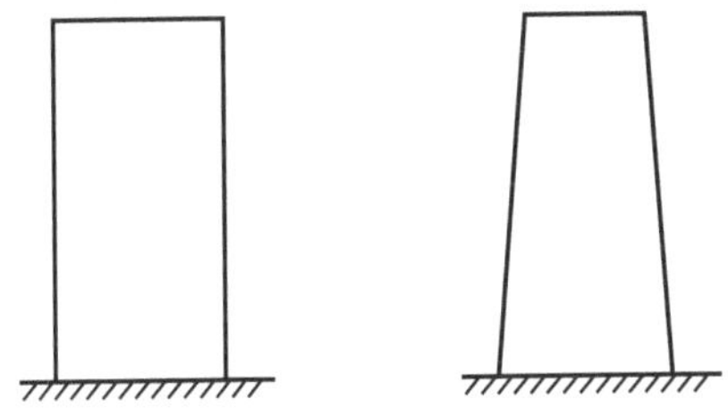

(a) Impression of unstability (b) Impression of stability

Fig. 9.10

**10. Style :**

- *Style* is a visual quality of product.
- Style makes a product attractive and appealing as compared to other products doing same function.

**11. Material and Surface Finish :**

- *Material* and *surface finish* of product contribute significantly to its appearance.
- Material like stainless steel gives better appearance than the cast iron, plain carbon steel or low alloy steel.
- Brass or bronze give richness to the appearance of product.
- Products with better surface finish are always aesthetically pleasing.
- Surface coating processes like spray painting, anodizing, electroplating etc. greatly enhances (or improves) aesthetic appeal of product.

**12. Tolerance :**

- Proper tolerance given to the mating parts improves aesthetic appeal of product.

**13. Noise :**

- Unwanted noise is disturbing and indicator of some malfunction within the product and hence, it greatly reduces the aesthetic appeal.

## 9.6 GUIDELINES IN AESTHETIC DESIGN

- For any product, there exists a relationship between its functional requirements and appearance.
- However, the aesthetic quality contributes to the sales volume of product in market. However, the extent of contribution may vary from product to product.
- For example : Chromium plating of automobile components improves corrosion resistance, in addition to aesthetic appearance. Similarly, aerodynamic shape of car improves the performance in terms of more mileage per litre due to decreased air resistance and at the same time, aerodynamic shape gives the pleasing appearance.
- The following guidelines may be used in aesthetic design (design for appearance) :

1. Appearance should contribute to the performance of product.

   **For example :** Aerodynamic shape of car will have lesser air resistance, resulting in lesser fuel consumption.

2. Appearance should reflect the function of product.

   **For example :** Aerodynamic shape of car increases its speed.

3. Appearance should reflect the quality of product.

   **For example :** Robust and heavy appearance of hydraulic press machine reflects its strength and rigidity.

4. Except prime requirements of any product, appearance should not involve high expenditures or extra costs. It must be economically feasible for producer as well as consumer.
5. Appearance should be achieved by effective and economical utilization of available resources and material.
6. Appearance should be suitable to the environment, in which, product is used.

## 9.7 RELATION BETWEEN AESTHETICS AND ERGONOMICS

- Ergonomics is the scientific study of relationship between men and their working environment. Since the appearance of a design (i) directly affects relationship between product and user and (ii) improves the efficiency of an operation, aesthetic engineering may be considered as an important part of ergonomics.
- Knowing the terms 'ergonomics', 'aesthetics' and their 'interrelation', we are able to understand more details about ergonomic factors influencing product design as well as man-machine relationship.

### *Theory Questions for Practice*

1. Define 'ergonomics'. State areas covered under ergonomics.
2. List ergonomic considerations in design of display and controls.
3. Illustrate the role played by 'colour' in ergonomics and aesthetic design of products.
4. Discuss the aesthetic considerations in machine design regarding shape and colour.
5. Explain the term "aesthetics". Give the important features.
6. State and describe in brief about any six ergonomic considerations in the designing of machine elements.
7. State four examples of ergonomic considerations in the design of a lathe machine.
8. What are the aesthetic considerations in design.
9. Describe the importance of aesthetic considerations in design related to shape, size, colour and surface finish.
10. State aesthetic consideration in design of office table context to following points:

    (i) Shape, (ii) Size, (iii) Color and (iv) Surface finish.
11. Explain the importance of Aesthetic considerations in design by giving any two examples.
12. State Morgan's colour code in Aesthetics.
13. State the meaning of following colour code in aesthetic consideration, while designing a product : (i) Red, (ii) Orange, (iii) Green, (iv) Blue.

❑❑❑

# Appendix : Standard Table for Square Threads

**Reference Table for Nominal Diameter and Pitch of Square Threads**

| Nominal diameter in mm | Pitch in mm | Nominal diameter in mm | Pitch in mm |
|---|---|---|---|
| 22<br>24<br>26<br>28 | 5 | 55<br>58<br>60<br>62 | 9 |
| 30<br>32<br>34<br>36 | 6 | 65<br>68<br>70<br>72 | 10 |
| 38<br>40<br>42<br>44 | 7 | 75<br>78<br>80<br>82 | 10 |
| 46<br>48<br>50<br>52 | 8 | 85<br>88<br>90<br>92 | 12 |